Urban Artificial Intelligence

This volume thoroughly explores the perceptions and ethical considerations surrounding urban artificial intelligence (AI). Tan Yigitcanlar delves into the complex public and professional views on AI, offering invaluable insights for policymakers, urban planners, and developers.

As the world rapidly advances technologically, the role of AI has become increasingly significant. AI's transformative power spans various sectors, revolutionising how we operate and innovate in fields such as healthcare, finance, agriculture, and space exploration. Despite its wide-reaching impact, the integration of AI into urban planning and development remains relatively underexplored. This is surprising given AI's immense potential to revolutionise urban design, management, and experience. Comprising eight comprehensive and insightful chapters, this book examines AI's role in urban contexts, including its applications, public perceptions, and ethical implications. The first part of the guidebook delves into varied perceptions of AI within different urban sectors, presenting detailed perception analyses on AI's role in urban planning, local government services, disaster management, and the construction industry. The second part shifts focus to the ethical implications and responsible implementation of AI in urban settings. It provides frameworks and strategies to ensure AI technologies contribute positively to urban development while mitigating potential risks and ethical concerns.

This volume, alongside its companion *Urban Artificial Intelligence: A Guidebook for Understanding Concepts and Technologies*, offers a holistic view of Urban Artificial Intelligence. Together, these books provide essential insights for urban planners, policymakers, researchers, and anyone interested in AI and urban development, guiding responsible AI integration to foster smarter, more sustainable, and equitable urban environments.

Tan Yigitcanlar is a globally renowned Australian researcher and author in urban studies, technology, and planning, and was acknowledged as one of the top 1% of scientists worldwide by Clarivate in 2023. He serves as a Professor of Urban Studies and Planning at Queensland University of Technology's School of Architecture and Built Environment in Brisbane, Australia.

Urban Artificial Intelligence

A Guidebook for Understanding Perceptions and Ethics

Tan Yigitcanlar

CRC Press
Taylor & Francis Group
Boca Raton London New York

CRC Press is an imprint of the
Taylor & Francis Group, an **informa** business

First edition published 2025
by CRC Press
2385 NW Executive Center Drive, Suite 320, Boca Raton FL 33431

and by CRC Press
4 Park Square, Milton Park, Abingdon, Oxon, OX14 4RN

CRC Press is an imprint of Taylor & Francis Group, LLC

Library of Congress Cataloging-in-Publication Data
Names: Yigitcanlar, Tan, author.
Title: Urban artificial intelligence : a guidebook for understanding
perceptions and ethics / Tan Yigitcanlar.
Description: First edition. | Boca Raton, FL : CRC Press, 2025. | Includes
bibliographical references and index. | Identifiers: LCCN 2024038036 (print) | LCCN
2024038037 (ebook) | ISBN 9781032861289 (hardback) | ISBN 9781032861234 (paperback) | ISBN
9781003521440 (ebook)
Subjects: LCSH: Smart cities. | City planning—Technological innovations. |
Artificial intelligence—Moral and ethical aspects.
Classification: LCC TD159.4 .Y5542 2025 (print) | LCC TD159.4 (ebook) |
DDC 307.1/416—dc23/eng/20240923
LC record available at https://lccn.loc.gov/2024038036
LC ebook record available at https://lccn.loc.gov/2024038037

ISBN: 9781032861289 (hbk)
ISBN: 9781032861234 (pbk)
ISBN: 9781003521440 (ebk)

DOI: 10.1201/9781003521440

Typeset in Minion
by codeMantra

This book is dedicated to the following four beloved, exceptional, and intellectually formidable women who have profoundly shaped my life:
Selin, Ela, Susan, and Cahide

Contents

Foreword

I magine a city where traffic patterns are analysed in real time, leading to optimised traffic flow and reduced congestion. Imagine urban environments where energy consumption is meticulously monitored and managed, waste disposal is streamlined, and public safety is significantly improved through artificial intelligence (AI)-driven insights. Imagine urban environments that can anticipate and prepare for disaster, saving lives and billions in damages. Such scenarios illustrate the profound impact AI can have on urban living, not only enhancing operational efficiency but also promoting long-term sustainability and resilience.

The integration of AI into urban planning and development processes presents both unprecedented opportunities and significant challenges. AI offers transformative potential to enhance efficiency, predict urban growth, and optimise resource allocation, yet it also necessitates a paradigm shift in methodologies. Those involved in planning, designing, and developing our cities must be ready to embrace a new generation of tools and techniques that leverage AI to analyse vast datasets, simulate dynamic urban systems, and make data-driven decisions. This transition promises to revolutionise urban design and management, making cities smarter, more resilient, and more responsive to the needs of their inhabitants. The importance of adopting AI in these processes cannot be overstated; it is a strategic imperative for creating sustainable urban futures in an increasingly complex world.

To be prepared for this technological change and innovation calls for a forward-thinking mindset and a commitment to embracing and leveraging these advancements. This involves continuous learning and adaptation, ensuring that planners, designers, and developers are equipped with the skills and knowledge required to harness AI effectively. A collaborative environment where interdisciplinary teams can work together to integrate AI solutions into urban development projects is critical.

By embracing these changes, we can enhance our abilities to address complex urban challenges, improve public services, and create more inclusive and equitable cities. Ultimately, the successful adoption of AI in urban planning and development will depend on the willingness to be proactive and be prepared to communicate this new paradigm to stakeholders and policymakers—especially in the face of fear and uncertainty. We need to be prepared.

From an urban planning perspective, today's urban areas face an array of complex challenges that include, but are not limited to, rapid population growth, climate change, traffic congestion, environmental sustainability, and the need for improved public services. Traditional urban planning methods often fall short in addressing these multifaceted issues. Herein lies the potential of AI. By leveraging AI's capabilities in advanced data analytics, predictive modelling, and automation, urban planners can devise innovative solutions to these challenges, enhancing the efficiency, sustainability, and resilience of urban systems.

However, the integration of AI into urban planning is not without its challenges. Public perception and acceptance of AI technologies play a critical role in their successful implementation. Concerns about privacy, security, job displacement, and the ethical implications of AI decisions are prevalent and must be addressed to build trust and ensure responsible development and deployment of these technologies.

Urban Artificial Intelligence: A Guidebook for Understanding Perceptions and Ethics is divided into two sections, each addressing critical aspects of AI in urban contexts. The first section, 'Perceptions on Urban Artificial Intelligence', focuses on how AI is perceived across various urban sectors. It examines public attitudes, the factors shaping these perceptions, and their implications for policymakers and urban planners. Understanding these perceptions is vital for creating policies that align with public expectations and facilitate the seamless integration of AI into urban environments. In this section, Yigitcanlar explores how different urban sectors, such as urban planning, local government services, disaster management, and the construction industry, perceive AI. By analysing social media discussions, survey responses, and case studies, he uncovers the prevailing sentiments and concerns surrounding AI. These insights are invaluable for policymakers and urban planners, helping them to design AI-driven initiatives that are both effective and publicly accepted.

The second section, 'Responsible Urban Artificial Intelligence', focuses on the ethical considerations and responsible implementation of AI technologies. Here, Yigitcanlar provides frameworks and strategies to balance the potential risks and benefits of AI, emphasising the importance of transparency, inclusivity, and robust ethical standards. This section offers practical guidance for ensuring that AI technologies contribute positively to urban development while mitigating ethical concerns. In exploring the ethical dimensions of AI, Yigitcanlar addresses key issues such as data privacy, algorithmic bias, accountability, and the social implications of AI-driven decisions. He presents a conceptual framework for responsible AI innovation, outlining principles and practices that promote ethical AI development. This framework serves as a critical tool for urban planners, developers, and policymakers, guiding them in implementing AI technologies that uphold ethical standards and foster public trust.

This book emphasises the importance of engaging diverse stakeholders in the AI implementation process. Inclusivity is crucial for ensuring that AI technologies address the needs and concerns of all community members, particularly marginalised and vulnerable groups. Yigitcanlar advocates for participatory approaches that involve the public in decision making, thereby enhancing the legitimacy and acceptance of AI-driven initiatives. This book provides readers with the knowledge and insights needed to navigate the complexities of AI in an array of urban circumstances and situations, ultimately contributing to the development of resilient and inclusive urban communities.

As a leading thinker on these issues, Tan Yigitcanlar's expertise and forward-looking perspective provide the foundation for this comprehensive guide. His work not only highlights the potential of AI in urban planning but also underscores the importance of addressing ethical and social considerations. *Urban Artificial Intelligence: A Guidebook for Understanding Perceptions and Ethics* is a vital contribution to the field of urban planning and development. It provides a roadmap for integrating AI technologies into urban contexts in a manner that is both effective and ethically sound. As we move forward into an era increasingly defined by AI, this book will serve as an essential guide for policymakers, urban planners, and developers, helping them to navigate the challenges and opportunities presented by AI and to build cities that are smarter, more sustainable, and more equitable for all.

Professor Thomas W. Sanchez
Texas A&M University, USA

Preface

As the world moves rapidly towards an era dominated by technological advancements, the role of artificial intelligence (AI) has become increasingly significant. AI's transformative power is evident across a myriad of sectors, revolutionising how we operate and innovate in fields such as healthcare, finance, agriculture, and space exploration. For instance, AI is used to diagnose diseases with unprecedented accuracy, optimise financial trading algorithms, enhance crop yields through precision agriculture, and even assist in navigating and analysing data from space missions. The pervasive influence of AI in these areas highlights its potential to drive efficiency, accuracy, and innovation.

Yet, despite its far-reaching impact, the integration of AI into urban planning and development remains relatively underexplored. This is surprising given the immense potential AI holds to revolutionise how we design, manage, and experience our cities. Urban areas are facing increasingly complex challenges, including rapid population growth, traffic congestion, environmental sustainability, and the need for enhanced public services. Traditional approaches to urban planning often struggle to address these multifaceted issues effectively, underscoring the need for innovative solutions.

AI can offer transformative approaches to urban planning and development by providing advanced data analytics, predictive modelling, and automation capabilities. For example, AI can analyse vast amounts of data from various sources, such as sensors, cameras, and social media, to offer real-time insights into traffic patterns, energy consumption, and public safety. This data-driven approach can lead to more efficient traffic management, reduced energy waste, and improved emergency response times.

Moreover, AI can facilitate more sustainable urban development. By optimising resource allocation and energy use, AI can help cities reduce their carbon footprint and promote environmental sustainability.

Smart grids, waste management systems, and water supply networks can all benefit from AI-driven optimisation, leading to more resilient and sustainable urban infrastructures. AI also holds promise for enhancing the quality of life in urban areas. Smart city initiatives, powered by AI, can provide residents with improved services, from personalised healthcare and education to enhanced public transportation and security. AI can enable more responsive and adaptive urban environments, where city services are tailored to the needs of the inhabitants, thus fostering a higher quality of life.

In summary, while AI has already made significant strides in various sectors, its application in urban planning and development is an area ripe with potential. By harnessing AI's capabilities, we can reimagine and reshape our cities to be more efficient, sustainable, and liveable, addressing the complex challenges of urbanisation in innovative and impactful ways.

Nevertheless, despite the rapid advancements and growing implementation of AI in urban settings, there remains a significant gap in understanding how urban AI is perceived by the public and professionals. This limited knowledge poses a challenge, as perceptions greatly influence the acceptance and effectiveness of AI technologies. Public concerns often revolve around privacy, security, and the potential loss of jobs, while professionals might grapple with the ethical implications and unintended consequences of deploying AI in urban environments. Therefore, it is critical to adopt an ethical and responsible approach to urban AI. This involves transparent communication, robust regulatory frameworks, and inclusive decision-making processes that consider the diverse perspectives and needs of all stakeholders. Ensuring ethical standards and fostering trust are essential for the successful integration of AI in urban development, ultimately leading to smarter, more equitable, and sustainable cities.

Against this backdrop, the second volume of the *Urban Artificial Intelligence* book, entitled *Urban Artificial Intelligence: A Guidebook for Understanding Perceptions and Ethics*, thoroughly explores and elucidates the key elements of the urban AI phenomenon. This volume delves into the perceptions and ethical considerations surrounding urban AI, providing invaluable insights and understanding. Comprising eight chapters, this volume offers a comprehensive examination of the crucial aspects of AI in urban contexts, including its applications, public perceptions, and ethical implications. By addressing these critical areas, this book aims to fill an essential gap in the existing literature, equipping readers with a well-rounded perspective on the role of AI in shaping modern urban environments. This second volume is accompanied with the first volume entitled, *Urban Artificial Intelligence: A Guidebook for Understanding Concepts and Technologies.*

Part 1: Perceptions on Urban Artificial Intelligence
The first part of this guidebook delves into the varied perceptions of AI within different urban sectors. It presents a detailed analysis of how people view AI's role in urban planning, local government services, disaster management, and the construction industry. These insights are crucial for policymakers, urban planners, and developers aiming to integrate AI into their projects and policies effectively.

Chapter 1: Perceptions on Urban Artificial Intelligence in Urban Planning and Development
This chapter explores how AI technologies are perceived and utilised within urban planning. By analysing Twitter (now rebranded as X) posts from Australia, this chapter highlights the leading AI technologies and their applications, revealing a strong community focus on digital transformation and sustainability. This chapter underscores the importance of big data, automation, and robotics in driving urban development and enhancing city sustainability.

Chapter 2: Perceptions on Artificial Intelligence in Local Government Services
This chapter examines public attitudes towards AI in local government operations. Through an online survey of respondents from major cities, this chapter identifies key factors influencing public perceptions, such as ease of use and perceived usefulness. The findings suggest that while Australians generally have a positive view of AI in local government, cultural and contextual differences influence these perceptions, as seen in the comparison with Hong Kong residents. This chapter provides valuable insights for local authorities to shape AI policies that meet public expectations and improve service delivery.

Chapter 3: Perceptions on Artificial Intelligence in Urban Disaster Management
This chapter addresses the role of AI in managing urban disasters. It highlights the varied perceptions among different demographic groups, revealing that younger people and those with higher education levels are more enthusiastic about AI-driven disaster management solutions.

By understanding these demographic differences, policymakers can tailor AI integration strategies to enhance disaster management practices effectively.

Chapter 4: Perceptions on Artificial Intelligence in the Construction Industry
This chapter focuses on the construction sector, where AI adoption has been slow despite its potential benefits. Using social media analytics, this chapter identifies prevalent AI technologies and public sentiments towards AI in the construction industry. The findings highlight opportunities for timesaving and innovation, as well as challenges such as project risk and data security. This chapter offers guidance for construction professionals to navigate AI adoption and leverage its benefits for industry advancement.

Part 2: Responsible Urban Artificial Intelligence
The second part of this book shifts focus to the ethical implications and responsible implementation of AI in urban settings. It provides frameworks and strategies to ensure that AI technologies contribute positively to urban development while mitigating potential risks and ethical concerns.

Chapter 5: Responsible Local Government Artificial Intelligence
This chapter discusses the principles of responsible urban innovation, emphasising the need for careful evaluation of AI systems deployed by local governments. It introduces a conceptual framework designed to balance the costs, benefits, risks, and impacts of AI technologies, guiding urban policymakers in promoting responsible and sustainable urban development.

Chapter 6: Conceptual Framework of Responsible Urban Artificial Intelligence
This chapter elaborates on the concept of responsible innovation and technology (RIT), which encompasses Responsible AI. This chapter provides a comprehensive review of RIT characteristics and introduces a framework to guide the design and implementation of AI technologies that align with societal values and human well-being. The framework serves as a valuable tool for governments, companies, and stakeholders to navigate the ethical challenges of technological advancement.

Chapter 7: Assessment Framework of Responsible Urban Artificial Intelligence
This chapter reviews the responsible innovation and technology (RIT) policies of leading tech companies, highlighting the central themes of trustworthiness and acceptability. It proposes an assessment framework to evaluate the commitment of tech companies to RIT practices, promoting the development of technology-driven societies that prioritise human and social well-being. This framework is essential for ensuring that AI technologies are implemented responsibly and sustainably.

Chapter 8: Artificial Intelligence and Sustainable Development Goals
This chapter explores AI's role in achieving sustainable development within the construction industry. This chapter maps AI applications, impacts, adoption challenges, and best practices, demonstrating how AI can advance specific Sustainable Development Goals (SDGs). It emphasises the importance of ethical considerations and specialised training to fully harness AI's potential and overcome adoption challenges, ensuring that AI contributes positively to sustainability in construction.

Given the wealth of knowledge it offers, *Urban Artificial Intelligence: A Guidebook for Understanding Perceptions and Ethics* is an essential resource for anyone involved in urban planning, development, and governance. It provides a comprehensive understanding of AI's potential applications, public perceptions, and ethical considerations in urban contexts. By bridging the knowledge gaps and offering practical frameworks, this book aims to guide policymakers, planners, and developers in integrating AI responsibly and effectively into urban environments, ultimately fostering sustainable and equitable urban development.

It is important to note that the second volume of this book, *Urban Artificial Intelligence: A Guidebook for Understanding Perceptions and Ethics*, is complemented by the first volume, *Urban Artificial Intelligence: A Guidebook for Understanding Concepts and Technologies*. To gain a comprehensive understanding of Urban Artificial Intelligence, it is recommended to read both volumes together. This will provide a complete picture, encompassing both the technical aspects and the ethical and perceptual dimensions of AI in urban settings.

Author Biography

Tan Yigitcanlar is a distinguished Australian researcher known for his significant contributions to urban studies and planning. He holds the position of Professor at the Queensland University of Technology's School of Architecture and Built Environment. In addition, he is an Honorary Professor at the Federal University of Santa Catarina in Brazil, Director of the Australia-Brazil Smart City Research and Practice Network, Lead of the QUT Smart City Research Group, Co-Director of the QUT City 4.0 Lab, and Director of the QUT Urban AI Hub. He is also a member of the Australian Research Council College of Experts.

With a career spanning over three decades, Tan's work has involved research, teaching, training, and capacity building in urban studies and planning at prestigious universities in Australia, Brazil, Finland, Japan, and Turkey. His research is dedicated to tackling contemporary challenges in urban planning and development, encompassing economic, societal, spatial, governance, and technology-related issues. At the heart of his work is a focus on smart and sustainable urbanisation.

Tan has provided research consultancy services to various levels of government, international corporations, and non-governmental organisations in Australia and abroad. His expertise has helped these entities develop strategies, enhance resilience, and prepare for emerging disruptive conditions. He serves as the lead Editor-in-Chief of the 'Elsevier Smart

Cities Book Series' and holds senior editorial positions in 12 high-impact journals. He chaired the annual 'Knowledge Cities World Summit' series from 2007 to 2019, organising 12 international conferences in various locations around the world. He has delivered over 75 keynote and invited talks at prestigious international academic conferences and national industry events.

Tan's research findings have been extensively disseminated, with over 300 articles published in high-impact journals and 30 key reference books published by esteemed international publishing houses. His work has significantly influenced urban policy, practice, and research internationally, with over 25,000 citations and an h-index of over 90. He is ranked #1 in Australia and top 10 worldwide in urban and regional planning according to the 2023 Science-wide Author Databases of Standardised Citation Indicators. He was also recognised as an 'Australian Research Superstar' in the Social Sciences Category by The Australian's 2020 Research Special Report and named Australia's Social Sciences 'Research Field Leader' for Urban Studies and Planning in the 2024 edition of The Australian Research Magazine. He is acknowledged as one of the top 1% of scientists worldwide by Clarivate in 2023.

PART 1

Perceptions on Urban Artificial Intelligence

This part of the book aims to provide an in-depth understanding of public perceptions regarding urban artificial intelligence. It captures and discusses a range of viewpoints on various applications of AI in urban settings. Topics covered include the role of urban artificial intelligence in planning and development, its integration into local government services, and its significance in urban disaster management. Additionally, it explores the impact of AI in the construction industry. Through thorough analysis and diverse perspectives, this part sheds light on how the public views and interacts with AI-driven innovations in their urban environments.

DOI: 10.1201/9781003521440-1

Perceptions on Urban Artificial Intelligence in Urban Planning and Development

1 INTRODUCTION AND BACKGROUND

Cities provide tangible and intangible infrastructures and platforms from which individuals can self-actualise, and consequently create goods and services that further enhance the standards of living of the broader population [1]. The city, therefore, has an overarching responsibility for the impact that its hard and soft attributes have on its inhabitants, rendering it an institution that must guarantee the efficiency and reliability of its urban matrix [2]. As part of the constant necessity to boost development and economic growth, cities are leveraging the benefits of technological advancements and implementing the latest artificial intelligence (AI) technologies; the aim is to exponentially increase sustainability through the efficient use of energy and resources [3–5].

Technologies that leverage AI are currently being utilised in many cities across the globe, for example, in Amsterdam, London, San Francisco, Stockholm, Singapore, Hong Kong, Vienna, and Toronto, to optimise their urban functionality and service efficiency [6,7]. For instance, the smart grid initiative acts as one of the foundations for the utilisation of AI in

DOI: 10.1201/9781003521440-2

cities; it facilitates spatial navigation in the form of interactive and automated systems that use data processing technology to reveal the dynamics of the urban grid. In this way, digitalisation has enabled cities to identify specific needs, leading to increased productivity and economic performance [2]. Subsequently, AI offers an opportunity for enhanced city governance [8]; AI concepts and technologies can influence and improve the way the city serves its citizens and provide everyone with the desired and responsible urban futures [9,10].

One of the most outstanding factors that have led cities to becoming smart is an inherent necessity to adapt to environmentally friendly initiatives [11]. The unprecedented reality of global warming has made it necessary to restructure the use of resources, where smart technologies are required to assist in the homogenous distribution of resources, resulting in the reduction in the carbon footprint of cities [12,13]. Accordingly, smart environment technologies are applied in cities—they are generally AI-driven systems, which come in the form of smart traffic lights, noise prediction, air quality prediction, and foot traffic as well as car traffic prediction faculties. All this is made possible thanks to the big data technology, which facilitates data capture [2]. This results in hyper-accurate urban data, which permits for highly productive interventions, enabling cities to use their resources sustainably [14].

It is evident that cities, so far, have been reaping significant benefits from utilising AI to design and implement city management strategies [15]. Nevertheless, it is estimated that a knowledge gap remains, specifically in the way the public perceives the implementation of those technologies, and how they feel about the extensive application of AI in their cities [16]. A solid understanding of the public's perceptions about AI concepts and technologies in their cities would inform policymakers of the public sentiment regarding the different aspects of AI [17,18]. Consequently, governing bodies would be better prepared to respond to the public's demands and to adopt urban AI technology and applications [19]. It is, hence, necessary to explore the ways in which AI directly interacts with individuals and to dissect the way the different AI instruments can potentially benefit or impair an individual or community.

The discourse on AI has become prevalent in Australia in recent times [20–23]. In particular, the arrival of autonomous vehicles (AV), robotics, machine learning (ML), Internet-of-Things (IoT), block chain, augmented reality (AR), and virtual reality (VR) technologies has resulted in a widespread debate on the future of AI in Australian cities; citizens have begun to contemplate how big data, 5G, surveillance, and cyber-security will impact their daily life. This chapter thus focuses on the public's perception of AI concepts and technologies in urban planning and development, in the context of Australian cities. The methodological approach of this study employs the social media analytics method and conducts sentiment and content analyses of location-based Twitter messages from Australia.

Following this introduction, Section 2 of the chapter provides a literature background on the topic of investigation. Then, Section 3 introduces the methodological approach of the study. Next, Section 4 presents the results of the analysis. Afterwards, Section 5 discusses the study findings and generated insights. Lastly, Section 6 concludes the chapter with the study highlights, final remarks and future research directions.

2 LITERATURE BACKGROUND

2.1 Artificial Intelligence

AI is one of the most disruptive technologies of our time [24]. AI can be defined as machines or computers that mimic cognitive functions that humans associate with the human mind, such as learning and problem solving [25]. AI is a branch of computer science that perceives its environment and acts to maximise its chances of success. Furthermore, AI is capable of learning from past experiences, making reasoned decisions, and responding rapidly [26]. The scientific goal of AI researchers, hence, is to understand intelligence by building computer programs that exhibit symbolic inference or reasoning. For instance, according to [27], the four main components of AI are:

- Expert system: Handles the situation under examination as an expert and yields the desired or expected performance.

- Heuristic problem solving: Consists of evaluating a small range of solutions and may involve some guesswork to find near-optimal solutions.

- Natural language processing: Enables communication between human and machine in natural language.

- Computer vision: Generates the ability to recognise shapes and features automatically.

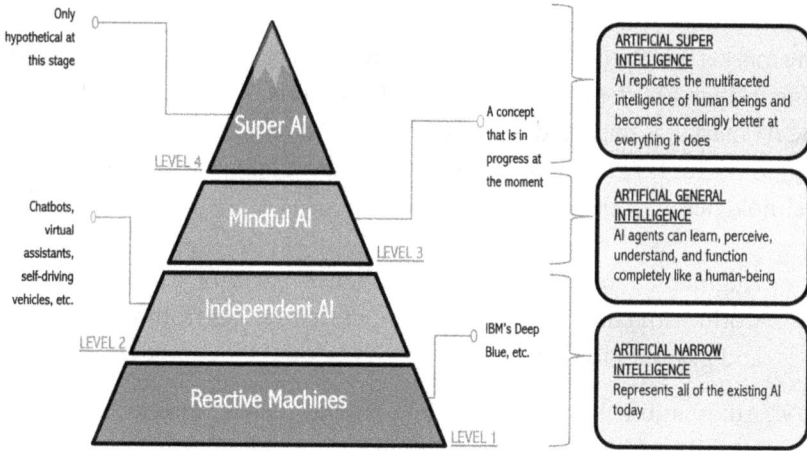

FIGURE 1.1   Levels of artificial intelligence, derived from [33].

AI is already being used today in numerous areas, including but not limited to marketing, finance, agriculture, healthcare, security, robotics, transport, and artificial creativity and manufacturing [28,29]. In recent years, AI has become an integral part of urban services, as it offers efficient and effective platforms and smart governance opportunities [30,31]. There are several types of AI hardware (e.g., machines/robots) and software (e.g., algorithms) that each have different capabilities at different levels of development [32]. These levels are illustrated in Figure 1.1 and described below.

Level 1 refers to 'Reactive machines', which are programmed to undertake a single task and carry it out perfectly. However, this type of machine cannot learn further as it reacts to human input, rather than planning and pursuing its own original agenda [34]. Level 2 is the 'Independent AI', as human actions do not dictate all the software actions; after some human lead, AI learns and improves its own ability to perform a given task. AI levels 1 and 2 are referred to as 'artificially narrow intelligence', and they are currently being used in practice and are commonly applied in urban planning and development. Level 3 AI is called 'Mindful AI' and is capable of thinking thanks to a consciousness of its own and multiple domains of knowledge. This level of AI is currently at the stage of conceptual progress. Lastly, Level 4 is called 'Super AI', as it does anything and everything better than humans do [35]. This level of AI is currently at the hypothetical stage.

2.2 Artificial Intelligence Technologies

The market for AI technologies is growing as large corporations are beginning to increase their investment in AI solutions. According to IDC [36], the AI market is expected to grow from USD 8 billion in 2016 to USD 57.6 billion by 2021. A study conducted by Cearley et al. [37] identified the AI technologies with the most growth so far:

- Augmented reality: Designers enhance parts of a user's physical world with computer generated input that ranges from sound, video, and graphics to GPS overlay.

- Automation: Software that follows the instructions or workflows established by individuals for simple and repetitive tasks.

- Big data: Structured and unstructured data that are collected by an organisation and can be mined for information extraction and used in ML projects, predictive modelling, and other advanced analytical applications.

- Biometrics: This enables natural interactions between humans and machines through image, touch recognition, speech, and body language.

- Block chain: This is a public electronic ledger that can openly share information with many disparate users to create an unalterable record of transactions.

- Deep learning platforms: These are ML systems that consist of artificial networks with multiple layers; deep learning can recognise and classifying patterns.

- Digital twins: These are digital representations that simulate a real-life object through the law of physics, material properties, virtualised sensors, and causality.

- Machine learning platforms: These provide algorithms, application programming interfaces (APIs), development and training toolkits.

- Natural language generation: This produces text from computer data, and it is currently being used in customer service, report generation, and summarising business intelligence insights.

- Robotics: This refers to the use of machines to perform tasks that are traditionally completed by humans.

- Virtual agents: These are advanced systems that can network with humans.

Thanks to the extensive research and advances in the field of AI, it is expected that by the end of 2035, our society and cities will transition from the current AI with complex machine language towards AI that will likely be fully understood by humans [38].

2.3 Artificial Intelligence Application Areas in Urban Planning and Development

The urban planning environment is increasingly turning to specialised technologies to address uses related to sustainability, society, security, transportation, infrastructure, and governance [39]. The term urban artificial intelligence refers to AI that is embodied in urban spaces and infrastructure. These technologies are turning cities into autonomous entities that operate in an unsupervised manner [40]. The emerging concept of smart cities has promoted the development of IoT and through it the incorporation of sensors and big data [41,42]. The surge in data brings new possibilities to design, management, and the economy. Furthermore, IoT supports increased connectivity that leads to the generation of data and its subsequent capture, analysis, and distribution, contributing to better smart city development [43,44].

AI can significantly contribute to planning by binding frameworks that encompass key dimensions, such as culture, metabolism, and governance, ensuring their achievement. Data can now be sourced from numerous neighbourhoods to gain a more holistic understanding of the urban fabric. This allows planners and policymakers to shift from closed systems (interlinked urban elements) to an open, fragmented, peri-urban fabric that has tangible impacts on density fragmentation, cohesion, and compactness [43].

AI-based data processing can help offer better prevision of livability, through the creation of a clean, healthy, and conducive environment for people to live and work in, overcoming the urban challenges of pollution and congestion [45]. Additionally, urban areas that leverage AI are enabling infrastructures that attract higher economic returns by offering connectivity, energy, and computing capabilities that support globally competitive jobs, as well as talented and knowledgeable workers [46].

3 RESEARCH DESIGN

3.1 Case Study

This chapter follows a case study approach. The conducted case study focused on the status of AI and on the identification of AI applications in urban planning and development activities in Australian cities. The reasons behind this selection include: (a) some of Australia's major cities are among the leading smart cities in the world. These cities are successfully adopting smart urban technologies that also include AI-related technologies and applications [47]; (b) Australia is among the countries that developed a national AI strategy and roadmap, meaning that AI uptake in cities is planned rather than done on an ad hoc basis [48]; and (c) social media use is highly popular in the country, making it a source of firsthand information regarding the Australian society, including its perception of urban technologies. About 66% of Australian Internet users use social media daily and around 34% use social media more than five times a day [49]. Among the 66%, 19% of them have accessed Twitter and one-third of them tweet daily [50]; (d) although large amounts of social media data are available regarding AI in Australia, very few studies have been conducted to draw conclusions from these big data. For instance, Yigitcanlar et al. [47] performed social media analytics on Twitter data to evaluate smart city concepts and technologies across Australia, such as AI, big data, 5G, IoT, AVs, and robotics. However, further investigations are required to capture and evaluate the public's perception of AI and its urban and social implications in Australia.

3.2 Methodology

Instead of using a traditional data collection method, the methodological approach applied in this study employs a contemporary method—i.e., social media analysis. As social media are ever-evolving platforms, where people can share thoughts and opinions, they have become a new source of qualitative data [51]. This data collection method has started to be used as the main data source in large number of studies. Social media have offered an opportunity to engage with larger groups of people, in an unbiased setting [52]. In addition, researchers can engage with people from broader geographic areas with the help of the location of social media users, which is tagged in their posts [53].

A geo-Twitter analysis has proven to be a very successful data collection method [54]; hence, this method was used in this study. A geo-Twitter analysis increases efficiency in analysing a large number of shared thoughts and opinions [55], and real-time information on ongoing social issues [56]. For instance, social media analytics has contributed to safeguarding Australian cities and their residents from the coronavirus outbreak (COVID-19) in 2020 [57].

Initially, sentiment and content analyses were completed for the total number of location-based Twitter messages—a.k.a. tweets. To do this, the original dataset obtained (from the QUT Digital Observatory—https:// www.qut.edu.au/institute-for-future-environments/facilities/digital-observatory/digital-observatory-databank) with 98,534 tweets was filtered down to 11,236 tweets. This was done using five data filtering processes, which included frequency analysis, location, date, bot, and relevance filters.

First, we selected the most recent one-year period for the analysis—hence, all tweets outside of Australia and not within the 10 June 2019 to 10 June 2020-time period were removed from this dataset. The reason for only selecting one-year period was two-fold. The first one is to capture the latest trends, as in the technology domain, the development is fast and public perceptions change rather rapidly. The second is easing the analysis tasks, as there have been around 50,000 to 100,000 tweets on AI shared annually in Australia during the last five years. The bot filter removed tweet repetitions with the program 'NVivo'—a content analysis automatic software system. For identifying the tweets on themes associated with AI applications in urban planning and development, NVivo was also used.

Second, word frequency analysis was conducted using NVivo, with the aim of identifying popular themes, concepts, and technologies.

Next, a word co-occurrence analysis identified the tweets that discussed both AI technologies and urban planning and development-related concepts (or AI application areas) in a single Twitter message. For this analysis, NVivo software was employed.

Fourth, a spatial analysis was conducted to complement the content analysis, which included the tweets being separated by location and collected to help categorise themes, concepts, and technologies based on these locations. This created an overview of the most popular themes, concepts, and technologies for each state/territory in Australia. ArcGIS Pro software was used for visualising the spatial information.

Then, the relevance filter was completed manually and was used to identify tweets that were related to or discussed AI technologies and urban planning and development-related concepts, noting key sentiment words. These words were then classified on a scale of one to three, to measure the sensitivity. This scale was read as: 1 = positive sentiment, 2 = negative sentiment, and 3 = neutral sentiment. These words were then pre-processed in the program 'Weka', which created a dataset that further analysed the word content. The sensitivity of these specific words was showcased in a 'Random tree' classification type.

Finally, a network analysis has been created to present the relationships between AI themes, concepts, and technologies, presenting the most popular relationships more centrally. In this analysis, nodes (themes, concepts, and technologies) and edges (relationships between these themes, concepts, and technologies) were used as the key elements of the network. These assist in understanding the network typology, which represents the arrangement of nodes and edges on the basis of the co-occurrence of the themes, concepts, and technologies found in the tweets. For this analysis Gephi software was employed.

4 ANALYSIS AND RESULTS

4.1 General Observations

From the final dataset of 11,262 tweets, 52% (n = 5850) were from NSW, 15% (n = 1704) were from VIC, 12% (n = 1349) were from SA, 10% (n = 1124) were from WA, 7% (n = 787) were from QLD, 2% (n = 260) were from TAS, and 1% (n = 133) were from ACT (Figure 1.2). Compared

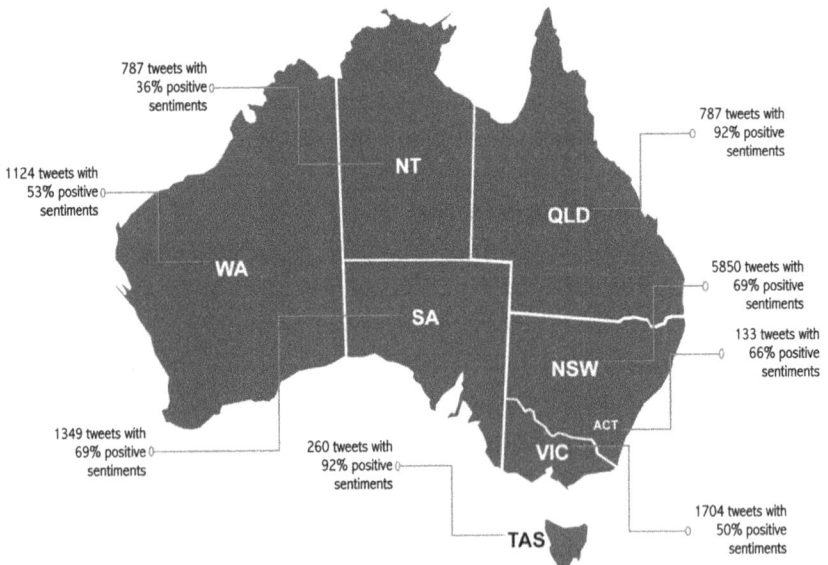

FIGURE 1.2 Tweet numbers and positive sentiment percentages by states and territories.

to the other states and territories, the number of tweets received from NT was recorded as 55, which represented a negligible percentage of 0%. This reveals the low interests among the NT community regarding AI-related applications. A wide range of hashtags were used in the circulated tweets. Among them, hashtags such as #AI, #IoT, #digital, #robotics, #future, #technology, #automation, #bigdata, #VR, #AR, #crypto, #bitcoin, #machine learning, and #ML were the most popular ones.

4.2 Community Sentiments

Out of analysed 11,262 tweets, 66% (n = 7475) of them carried positive sentiments related to AI technologies and applications within the context of urban planning and development. About 17% (n = 1935) had negative sentiments towards AI technologies. Around 16% (n = 1852) of the tweets had neutral sentiments, where such tweets used only a set of hashtags to express their ideas rather than elaborative comments (Table 1.1).

From the tweets originating from NSW and SA, n = 5850 and 1349, respectively, 69% of them contained positive sentiments—4022 and 932 tweets from NSW and SA were positive in nature, respectively. While 14% of tweets were negative in NSW, this figure was only 4% in SA. Out of 133 tweets originating from ACT, 66% (n = 88) were positive and 17% (n = 22) were negative. VIC had the second-highest number of tweets (n = 1704), and among them 50% (n = 854) were positive and 42% (n = 715) were negative. From the 1124 tweets originating from WA, 53% (n = 599) were positive, and 22% (n = 251) were negative in nature. NT had the lower number of tweets related to AI; among them, 36% (n = 20) were positive and only 11% (n = 6) were negative. Significantly, 53% (n = 29) of tweets from NT were neutral. Example tweets for each sentiment category are given in Table 1.2.

4.3 Artificial Intelligence Technologies

Using a word frequency analysis technique, 15 key AI-related technologies were derived from the collected tweets (Figure 1.3 and Table 1.3). These technologies are 'robotics' (n = 3055), 'drones' (n = 1943), 'automation' (n = 717), 'digital twins' (n = 337), 'block chain' (n = 263), 'machine learning' (n = 236), 'digital networks' (n = 207), 'digital currency' (n = 192), '5G technology' (n = 178), 'big data' (n = 154), 'augmented reality' (n = 124), '3D printing' (n = 101), 'virtual reality' (n = 86), 'telephony' (n = 13), and 'chatbots' (n = 11).

TABLE 1.1 Tweet sentiments in percentages per state/territory

	Queensland (QLD)	Tasmania (TAS)	New South Wales (NSW)	South Australia (SA)	Australian Capital Territory (ACT)	Victoria (VIC)	Western Australia (WA)	Northern Territory (NT)	Australia
Positive sentiments	92%	92%	69%	69%	66%	50%	53%	36%	66%
Negative sentiments	5%	3%	14%	4%	17%	42%	23%	11%	15%
Neutral sentiments	3%	5%	17%	27%	17%	8%	24%	53%	19%
Total	100%	100%	100%	100%	100%	100%	100%	100%	100%

TABLE 1.2 Example tweets for three sentiment categories

Date and Time	State	Tweet	Sentiment
12 August 2019 21:03	NSW	#drones are changing the meaning of "many hands make light work" for these farmers. The farm of the future will be a technology-enabled farm. We should be happy with that.	Positive
23 July 2019 8:01	VIC	In other words, just move to a city or large town. People are losing their jobs daily. Robots and technology aren't consumers of goods and services.	Negative
15 March 2019 9:42	QLD	Automation has much to offer #IoT #AutonomousVehicles #IoT #SmartCity #smartgrid #Healthcare, but also much to take away from us #Cyberecurity #jobloss #Disruption	Neutral

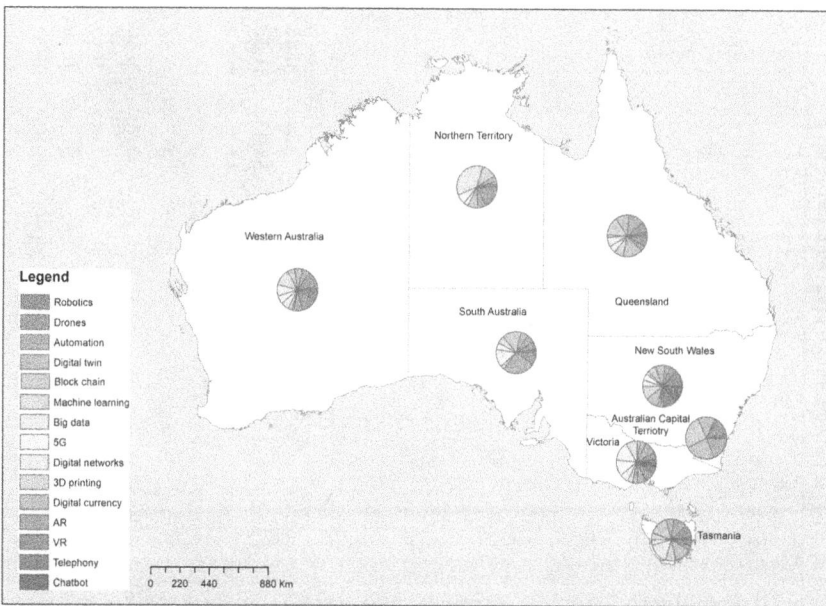

FIGURE 1.3   Distribution of tweets by artificial intelligence (AI)-related technologies per state/territory.

Nonetheless, the popularity of each technology differs from one state or territory to another. For instance, in VIC, there are more tweets about 'automation' ($n = 317$) compared to NSW ($n = 180$). Conversely, 'robotics' is around seven times more popular ($n = 2,328$) in NSW than in VIC ($n = 340$). Likewise, tweets from different states and territories had more

TABLE 1.3 Distribution of tweets by AI-related technologies per state/territory

	Robotics	Drones	Automation	Digital Twins	Block Chain	Machine Learning	Big Data	5G	Digital Networks	3D Printing	Digital Currency	AR	VR	Telephony	Chatbot
NSW	2328	694	180	164	104	40	0	54	63	63	116	18	14	8	9
VIC	340	499	317	72	31	105	107	104	65	21	36	40	34	3	2
WA	157	433	74	40	28	55	25	20	24	6	5	16	19	2	0
SA	128	146	62	26	54	19	11	0	38	4	2	43	11	0	0
QLD	77	96	62	24	26	3	8	0	8	5	21	3	5	0	0
TAS	13	48	19	5	10	3	2	0	8	2	6	1	1	0	0
ACT	9	22	1	4	6	0	0	0	0	0	4	2	0	0	0
NT	3	5	2	2	4	11	1	0	1	0	2	1	2	0	0
Total	3055	1943	717	337	263	236	154	178	207	101	192	124	86	13	11
NSW	76.2	35.72	25.1	48.66	39.54	16.95	0	30.34	30.43	62.38	60.42	14.52	16.28	61.54	81.82
VIC	11.13	25.68	44.21	21.36	11.79	44.49	69.48	58.43	31.42	20.79	18.75	32.26	39.53	23.08	18.18
WA	5.14	22.29	10.32	11.87	10.65	23.31	16.23	11.23	11.59	5.94	2.6	12.89	22.09	15.38	0
SA	4.19	7.51	8.65	7.72	20.53	8.05	7.15	0	18.36	3.96	1.04	34.68	12.79	0	0
QLD	2.52	4.94	8.65	7.13	9.89	1.27	5.19	0	3.86	4.95	10.94	2.42	5.82	0	0
TAS	0.43	2.47	2.65	1.48	3.8	1.27	1.3	0	3.86	1.98	3.13	0.81	1.16	0	0
ACT	0.29	1.13	0.14	1.19	2.28	0	0	0	0	0	2.08	1.61	0	0	0
NT	0.1	0.26	0.28	0.59	1.52	4.66	0.65	0	0.48	0	1.04	0.81	2.33	0	0
Total	100%	100%	100%	100%	100%	100%	100%	100%	100%	100%	100%	100%	100%	100%	100%

TABLE 1.4 Example tweets for AI-related technologies

Technology	Data and Time	State	Tweet	Sentiment
Robotics	18 June 2020 2:14 p.m.	NSW	Boston Dynamics starts selling its Spot robot your own pet robot dog for $74,500 #Robotics #ArtificialIntelligence #robotpetdog #exciting	Positive
Drones	10 January 2020 7:22 p.m.	QLD	These drones plant thousands of trees ðŸCE³ every day. Shooting the seeds into the ground. Huge opportunity for massive global tree planting!	Positive
Automation	18 December 2019 9:18 p.m.	VIC	As automation technology becomes more ingrained into the workplace, employee training becomes critical to direct employees' time towards higher-value work. The results of this survey are fascinating #SkillsGap #DigitalSkills	Positive
Digital twins	5 February 2020 1:38 p.m.	VIC	See my virtual replica. Experience the difference and the excitements #digitaltwin	Positive
Block chain	16 July 2019 8:35 p.m.	TAS	A place with abundant renewable generation such as wind, pumped hydro and cool climate would be perfect. #TAS and @HydroTasmania has all three, combined with a block chain based electricity marketplace and we can use the exist #futuristic #sustainableworld	Positive
Machine learning	5 February 2020 11:03 a.m.	ACT	Really excited for this one—our contribution to the discussion on predicting performances based on training load. Plus, an extra section using machine learning to combine the data from multiple athletes to predict outcomes for one. All done using #rstats h	Positive
Network	22 January 2019 5:34 a.m.	SA	A Hacker-Proof Quantum Network Is Hiding In This City Tunnel. We all are at a big risk	Negative

(Continued)

TABLE 1.4 Continued

Technology	Data and Time	State	Tweet	Sentiment
Digital currency	9 January 2020 11:49 a.m.	NSW	The latest The Bitcoin Profits Daily! https://t.co/qTJJaBkGEo Thanks to @EllenDibble @linasantlinijos #cryptocurrency #cryptocurrency #enjoytheprofit	Positive
5G technology	9 December 2019 2:27 a.m.	TAS	What has #5G got to do with helping reduce road traffic accidents? #EmergingTech #AI #ML #IoT #SelfDrivingCars #SmartCities #SelfDriving #Driverless #AutonomousVehicles #SelfDrivingCars #autonomousdriving #Automotive #selfDrivingCar #4IR #safercities	Positive
Big data	2 April 2019 11:09 a.m.	WA	Training doctors while using #AugmentedReality via @futurism|#AR #VR #Healthcare #InternetofThings #IoT #SmartCity #SmartPhones #ArtificialIntelligence #AI #BigData #DataAnalytics #Data #Video	Positive
Augmented reality	17 January 2019 10:08 a.m.	WA	The AR market today is similar to where the IoT market was in 2010. AR's capacity to visualise, instruct, and interact can transform the way we work with data #success #newtech	Positive
3D printing	17 January 2019 12:33 a.m.	WA	Did my first 3D printing? It's amazing super-duper excited to share it with you	Positive
Virtual reality	29 March 2019 8:34 a.m.	SA	How exciting to see what is possible when AI meets virtual reality in the treatment of mental health conditions	Positive
Telephony	29 August 2019 5:12 p.m.	NT	Telephony technology has evolved rapidly keeping people distant emotionally and physically	Negative
Chatbots	19 June 2020 3:26 a.m.	NSW	How can I find screenshots or scripts from the CyberLover chatbot (the bot used to flirt with people in order to steal their data)? I would like to see some of the conversations it held. #wrongexamples	Negative

tweets related to different AI-related technologies. 'Robotics' ($n = 2348$) was the most tweeted about AI technology in NSW. In contrast, 'drones' were the most tweeted technology in VIC ($n = 499$), WA ($n = 433$), SA ($n = 146$), QLD ($n = 96$), TAS ($n = 48$), and ACT ($n = 22$). 'Machine learning' ($n = 11$) was comparatively high among the tweets circulated in NT. Table 1.4 provides exemplar tweets related to each technology.

4.4 Artificial Intelligence-Related Urban Planning and Development Concepts

Based on a word frequency analysis, 16 key urban planning and development-related concepts were derived from the AI-related tweets (Figure 1.4 and Table 1.5). These concepts are 'sustainability' ($n=774$), 'cybersecurity' ($n = 741$), 'innovation' ($n = 734$), 'construction' ($n = 644$), 'governance' ($n=585$), 'transportation' ($n=275$), 'health' ($n=263$), 'communication' ($n = 241$), 'digital transformation' ($n = 203$), 'mobility' ($n = 190$), 'energy' ($n = 184$), 'infrastructure' ($n = 156$), 'waste' ($n = 144$), 'economy' ($n = 124$), 'environment' ($n = 118$), and 'tourism' ($n = 20$).

Sustainability was the most discussed urban planning and development concept, but its usability differed from one state/territory to another.

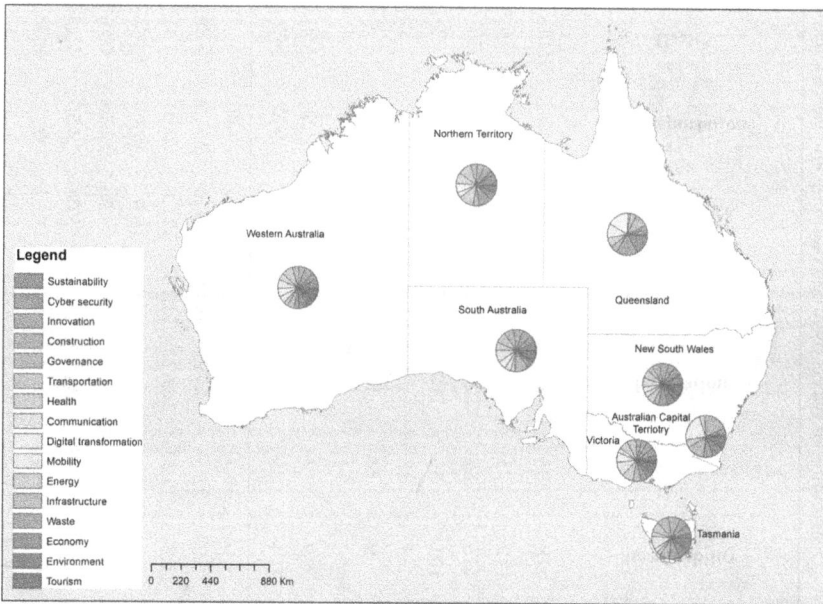

FIGURE 1.4 Distribution of tweets by urban planning and development concepts per state/territory.

TABLE 1.5 Distribution of tweets by urban planning and development concepts per state/territory

	Sustainability	Cybersecurity	Innovation	Construction	Governance	Transportation	Health	Communication	Digital Transformation	Mobility	Energy	Infrastructure	Waste	Economy	Environment	Tourism
NSW	418	358	285	324	264	78	115	116	63	54	79	48	72	61	37	13
VIC	119	173	182	123	101	78	76	35	61	82	67	42	30	24	38	2
WA	115	86	122	72	66	55	30	36	29	24	13	20	12	20	15	5
SA	101	89	110	97	93	47	33	46	26	19	13	35	11	6	21	0
QLD	3	13	11	1	29	6	2	0	16	10	6	4	12	8	1	0
TAS	9	18	16	20	18	7	3	8	2	0	2	4	4	3	4	0
ACT	0	1	1	2	10	0	1	0	4	0	1	2	1	1	1	0
NT	9	3	7	5	4	4	3	0	2	1	3	1	2	1	1	0
Total	774	741	734	644	585	275	263	241	203	190	184	156	144	124	118	20
NSW	54.01	48.31	38.83	50.31	45.13	28.36	43.73	48.13	31.03	28.42	42.93	30.77	50.01	49.19	31.36	65
VIC	15.37	23.35	24.79	19.09	17.26	28.36	28.89	14.52	30.05	43.16	36.41	26.92	20.83	19.35	32.19	10
WA	14.86	11.61	16.62	11.18	11.28	20.01	11.41	14.94	14.29	12.63	7.07	12.82	8.33	16.13	12.71	25
SA	13.05	12.01	14.99	15.06	15.9	17.09	12.55	19.09	12.81	10	7.07	22.44	7.64	4.84	17.8	0
QLD	0.39	1.75	1.5	0.16	4.96	2.18	0.76	0	7.88	5.26	3.26	2.56	8.33	6.45	0.85	0
TAS	1.16	2.43	2.18	3.11	3.08	2.55	1.14	3.32	0.99	0	1.09	2.56	2.78	2.42	3.39	0
ACT	0	0.13	0.14	0.31	1.71	0	0.38	0	1.97	0	0.54	1.28	0.69	0.81	0.85	0
NT	1.16	0.41	0.95	0.78	0.68	1.45	1.14	0	0.98	0.53	1.63	0.65	1.39	0.81	0.85	0
Total	100%	100%	100%	100%	100%	100%	100%	100%	100%	100%	100%	100%	100%	100%	100%	100%

While 'sustainability' ($n = 418$) was the most tweeted urban planning and development concept in NSW, 'innovation' was most popular in VIC ($n = 182$), WA ($n = 122$) and SA ($n = 110$). The use of AI-related technologies in 'governance' was the most popular concept in QLD ($n = 29$) and ACT ($n = 10$). Tweets from TAS had more discussions related to use of AI in 'construction' ($n = 20$). Although there was a lower number of tweets in NT, most of them were related to use of AI for 'sustainability' ($n = 9$).

As shown in Table 1.5, the use of AI technologies in relation to transportation, health, communication and digital transformation were also some of the frequently used concepts (or AI application areas). Furthermore, concepts such as 'mobility', 'energy', 'waste', 'economy', 'environment', and 'tourism' did not receive much attention, and thus they can be identified as emerging topics within the research contexts of novel AI applications. Table 1.6 provides exemplar tweets related to each urban planning and development concept.

4.5 Relationships between Artificial Intelligence Technologies and Urban Planning and Development Concepts

The objective has been to understand AI-related technologies and the public perception of their application in the urban context. To this end, this study conducted a word co-occurrence analysis, which identified the number of tweets that mentioned both technology and urban planning and development concepts (Table 1.7).

Figure 1.5 represents the network topology, developed based on the word co-occurrence analysis. This network typology was initially generated by using the Gephi software. Nonetheless, due to the crowdedness of the original figure—shown in the lower left side of Figure 1.5—a less complex version was recreated by only showing the stronger relationships that occurred between AI technologies and urban planning and development concepts. For that, we identified connections less than 50 as weak or mid-strength and removed them from the figure. Connection counts between 50 and 99 are determined as semi-strong, connections between 100 and 199 are categorised as strong, and connections over 200 are labeled as very strong. Figure 1.5 illustrates these connections, where only the prominent connections are shown in the main part of the figure, and the full connections are given at the lower left side of the figure.

As shown in Figure 1.5, robotics has a close relationship with urban planning and development concepts such as 'innovation' ($n = 337$), 'sustainability' ($n = 450$), 'cybersecurity' ($n = 376$), 'construction' ($n = 242$),

TABLE 1.6 Example tweets for urban planning and development concepts

Concepts	Data and Time	State	Tweet	Sentiment
Sustainability	17 June 2020 5:26 a.m.	NSW	3Ai Director @feraldata and @anucecs Dean @profElanor join the world-first Global Partnership on Artificial Intelligence. An exciting opportunity for Australia to contribute to global work on AI and to shape a safe, responsible and sustainable future.	Positive
Cybersecurity	18 June 2020 10:07 a.m.	NSW	Digital human rights issues such as data privacy, cybersecurity and social impacts of AI can pose risks to companies, and protection of digital human rights take on new considerations in the post-COVID-19 era, according to @ Robeco https://t.co/9zvXiVIDdJ	Negative
Innovation	9 January 2020 10:29 p.m.	QLD	We are thrilled to be featured in an @AllianceQQ Mag Dec/Jan issue article focusing on new #technology impacting #mining. "The industry is now seeing a second wave of technological #innovation based on #digitisation and #IoT"	Positive
Construction	4 February 2020 2:17 p.m.	SA	I'm working on some amazing #hightech projects with the awesome team. #IOT #industrialiot #meshnetworks #smartmine #miningsolutions #miningtechnology #agriculture #agritech #agribusiness #construction #smartcity	Positive
Governance	29 October 2019 2:14 p.m.	ACT	How can governments earn trust in the next generation of AI; bot powered digital services? @piawaugh introduces our new fave term Citizen's Ledger in this A+ read on trust infrastructure for the future of democratic government #fake&fraud	Negative
Transportation	29 May 2019 10:13 a.m.		One of my favourite PBLs that my Ts do is #smartreynoldsburg. Based on what Ss learn about our city's past; the future of transportation; energy, Ss create a 3D model of what Reynoldsburg will look like in 50 years, complete with an autonomous car. #teachingland	Positive
Health	16 June 2020 5:56 p.m.	NSW	Big day today—I have now performed more than 300 transoral robotic surgeries on the da Vinci platform. Thank you to my surgical team and @SVHSydney for the cake! #TORS #HNC #HeadandNeckCancer #roboticsurgery @device_ robotics https://t.co/KNhGvlYF4n	Positive

Category	Date	State	Text	Sentiment
Communication	3 September 2019 11:50 a.m.	SA	We are excited to announce the new research initiative: Information, Communication the Data Society. ICDS is an interdisciplinary research initiative on the way AI and algorithms affect the role, impact and regulation of information	Positive
Digital transformation	1 April 2019 5:29 a.m.	NSW	What is the #InternetOfThings? Why is it so important? #IoT #DigitalTransformation #Automation #SmartCity #AutonomousVehicles #Driverless #SmartCars #SmartHome #CyberSecurity #SmartTech	Neutral
Mobility	31 March 2019 4:51 p.m.	SA	Should the AIUS SA focus on the Future of Mobility such as driverless shuttles and other autonomous vehicles? Let us know by completing our 5 min survey!	Neutral
Energy	20 August 2019 6:27 a.m.	WA	Australian @PowerLedger_io successfully trialled its block chain platform's use in P2P trading of renewable electricity in Japan	Positive
Infrastructure	16 April 2020 1:38 p.m.	QLD	#Virtual presence for physical one could have taken at least a generation #Coronivrus #Covid19 did it in months To #sustain it with #reliability #security & #capacity strong #Telecom infrastructure like #5G is important than ever #ICT #VR #AR #AI #Cloud #Data #IoT #CyberSecurity #safercities	Positive
Waste	17 June 2020 5:43 a.m.	NSW	As part of a new partnership with @Microsoft, we're using artificial intelligence (AI) and other digital technologies to boost farming and tackle global challenges including illegal fishing and plastic waste	Positive
Economy	1 May 2020 2:08 a.m.	TAS	Excited to introduce the AI Economist: Extends ideas from Reinforcement Learning for tackling inequality through learned tax policy design. The framework optimises productivity and equality.	Positive
Environment	10 September 2019 6:57 p.m.	TAS	Going digital will save the environment. Go digital!!!	Positive
Tourism	25 June 2019 7:00 a.m.	QLD	Autonomous regions, have been well prepared for the peak #tourism season #easytravel #easyapps	Positive

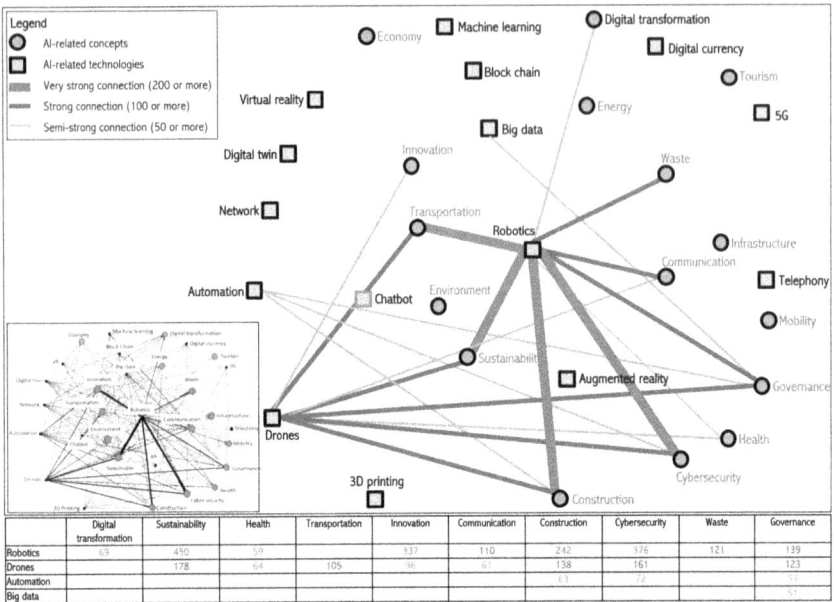

FIGURE 1.5 Relationships between AI technologies and urban planning and development concepts.

	Digital transformation	Sustainability	Health	Transportation	Innovation	Communication	Construction	Cybersecurity	Waste	Governance
Robotics	63	450	59		337	110	242	376	121	139
Drones		178	64	105	98	61	138	161		123
Automation							63	72		55
Big data										51

'governance' (n = 139), and 'waste management' (n = 121). Second, the relationships between 'drone' technology and 'sustainability' (n = 178), 'cybersecurity' (n = 161), and 'construction' (n = 138) were very pronounced. The third popular relationship was the use of 'autonomation' in 'cybersecurity' (n = 72), 'construction' (n = 63), 'innovation' (n = 61), and 'transportation' (n = 48). Fourth, the relationship between 'big data' and 'governance' (n = 51), and 'cybersecurity' (n = 42) was visible.

Although the relationships among the AI-related technologies such as 'digital twin', 'digital networking', 'machine learning', '3D printing', and '5G' were not as frequently used in relation to the derived urban planning and development concepts, they can be identified as emerging discussions within the Twitter hemisphere. The existence of tweets related to the 'digital twin' and 'governance' (n = 28), 'digital transformation' (n = 28), and 'innovation' (n = 26) highlighted the increasing importance of 'digital twin' technology in our society—both the public and private sectors. Nevertheless, tweets related to technologies such as 'block chain', 'AR', 'VR', 'digital currency', 'chatbot', and 'telephony' were comparatively low. Table 1.8 shows the example tweets that discuss AI technologies in the derived urban planning and development concepts (or, in other words, AI application areas).

TABLE 1.7 Technology-concept relationship word co-occurrence analysis

	Digital Transformation	Sustainability	Health	Environment	Economy	Transportation	Innovation	Communication	Construction	Tourism	Infrastructure	Energy	Cybersecurity	Mobility	Waste	Governance	Total
Robotics	69	450	59	13	22	40	337	110	242	2	14	16	376	44	121	139	2054
Drones	10	178	64	37	17	105	96	61	138	2	29	16	161	13	26	123	1076
Automation	28	48	8	0	18	16	61	26	63	2	4	12	72	8	9	53	428
Big data	11	2	13	12	6	12	25	19	33	0	10	14	42	6	2	51	258
Digital twin	28	23	12	2	8	6	26	4	20	0	12	9	15	6	6	28	205
Network	28	23	12	2	0	8	16	10	23	0	11	20	22	4	2	17	198
Machine learning	4	12	2	2	2	4	16	41	2	0	0	6	6	0	0	8	105
3D printing	2	19	4	0	0	2	13	0	14	0	8	6	6	2	2	6	84
5G	5	14	4	0	2	4	15	2	13	0	2	2	2	8	0	4	77
Block chain	4	10	5	2	0	2	10	6	0	0	0	0	11	2	4	2	58
AR	10	10	2	2	2	2	10	0	7	0	2	2	3	6	0	0	58
VR	11	9	0	0	4	2	13	0	4	0	0	4	0	4	0	4	55
Digital currency	2	10	2	0	0	5	9	0	2	0	0	3	5	0	0	8	46
Chatbot	0	0	0	0	0	0	0	4	6	0	0	0	1	0	0	2	13
Telephony	0	0	0	0	0	0	0	6	0	0	0	0	0	0	0	0	6
Total	212	808	187	72	81	208	647	289	567	6	92	110	722	103	172	445	4721

TABLE 1.8 Example tweets showing the relationship between AI technologies and urban planning and development concepts

Data and Time	State	Tweet	AI Technology	Urban Planning and Development Concept	Sentiment
14 November 2019 5:57 p.m.	VIC	Great to see @cserAdelaide Lending Library #sphero kit in action with classes designing and building a Sustainable City and then coding robots through the streets of the city.	Robotics	Sustainability	Positive
3 January 2019 7:26 a.m.	NSW	Building #Sustainable #transport platforms will provide a more efficient #smartcity and cheaper than autonomous and electric vehicles	Automation	Transportation	Positive
17 June 2019 10:36 a.m.	QLD	City Loses $500,000 to Phishing Attack #CyberSecurity #Databreach #Ransomware #Hackers #infosec @reach2ratan #AI #bots #malware #DDoS #Digitaltransformation #Fintech #Blockchain #Chatbots #Bigdata #datascience #Dgital	Chatbot, Big data	Cybersecurity, Digital transformation	Negative
8 August 2019 11:35 a.m.	TAS	@UTAS_ @DeformedEarth @CityByrne @ homehillwines Drone video of @homehillwines landslide and @UTAS_ #UTAS_GSS student at work collecting 3D spatial data. Thanks @ homehillwines for your fantastic hospitality!	Drone	Environment	Positive
5 August 2019 4:24 p.m.	NSW	Humanity must now accept that a digital economy implemented by global governance w/AI world systems for ppl and planet is the way forward from 2020 #bitcoins	Digital currency	Economy, Governance	Positive
29 March 2019 8:34 a.m.	SA	How exciting to see what is possible when AI meets virtual reality in the treatment of mental health conditions	VR	Health	Positive

5 FINDINGS AND DISCUSSION

AI is a widely used technology in Australia across many urban planning and development areas, including, but not limited to, health, safety, environment, energy, infrastructure, transport, education, and urban services [58]. Nonetheless, public perceptions regarding the use of AI are an understudied line of research [59]. The study at hand focused on addressing this limitation. Accordingly, the community's positive perceptions regarding the use of AI are evident in the presented findings. This is to say, overall, the Australian public has a positive perception of AI and its use to make cities more sustainable, innovative, accessible, healthy, and livable [60,61].

Nonetheless, in the analysed tweets, the public has also raised concerns about the use of AI, particularly in terms of cybersecurity breaches. Especially during the pandemic, a large part of Australian society was intending to move towards digital transformation. Due to the cyber-attack boom in 2020 in Australia, the discussions on cybersecurity and the ethical concerns associated with using AI technologies become highly prominent [62,63]. Australian researchers have also highlighted the importance of understanding the loopholes in the present AI systems [64]. Furthermore, the digital transformation has had a negative impact on the elderly population, as they need more assistance to use the technology [65,66].

The Australian government has already drafted an 'AI Action Plan' for all Australians and is currently seeking feedback from the wider community [67]. Through this plan, the Australian government has attempted to address the issue of cybersecurity by preparing and publishing an 'AI Ethics Framework'. This framework addresses the following issues: (a) human, social and environmental wellbeing; (b) human-centred values; (c) fairness; (d) privacy protection and security; (e) reliability and safety; (f) transparency and explainability; (g) contestability; and (h) accountability [68]. Moreover, the Australian government has identified the importance of using AI in the ageing and disability sector to reduce costs while making quality care accessible to adult Australians [48]. Nevertheless, it is important to pay attention to the user-friendliness and affordability of AI technologies, particularly concerning the disadvantaged populations [69].

In 2019, the Australian government released an AI roadmap that recognises the current global shift towards smart cities and smart urban infrastructure [48]. The roadmap suggests that government institutions should work with private organisations to develop, advance, and deploy AI solutions that will improve the urban environment and will help shape sustainable urban futures.

The roadmap has pointed to the potential benefits of AI, such as economic growth (Australia could become a key player on the global AI market in 2030, reaching a value of AUD 22.17 trillion), improved quality of life, environmental sustainability, and solution of the problems experienced by the ageing society [48]. The roadmap involves the use of AI to decrease the costs and improve the effectiveness of built infrastructure planning, design, construction, operation, and maintenance. This is significant in the built environment as there is a shortcoming of built infrastructure, as it is already impacting the operations of towns and cities around Australia [48]. The following are urban planning and development concepts that were mentioned in the roadmap:

- Improve the digital infrastructure (for data transmission storage, analysis and acquisition) so that AI can safely and effectively be used across Australian cities.

- Develop AI for better towns, cities, and infrastructure, to improve the safety, efficiency, cost-effectiveness, and quality of the built environment.

- Improve design, planning, construction, operation, and maintenance of infrastructure and building with AI.

- Utilise AI to improve the efficiency and safety of transportation, electricity, and water services throughout the urban environment.

- Improve AI technology that reduces high construction costs and unplanned cost overruns as it is limiting the ability to improve cities and infrastructure.

Since 2010, The Australian Research Council (ARC) has awarded over AUD 243 million to research regarding AI and data processing. Significant investment went towards block chain, AR/VR, robotic process automation (RPA), natural language processing (NLP), and computer vision. These technologies represent the functionality of a digital co-worker, as it encompasses both rule-based activities and judgement-based activities [70]. The following are the key concepts that were developed in the funded research and have influenced the perception of AI-related urban planning and development:

- Data analytics: Real-time or historical data that can provide insights into an urban environment. A key example is intelligent traffic lights that use data analytics to coordinate and make time-based changes in the traffic lights.

- Machine learning: Computer vision techniques to collect and annotate datasets. The model can be applied to predict the roads that will undergo more 'Wear and Tear', allowing maintenance crews to focus their energy on repairing potholes, instead of looking for them.

- Deep learning: Complex algorithm that analyses large datasets to give planners a predictive insight into data. This provides urban planners with an insight into the nature of traffic, management of traffic flows, and the design of new public transportation.

Perhaps the most important digital infrastructure in the future would be to provide distributed AI services to support the development and operations of ubiquitous urban, rural, industrial, and other applications [71]. The 5G networks that promise us unprecedented mobile internet speed have started to appear around the world. However, AI infrastructure requires more than mere fast networks. The work on the sixth generation (6G) networks has begun. The 6G networks are expected to support extreme-scale ubiquitous AI services through next-generation softwarisation, platformisation, heterogeneity, and configurability of networks [71]. This is a technology that will likely have unimaginable impacts on urban planning and development [72].

6 CONCLUSION

AI is undoubtfully a powerful technology and has already started to reshape and disrupt our economy, society, cities, and urban management systems [73,74]. Today, there is limited understanding of the trending AI technologies and their application areas—or concepts—in the urban planning and development fields [75]. Moreover, there is a knowledge gap in how the public perceives AI technologies, their application areas, and the AI-related policies and practices of our cities [76,77]. Hence, the study at hand aimed at advancing our understanding of the relationship between the key AI technologies and their key application areas in urban planning and development.

The social media analytics undertaken in this study has important findings. Overall, the location-based Twitter analysis throughout this study has identified that: (a) 'Sustainability' ($n = 774$ tweets); (b) 'Cybersecurity' ($n = 741$); (c) 'Innovation' ($n = 734$); and (d) 'Construction' ($n = 644$) are generally the mostly discussed urban planning and development concepts across the entirety of Australia, although the popularity differs by states and territories. To accomplish the listed concepts, the following AI-related technologies are the most popularly discussed ones: (a) 'Robotics' ($n = 3,055$ out of 11,262, 27%); (b) 'Drones' ($n = 1,943$, 17%); and (c) 'Automation' ($n = 717$, 23%). The sentiment analysis has also defined that the degree of satisfaction across Australian communities is relatively high. It has been demonstrated that 'robotics', 'drones', and 'automation' are the AI fields that have a close relationship with the urban planning and development concepts of 'sustainability', 'cybersecurity', 'innovation', and 'construction'.

This study has also disclosed that QLD and TAS have the highest degree of satisfaction (92% of positive sentiments) among the other states and territories. In contrast, given that most states and territories gained dominant positive sentiments, NT has had the lowest degree of satisfaction, as demonstrated by a higher level of neutral sentiments (53%), as well as low interests in sharing their views on social media channels (i.e., Twitter), comprising slightly over 0% of the tweets studied. Meanwhile, NSW and VIC, to which the highest percentage of the total tweets belonged, had a lower degree of satisfaction than QLD and TAS. However, their degrees of satisfaction are also relatively high. The close relationship between popular technologies and concepts was also justified in number of analysis procedures—i.e., sentiment and content analyses, frequency analysis, content analysis, co-occurrence analysis, and spatial analysis.

In addition, concepts and technologies that have received less attention on Twitter are emerging topics and thus are deemed important to keep track of. This study also addressed the significance of improving all the identified AI-related technologies for their safety, effectiveness, efficiency, and affordances throughout the Australian AI Roadmap. Further empirical studies and analyses are needed to make concerted consolidations of AI across Australia to better understand the public perceptions with improved ethics, regulation, design, planning, construction, operation, and maintenance towards better Australian towns, cities, infrastructure, and buildings. In this prospective research, a particular attention should also be paid to further consolidate the understanding and relation between AI and responsible urban innovation [78–80].

ACKNOWLEDGEMENTS

This chapter, with permission from the copyright holder, is a reproduced version of the following journal article: Yigitcanlar, T., Kankanamge, N., Regona, M., Ruiz Maldonado, A., Rowan, B., Ryu, A., Desouza, K., Corchado, J., Mehmood, R., & Li, R. (2020). Artificial intelligence technologies and related urban planning and development concepts: how are they perceived and utilised in Australia? *Journal of Open Innovation: Technology, Market, and Complexity*, 6(4), 187.

REFERENCES

1. Dyer, M.; Dyer, R.; Weng, M.H.; Wu, S.; Grey, T.; Gleeson, R.; Ferrari, T.G. Framework for soft and hard city infrastructures. In *Urban Design and Planning*; ICE Virtual Library: London, 2019; Volume 172, pp. 219–227.
2. Liu, H. *Smart Cities: Big Data Prediction Methods and Applications*; Springer: Singapore, 2020; pp. 1–314.
3. Arbolino, R.; Carlucci, F.; Cirà, A.; Ioppolo, G.; Yigitcanlar, T. Efficiency of the EU regulation on greenhouse gas emissions in Italy: The hierarchical cluster analysis approach. *Ecol. Indic.* 2017, 81, 115–123.
4. Abduljabbar, R.; Dia, H.; Liyanage, S.; Bagloee, S.A. Applications of artificial intelligence in transport: An overview. *Sustainability* 2019, 11, 189.
5. Yigitcanlar, T.; Butler, L.; Windle, E.; Desouza, K.C.; Mehmood, R.; Corchado, J.M. Can building "artificially intelligent cities" safeguard humanity from natural disasters, pandemics, and other catastrophes? An urban scholar's perspective. *Sensors* 2020, 20, 2988.
6. Kassens-Noor, E.; Hintze, A. Cities of the future? The potential impact of artificial intelligence. *AI* 2020, 1, 192–197.
7. Kirwan, C.G.; Zhiyong, F. *Smart Cities and Artificial Intelligence*; Elsevier: London, 2020.
8. Ortega-Fernández, A.; Martín-Rojas, R.; García-Morales, V.J. Artificial intelligence in the urban environment: Smart cities as models for developing innovation and sustainability. *Sustainability* 2020, 12, 7860.
9. Zhang, F.; Zhou, B.; Liu, L.; Liu, Y.; Fung, H.H.; Lin, H.; Ratti, C. Measuring human perceptions of a large-scale urban region using machine learning. *Landsc. Urban Plan.* 2018, 180, 148–160.
10. Yigitcanlar, T.; Desouza, K.C.; Butler, L.; Roozkhosh, F. Contributions and risks of artificial intelligence (AI) in building smarter cities: Insights from a systematic review of the literature. *Energies* 2020, 13, 1473.
11. Mah, D.N.; van der Vleuten, J.M.; Hills, P.; Tao, J. Consumer perceptions of smart grid development: Results of a Hong Kong survey and policy implications. *Energy Policy* 2012, 49, 204–216.
12. Chang, D.L.; Sabatini-Marques, J.; Da Costa, E.M.; Selig, P.M.; Yigitcanlar, T. Knowledge-based, smart and sustainable cities: A provocation for a conceptual framework. *J. Open Innov. Technol. Mark. Complex.* 2018, 4, 5.
13. Quan, S.J.; Park, J.; Economou, A.; Lee, S. Artificial intelligence-aided design: Smart design for sustainable city development. *Environ. Plan. B* 2019, 46, 1581–1599.
14. Pan, Y.; Tian, Y.; Liu, X.; Gu, D.; Hua, G. Urban big data and the development of city intelligence. *Engineering* 2016, 2, 171–178.
15. Zhou, J.; Liu, T.; Zou, L. Design of machine learning model for urban planning and management improvement. *Int. J. Perform. Eng.* 2020, 16, 958.

16. Adikari, A.; Alahakoon, D. Understanding citizens emotional pulse in a smart city using artificial intelligence. *IEEE Trans. Ind. Inform.* 2020, doi:10.1109/TII.2020.3009277.

17. Fast, E.; Horvitz, E. Long-Term Trends in the Public Perception of Artificial Intelligence. In *Proceedings of the AAAI conference on artificial intelligence.* 2017 February, (Vol. 31, No. 1).

18. Neri, H.; Cozman, F. The role of experts in the public perception of risk of artificial intelligence. *AI Soc.* 2019, 35, 663–673.

19. Wirtz, B.W.; Weyerer, J.C.; Geyer, C. Artificial intelligence and the public sector: Applications and challenges. *Int. J. Public Adm.* 2019, 42, 596–615.

20. Abbot, J.; Marohasy, J. Application of artificial neural networks to rainfall forecasting in Queensland, Australia. *Adv. Atmos. Sci.* 2019, 29, 717–730.

21. Aziz, K.; Haque, M.M.; Rahman, A.; Shamseldin, A.Y.; Shoaib, M. Flood estimation in ungauged catchments: Application of artificial intelligence-based methods for Eastern Australia. *Stoch. Environ. Res. Risk Assess.* 2017, 31, 1499–1514.

22. Williams, M.A. The artificial intelligence race: Will Australia lead or lose? *J. Proc. R. Soc. New South Wales* 2019, 152, 105–114.

23. Rahmati, O.; Falah, F.; Dayal, K.S.; Deo, R.C.; Mohammadi, F.; Biggs, T.; Bui, D.T. Machine learning approaches for spatial modeling of agricultural droughts in the south-east region of Queensland Australia. *Sci. Total Environ.* 2020, 699, 134230.

24. Donald, M. *Leading and Managing Change in the Age of Disruption and Artificial Intelligence*; Emerald Group Publishing: London, 2019.

25. Schalkoff, R.J. *Artificial Intelligence: An Engineering Approach*; McGraw-Hill: New York, NY, 1990; pp. 529–533.

26. Jackson, P.C. *Introduction to Artificial Intelligence*; Courier Dover Publications: New York, NY, 2019.

27. Wah, B.W.; Huang, T.S.; Joshi, A.K.; Moldovan, D.; Aloimonos, J.; Bajcsy, R.K.; Fahlman, S.E. Report on workshop on high performance computing and communications for grand challenge applications: Computer vision, speech and natural language processing, and artificial intelligence. *IEEE Trans. Knowl.* Data Eng. 1993, 5, 138–154.

28. Yun, J.J.; Lee, D.; Ahn, H.; Park, K.; Yigitcanlar, T. Not deep learning but autonomous learning of open innovation for sustainable artificial intelligence. *Sustainability* 2016, 8, 797.

29. Kankanamge, N.; Yigitcanlar, T.; Goonetilleke, A.; Kamruzzaman, M. How can gamification be incorporated into disaster emergency planning? A systematic review of the literature. *Int. J. Disaster Resil. Built Environ.* 2020, 11, 481–506.

30. Paulin, A. *Smart City Governance*; Elsevier: London, 2018.

31. Caprotti, F.; Liu, D. Emerging platform urbanism in China: Reconfigurations of data, citizenship and materialities. *Technol. Forecast. Soc. Chang.* 2020, 151, 119690.

32. Bach, J. When artificial intelligence becomes general enough to understand itself. Commentary on Pei Wang's paper "on defining artificial intelligence". *J. Artif. Gen. Intell.* 2020, 11, 15–18.

33. Yigitcanlar, T.; Cugurullo, F. The sustainability of artificial intelligence: An urbanistic viewpoint from the lens of smart and sustainable cities. *Sustainability* 2020, 12, 8548.

34. Girasa, R. *AI as a Disruptive Technology*; Springer: Cham, Switzerland, 2020.

35. Pueyo, S. Growth, degrowth, and the challenge of artificial superintelligence. *J. Clean. Prod.* 2018, 197, 1731–1736.

36. IDC. The Next Generation of Intelligence. Available online: https://www.idc.com/itexecutive/research/topics/ai (accessed on 15 November 2020).

37. Cearley, D.; Burke, B.; Searle, S.; Walker, M.J. Top 10 Strategic Technology Trends for 2018. Available online: https://brilliantdude.com/solves/content/GartnerTrends2018.pdf (accessed on 16 November 2020).

38. Press, G. Top 10 Hot Artificial Intelligence (AI) Technologies. Available online: https://www.forbes.com/sites/gilpress/2017/01/23/top-10-hot-artificial-intelligence-ai-technologies (accessed on 10 November 2020).

39. Audirac, I. Information technology and urban form. *J. Plan. Lit.* 2002, 17, 212–226.

40. Cugurullo, F. Urban artificial intelligence: From automation to autonomy in the smart city. *Front. Sustain. Cities* 2020, 2, 38.

41. Vilajosana, I.; Llosa, J.; Martinez, B.; Domingo-Prieto, M.; Angles, A.; Vilajosana, X. Bootstrapping smart cities through a self-sustainable model based on big data flows. *IEEE Commun. Mag.* 2013, 51, 128–134.

42. Rathore, M.M.; Ahmad, A.; Paul, A.; Rho, S. Urban planning and building smart cities based on the internet of things using big data analytics. *Comput. Netw.* 2016, 101, 63–80.

43. Batty, M. Artificial intelligence and smart cities. *Environ. Plan. B* 2018, doi:10.1177/2399808317751169.

44. Ullah, Z.; Al-Turjman, F.; Mostarda, L.; Gagliardi, R. Applications of artificial intelligence and machine learning in smart cities. *Comput. Commun.* 2020, 154, 313–323.

45. Allam, Z.; Dhunny, Z. On big data, artificial intelligence and smart cities. *Cities* 2019, 89, 80–91.

46. Davenport, T.H. *The AI Advantage: How to Put the Artificial Intelligence Revolution to Work*; MIT Press: Boston, MA, 2018.

47. Yigitcanlar, T.; Kankanamge, N.; Vella, K. How are smart city concepts and technologies perceived and utilized? A systematic geo-Twitter analysis of smart cities in Australia. *J. Urban Technol.* 2020, doi:10.1080/10630732.2020.1753483.

48. CSIRO. Australia's AI Roadmap. Available online: https://research.csiro.au/robotics/australias-ai-roadmap-launched-solving-problems-growing-the-economy-and-improving-our-quality-of-life (accessed on 10 November 2020).

49. Yellow. Yellow Social Media Report 2018: Part One–Consumers. Available online: https://www.yellow.com.au/wp-content/uploads/2018/06/Yellow-Social-Media-Report-2018-Consumer.pdf (accessed on 10 November 2020).

50. Business Queensland. Who Uses Twitter? Available online: https://www.business.qld.gov.au/running-business/marketing-sales/marketing-promotion/online-marketing/twitter/who (accessed on 10 November 2020).

51. Kankanamge, N.; Yigitcanlar, T.; Goonetilleke, A.; Kamruzzaman, M. Can volunteer crowdsourcing reduce disaster risk? A systematic review of the literature. *Int. J. Disaster Risk Reduct.* 2019, 35, 101097.

52. Kankanamge, N.; Yigitcanlar, T.; Goonetilleke, A.; Kamruzzaman, M. Determining disaster severity through social media analysis: Testing the methodology with South East Queensland Flood tweets. *Int. J. Disaster Risk Reduct.* 2020, 42, 101360.

53. Kankanamge, N.; Yigitcanlar, T.; Goonetilleke, A. How engaging are disaster management related social media channels? The case of Australian state emergency organisations. *Int. J. Disaster Risk Reduct.* 2020, 48, 101571.

54. Alomari, E.; Katib, I.; Mehmood, R. Iktishaf: A big data road-traffic event detection tool using Twitter and spark machine learning. *Mob. Netw. Appl.* 2020, doi:10.1007/s11036-020-01635-y.

55. Fan, W.; Gordon, M.D. The power of social media analytics. *Commun. ACM* 2014, 57, 74–81.

56. Gu, Y.; Qian, Z.; Chen, F. From Twitter to detector: Real-time traffic incident detection using social media data. *Transp. Res. Part C* 2016, 67, 321–342.

57. Yigitcanlar, T.; Kankanamge, N.; Preston, A.; Gill, P.S.; Rezayee, M.; Ostadnia, M.; Ioppolo, G. How can social media analytics assist authorities in pandemic-related policy decisions? Insights from Australian states and territories. *Health Inf. Sci. Syst.* 2020, 8, 37.

58. Australian Government. Artificial Intelligence. Available online: https://www.industry.gov.au/policies-and-initiatives/artificial-intelligence (accessed on 15 November 2020).

59. Gao, S.; He, L.; Chen, Y.; Li, D.; Lai, K. Public perception of artificial intelligence in medical care: Content analysis of social media. *J. Med. Internet Res.* 2020, 22, e16649.

60. Yigitcanlar, T.; Kamruzzaman, M. Planning, development and management of sustainable cities: A commentary from the guest editors. *Sustainability* 2015, 7, 14677–14688.

61. Yigitcanlar, T. *Rethinking Sustainable Development: Urban Management, Engineering, and Design;* IGI Global: Hersey, PA, 2010.

62. Webb, T.; Dayal, S. Building the wall: Addressing cybersecurity risks in medical devices in the USA and Australia. *Comput. Law Secur. Rev.* 2020, 33, 559–563.

63. Taddeo, M.; McCutcheon, T.; Floridi, L. Trusting artificial intelligence in cybersecurity is a double-edged sword. *Nat. Mach. Intell.* 2019, 1, 557–560.

64. Chanthadavong, A. Australian and Korean Researchers Warn of Loopholes in AI Security Systems. Available online: https://www.zdnet.com/article/australian-and-korean-researchers-warn-of-loopholes-in-ai-security-systems (accessed on 15 November 2020).

65. Datta, A.; Bhatia, V.; Noll, J.; Dixit, S. Bridging the digital divide: Challenges in opening the digital world to the elderly, poor, and digitally illiterate. *IEEE Consum. Electron. Mag.* 2018, 8, 78–81.
66. Hu, S.H. Analysis of the effect of the digital divide on the digital daily life of the elderly. *J. Digit. Converg.* 2020, 18, 9–15.
67. Australian Government. Australia's AI Action Plan. Available online: https://www.industry.gov.au/news/australias-ai-action-plan-have-your-say (accessed on 15 November 2020).
68. Australian Government. AI Ethics Principles. Available online: https://www.industry.gov.au/data-and-publications/building-australias-artificial-intelligence-capability/ai-ethics-framework/ai-ethics-principles (accessed on 15 November 2020).
69. Lutz, C. Digital inequalities in the age of artificial intelligence and big data. *Hum. Behav. Emerg. Technol.* 2019, 1, 141–148.
70. Chaudhry, S.; Dhawan, S. AI-based recommendation system for social networking. In *Soft Computing: Theories and Applications: Proceedings of SoCTA 2018*; Springer: Singapore, 2020; pp. 617–629.
71. Janbi, N.; Katib, I.; Albeshri, A.; Mehmood, R. Distributed artificial intelligence-as-a-service (DAIaaS) for smarter IoE and 6G environments. *Sensors* 2020, 20, 5796.
72. Allam, Z.; Jones, D.S. (Future (post-COVID) digital, smart and sustainable cities in the wake of 6G: Digital twins, immersive realities and new urban economies. *Land Use Policy* 2021, 101, 105201.
73. Thirgood, J.; Johal, S. Digital disruption. *Econ. Dev. J.* 2017, 16, 25–32.
74. Panda, G.; Upadhyay, A.K.; Khandelwal, K. Artificial intelligence: A strategic disruption in public relations. *J. Creat. Commun.* 2019, 14, 196–213.
75. Wu, N.; Silva, E.A. Artificial intelligence solutions for urban land dynamics: A review. *J. Plan. Lit.* 2010, 24, 246–265.
76. Hengstler, M.; Enkel, E.; Duelli, S. Applied artificial intelligence and trust: The case of autonomous vehicles and medical assistance devices. *Technol. Forecast. Soc. Change* 2016, 105, 105–120.
77. Musikanski, L.; Rakova, B.; Bradbury, J.; Phillips, R.; Manson, M. Artificial intelligence and community well-being: A proposal for an emerging area of research. *Int. J. Community Well-Being* 2020, 3, 39–55.
78. Nagenborg, M. Urban robotics and responsible urban innovation. *Ethics Inf. Technol.* 2018, 22, 345–355.
79. Alami, H.; Rivard, L.; Lehoux, P.; Hoffman, S.J.; Cadeddu, S.B.; Savoldelli, M.; Fortin, J.P. Artificial intelligence in health care: Laying the Foundation for Responsible, sustainable, and inclusive innovation in low-and middle-income countries. *Glob. Health* 2020, 16, 52.
80. Theodorou, A.; Dignum, V. Towards ethical and socio-legal governance in AI. *Nat. Mach. Intell.* 2020, 2, 10–12.

CHAPTER 2

Perceptions on Artificial Intelligence in Local Government Services

1 INTRODUCTION

Artificial intelligence (AI) applications have become an important part of our daily lives (Yigitcanlar et al., 2020a; Son et al., 2023). AI improves health services, safety, and cities' cleanliness and reduces problems, including congestion and pollution (Arteaga et al., 2020; Liu et al., 2021). Robots run restaurants and shops and repair urban infrastructure. AI manages transport systems and autonomous vehicles. Intelligent platforms govern multiple urban domains, such as garbage collection and air quality monitoring. Indeed, urban AI, where AI is embodied in urban spaces, infrastructures, and technologies, turns our cities into autonomous entities functioning unsupervised (Yigitcanlar & Cugurullo, 2020). Digitally supporting intelligent and responsive services can be achieved conveniently in real time. Many cities take the initiative now to leverage big data with AI to increase economic returns by enabling our infrastructures with better energy, computing competency, and connectivity (Allam & Dhunny, 2019; Dwivedi et al., 2021).

Recently, many governments started to utilise AI for various public services as AI reduces administrative costs and time. For example, the robotic automation of immigration processes reduces processing time and improves efficiency (Alshahrani et al., 2021). AI brings a technological

 DOI: 10.1201/9781003521440-3

breakthrough in local government services. AI agents assist urban planners in scenario planning based on a goal-oriented Monte Carlo tree search. The goal-reasoning AI agents provide the optimum land-use solutions and help us make democratic urban land-use planning (Chen et al., 2020). AI utilises online data to monitor and amend policies for environmental threats. During the Chennai water crisis in 2019, the Latent Dirichlet Allocation approach identified the most discussed topics on Twitter, a naïve Tweet classification method classified topics such as the impact and causes of drought, the government response, and potential solutions (Xiong et al., 2020). AI tools complement human judges in the judicial sector to provide objective and consistent risk assessments (McKay, 2019).

Rapid advances in AI significantly raise our productivity and change the way we work (Yigitcanlar, 2021). Governments adopt AI for enhancing administrative efficiency and public service delivery due to its economic and societal benefits (Wilson & Van Der Velden, 2022). Within the next 15 years, AI is expected to increase the annual economic growth rates in Japan, the US, and Germany by 2%. Accordingly, AI can change and bring benefits to the public sector. AI can complement the skills of existing and virtual workforces that raise savings and cost-efficiency. Thus, there is an increase in investments in AI around the world. While Europe spends around $800 million on public-private partnerships for AI and robotics, the US allocates about $1.2 billion on research and development (R&D) of AI-related technologies. Likewise, China invests around $148 billion with the hope to become a global innovator in AI by 2030 (Wirtz et al., 2019).

In Australia, the Commonwealth Government is strengthening Australia's capabilities through the $124.1 million Artificial Intelligence Action Plan to position Australia as a global leader in AI technology. The plan sets out the Government's plan to build Australia's AI capability to grow the economy, support industry competitiveness, create jobs, and improve lives such as modernising manufacturing and farming activities, improving diagnosis and treatment of diseases, and enhancing our defence capabilities (Australian Government, 2022).

In Hong Kong, the government introduced AI and chatbot functions to the GovHK portal in 2019 to facilitate access of e-Government services, searching by the public; and handling public enquiries (News.gov.hk, 2022). Hong Kong Department of Transport has used the access control system and vehicle recognition system for the car parks for over a decades' time. It evaluates the cyber security information by using AI for collecting, classifying, and associating the cybercrime data (The Hong Kong Special Administrative Region, 2017).

In stark contrast to the fast adoption and implementation of AI for government services, implementation of these tools also leads to different kinds of concerns. Feldstein (2019) reported that at least 75 countries actively use AI technologies for surveillance purposes, such as facial recognition, which may cause privacy and cybersecurity threats (Shao et al., 2021). The risks are often articulated as bias, fairness, and privacy (Wilson & Van Der Velden, 2022). For example, Russia installed a video surveillance system, 'Safe City', which allows facial and moving objects recognition and is automatically sent to the government authorities (Polyakova & Meserole, 2019). Yet, as all individuals can be tracked daily, video surveillance raises significant privacy concerns. Analysing such data might reveal sensitive personal information, such as home and work addresses, health conditions, and religious affiliations. Even if we trust the law enforcement authorities to protect citizens' location privacy, the stored location data may still be accessed by malicious users and hackers (Bentafat et al., 2021). Others raised concerns such as user security and cybersecurity of intelligent transportation (Ullah et al., 2021).

Despite discussions regarding the problems and benefits of applied AI for governments, perceptions of AI technologies differ in different groups, owing to factors like prior experience on technologies, culture, perceived usefulness, ease-to-use, and social influence and risks of AI (Gursoy et al., 2019; Young et al., 2021). A study found that, in the US, government employees with experience in AI tend to support AI technologies more (Ahn & Chen, 2022). Perceived difficulty of the AI technologies significantly affected customer emotions, which impacted customers' acceptance of AI tools use in the service context (Gursoy et al., 2019).

All in all, the existing literature sheds light on the usefulness of AI in improving government services, while keeping the negative externalities in mind. Nevertheless, the impact of cultural differences on the perception of AI in local government services and the attitude towards AI impacts perceived ease of use and AI usefulness is still unknown. To address this issue, we aimed to tackle the following research questions:

> RQ1: Do culture differences affect people's perceived usefulness of AI in improving local government services?
>
> RQ2: Does the attitude towards AI technology affect perceived ease of use and perceived usefulness of AI in improving local government services?

This chapter aims to fill the knowledge gap by addressing these research questions and shedding light on the similarities and differences of Australians and Hong Kongers' perceptions on AI, where these two places displayed vast differences in cultures (see Section 2.2). Following this introduction, Section 2 reviews literature, Section 3 describes data and methods used in this research, Section 4 presents analysis and results, and Section 5 offers a discussion and concludes this chapter.

2 LITERATURE BACKGROUND

2.1 Useful Government AI Applications

Over the past decade, we experienced remarkable advances in AI technologies such as computer vision, machine learning, natural language processing, and virtual agents (Yun et al., 2016; Yigitcanlar et al., 2020b). AI allows computers to learn from the past, understand the world via different concepts, and automate tasks (Pencheva et al., 2020). It frees up government labour by automating tasks, fastening government services, and increasing accuracy in assessing policy outcomes. Thus, AI has a vast potential in almost all government sectors, including the legal system, construction, R&D, education, telecommunication, transportation, data security and management, policymaking, finance, and healthcare, among many others (Assemi et al., 2021). While increasing governance efficiency, AI can also improve quality of life. The government may utilise AI to optimise and predict energy utilisation, accurate forecasts and simulate complex systems, make decisions, and formulate policies for environment monitoring and management, urban planning, motor vehicle navigation, distribution logistics, and so on (Sharma et al., 2020; Zuiderwijk et al., 2021).

Similarly, other government domains such as telecommunication have room to adopt AI tools that may result in immense benefits in public services. For instance, built-in GPS and voice assistants can provide real-time traffic information for planning road trips and avoid road crashes (Dennis et al., 2021). Virtual agents significantly reduce the administrative burden of the public sector organisation and can help in government-to-citizen (G2C) cognitive communication. Citizens using such services prompt answers to their queries concerning public service and can easily navigate through the jungle of public data. The officials can be liberated from spending endless hours on repetitive work and can focus their attention on delivering a personalised and high-level service experience to citizens (Chohan et al., 2020).

To enhance public safety, networked cameras are installed on roads, hotels, shopping malls, or other public areas to detect theft and track criminals. Police force can then track the location of offenders if GPS is installed in these cameras. Along with these, autonomous vehicles are very popular AI applications for transportation systems (Dennis et al., 2021). Likewise, in the healthcare sector, health monitoring devices for blood pressure, oxygen level, and heart rate can be linked to remote AI systems to enable health surveillance for older people at homes or in hospitals. Moreover, AI can be applied in governance, data management, and security due to increased data transparency and internet usage. Government can also utilise data to enhance people's awareness and active participation in public administration (Sharma et al., 2020).

Given AI's potential for improving work efficiency and user experience, reducing service costs, and relieving human workloads, public sectors are seizing the opportunities that AI can bring and utilising AI-based self-service technology at an accelerating rate (D'Amico et al., 2020). The recent successful applications represent the third wave of AI that began in 2006. It is expected that AI will profoundly impact worldwide government services (Pencheva et al., 2020). Table 2.1 summarises some of the useful applications of AI.

2.2 Downside of Government AI Applications

Parallel to the abovementioned views, critics argued that there is a lack of fairness, transparency, unclear responsibility, and accountability in using AI for public governance (Cui & Wu, 2021). Failures due to AI use in government might negatively impact governments and society (Zuiderwijk et al., 2021). Wirtz et al. (2019) proposed that the challenges of AI include ethics, technology implementation, regulation, and society. For example, the UK, Australia, and China residents did not trust their governments to use artificial facial recognition in a responsible way (Ritchie et al., 2021). The lack of trust may affect people's attitude on the government.

Roski et al. (2021) proposed that the rise of AI has primarily occurred in a regulatory vacuum. Apart from the US legislation regarding autonomous drones and vehicles, laws that specifically address AI are scarce. While our digital urban infrastructure has been supported by AI, providing incredible opportunities for urban innovation, a responsible approach is needed to prevent problems such as an increase in bias and negative environmental externalities (Yigitcanlar et al., 2021a). Sharma et al. (2020) suggested

TABLE 2.1 Examples of useful government AI applications

Literature	AI's Benefit	Elaboration
Assemi et al. (2021), Chen et al (2021)	Economic	• AI frees up government labour by automating tasks, fastening services, and increasing policy outcomes assessment. • Improve work efficiency and user experience, reduce costs, and relieve human workloads.
Chohan et al. (2020), Sharma et al. (2020), Dennis et al. (2021)	Social	• Built-in GPS and voice assistants provide real-time traffic information for planning road trips and avoid car crashes. • Virtual agents reduce the administrative burden of the public sector and help G2C communication. • The public officials can then focus on a personalised and high-level service experience for citizens • Networked cameras with GPS track criminals' location. • In the healthcare sector, health monitoring devices for blood pressure, oxygen level, heart rate, can be linked to remote AI systems to enable health surveillance. • AI can be applied in governance, data management, and cybersecurity. The government can utilise data to enhance people's awareness and active participation in public administration.
Sharma et al. (2020), Zuiderwijk et al. (2021), Yigitcanlar et al. (2022a)	Environmental	• AI optimises and predicts energy use accurately, simulates complex systems, makes decisions, and formulates policies for environment monitoring and management, urban planning, motor vehicle navigation, and distribution logistics.

that government faces numerous administrative, technological, fiscal, and policy challenges when it applies AI in different fields.

Desouza et al. (2020) advocated the view that the public sector must contend with complex economic, societal, policy, and legal elements that their private-sector counterparts might skirt. The applied AI projects must attain more than simple cost and efficiency gains to satisfy various stake-holders who may have conflicting needs. The need for transparency and fairness in decision making. As taxpayers fund the public-sector projects and systems, these efforts face regular scrutiny which does not generally happen in the private sector. In addition to these, Yigitcanlar et al. (2021b) raised the sustainability of AI applications under the concept of 'green AI'.

Wirtz et al. (2019) pinpointed that AI needs policies and regulations in response to core societal values and principles. Nonetheless, research on AI in the government sector remains an emerging field (Sharma et al., 2020). In sum, while many studies proposed the merits and drawbacks or even threats that AI may bring to society, little is known as per how the public perceive AI for local government service delivery (Yigitcanlar et al., 2020c; Kankanamge et al., 2021; Selwyn & Gallo-Cordoba, 2021), not to mention international comparative studies considering cultural and con-textual differences that impacts their perceptions.

2.3 Impact of Cultural Factors on Perceived Usefulness and Ease of Use of Technologies

While we may consider the prospects and constraints clear as crystal, atti-tudes towards AI applications, such as perceived usefulness, ease to use, and risks, may be affected by our cultural background. Perceived useful-ness refers to the extent to which an individual believes using technology will make a job easier to perform and improve productivity and effective-ness. It influences users' technology acceptance due to the reinforcement outcomes' value. According to the technology acceptance model (TAM), ease of technological use enhances the perceptions of usefulness (Akour, 2006), and an individual's level of acceptance of technology depends on their perceived usefulness of the technology and the perceived ease of use (Parboteeah & Parboteeah, 2005).

Regarding cultural impacts on perceived ease of use, Dong (2011) sur-veys 534 first-year students using e-school systems. It is found that ease of use is more critical for users who espouse feminine values. Given the low level of masculinity, Chinese culture increases the ease of use of informa-tion technology innovations and the early stage of using e-school systems. Abdelhakim et al. (2023) note that uncertainty avoidance represents the extent to which society's intolerance for unknown or uncommon condi-tions. People with high uncertainty avoidance culture are less likely to adopt innovation. Because they experience less ease of use, people in a high uncertainty avoidance culture make more effort to adopt the technology, learn, and adopt new technology. Their study finds that uncertainty avoidance significantly moderates the behavioural intention of 428 Egyptian and Malaysian employees in using robots for fast food shops. Using a meta-analysis of 20 articles with 6,128 research participants, Zhang et al. (2022) evidence that uncertainty avoidance modifies the link between perceived usefulness and perceived ease of use and adoption intention of wearable health devices in a negative way. It suggests that people in a high uncertainty avoidance culture are more likely to acquire technology impulsively, ignoring perceived ease of use.

Regarding perceived usefulness, personal beliefs that result from an individual's socio-cultural background could affect the perceived usefulness of social media according to mean and standard deviation analysis of 184 surveys (Izuagbe et al., 2019). In another study, Yoon (2009) analyses 270 questionnaires using partial least square graph. The results show that masculinity moderately impacts the perceived ease of use, usefulness, and intention to use e-commerce. Nonetheless, individualism and power distance do not have a significant impact. In sharp contrast, Mahomed et al.'s (2018) findings indicate the perceived usefulness of email is negatively impacted by collectivism and considerable power distance.

Zhang et al. (2022) consider that performance expectation under the unified theory of acceptance and use of technology (UTAUT) is equivalent to the perceived usefulness of TAM, and effort expectation of UTAUT is equivalent to the perceived ease of use under TAM. They conduct a meta-analysis and find that the perceived usefulness of wearable health-care devices is affected negatively by uncertainty avoidance but positively moderated by individualism/collectivism, indulgence/restraint, and masculinity/femininity. On the other hand, the relationship between perceived ease of use and adoption intention of wearable healthcare devices is negatively moderated by uncertainty avoidance and positively moderated by masculinity/femininity, indulgence/restraint, and individualism/collectivism.

Chopdar et al. (2018) survey 366 Indian and American, and they find that individuals from a higher power distance culture perceived more risks with mobile shopping apps than those from a lower power distance culture under the UTAUT framework. This is because higher power distance societies raise distrust among individuals, which enhances their risk perception. Huang et al. (2020) survey students from 16 universities in China and reveal that perceived usefulness does not affect Chinese students' intents to use Internet-based technologies. Although students view

Internet-based technology as helpful, their intention to utilise it for learning was not directly related to their judgments of its usefulness. Because the Chinese have a collectivist value orientation, they place great importance on collective requirements, needs, and harmony. Hence, conformity is more important.

Despite studies on cultural impacts on perceived ease of use and usefulness, it is interesting to note that many studies that study the effects of culture on technology throw light on one country only. For example, Dasgupta and Gupta (2019) utilise Partial Least Squares Regression and find that users' acceptance and use of Internet technology in India's government organisation is influenced by professed cultural characteristics. Research comparing Eastern and Western cultures, like Hong Kong and Australia, is absent to the best of our knowledge, not to mention AI for local government services.

While previous research shed light on the benefits that AI may bring to government services and the downside of AI, research on the effects of cultural factors that influence the perceived usefulness and ease of use for local government services is absent, to the best of our knowledge.

2.4 Cultural Differences Between Australia and Hong Kong

As per new institutional economics, informal institutions like culture affect people's perceptions and behaviours (Li et al., 2019). While cultural values influence and (re)shape technology, introducing new technologies leads to a new scene that requires social acceptance and social, political, and cultural changes (Amershi, 2019). Hong Kong and Australia represent two different Eastern and Western cultures. Australia is a Western society despite being in the Asia-Pacific geography due to its historical close ties with the UK and the Western world (Phau & Kea, 2007). The British drastically decreased the local Indigenous Aboriginal population in a few decades through colonisation, introduction of smallpox and other diseases, invasion of Indigenous land, and several violent conflicts with Indigenous peoples. This consolidated British dominance and Australia was founded as a semi-autonomous British Dominion (Bogais et al., 2022). Despite the recent increase in migration to Australia, notably from Asia, most Australians are still English, Irish, and Scottish descent (Australian Bureau of Statistics, 2022). Therefore, its ideology and culture have not diverged significantly from Western ones. Even before, according to a study, there was no discernible difference between the ethical perspectives of Americans and Western Australians (Phau & Kea, 2007).

Hong Kong's population was very small in the early colonial days. During the 1937–1941 Chinese Civil War, the migration of mainland Chinese nearly doubled Hong Kong's population. Although the Immigration Control Ordinance 1940 was passed by the Hong Kong government in 1940, it limited the number of Chinese citizens to enter and remain in the colony, which marked a significant departure from earlier customs that permitted Chinese citizens 'to enter and stay in the colony without restrictions', these were 'not effectively enforced' in practice (Lopes, 2022). On the other hand, despite Hong Kong was also once upon a time a British colony, Hong Kongers manifests with Eastern culture. Chinese provinces of Fujian and Guangdong immigrant investors fuelled the economic boom in the 1960s (Lin et al., 2022). Although immigrants in Hong Kong include both Chinese and non-Chinese background people, the former comprises over 90% of the immigrant population. The remaining 10% non-Chinese immigrants are East Asian, South Asian, Southeast Asian, white and others (Sun & Eric Fong, 2022). In sum, historical reasons explain the following distinct cultural difference: the UK strengthened the colonisation by removing the indigenous Australians, while the migrants from the mainland boomed in colonial Hong Kong. Despite both having migrants in recent decades, the proportions of migrants remained small compared to the population. Second- and third-generation migrants in Hong Kong and Australia usually share similar cultures with locals.

Given the historical background differences, Hong Kong and Australia score significantly different power distance index, individualism vs. collectivism, long-term orientation, and indulgence vs. restraint. Hong Kongers' culture has high power distance, they are less individualistic, long-term oriented, and less indulgent. Australian's culture has low power distance (Huang & Crotts, 2019), evidenced in Hofstede-insights (2022)'s study. They are highly individualistic, short-term oriented, and more tolerant. Besides, Australia and Hong Kong have different organisational structures due to power distance differences. The low power distance is often associated with a flatter organisational structure in Western firms, where Australia is one example. On the other hand, Asian firms, including those in Hong Kong, tend to have central decision-making and are policy-driven. Leadership tends to be based on authority, seniority, and position (Lok & Crawford, 2004). Figure 2.1 presents six-dimension culture differences between Hong Kong and Australia (Hofstede-insights, 2022).

Some cultural studies have placed Australia and Hong Kong as contrasting cultural contexts, despite some similarities. For example, Landman

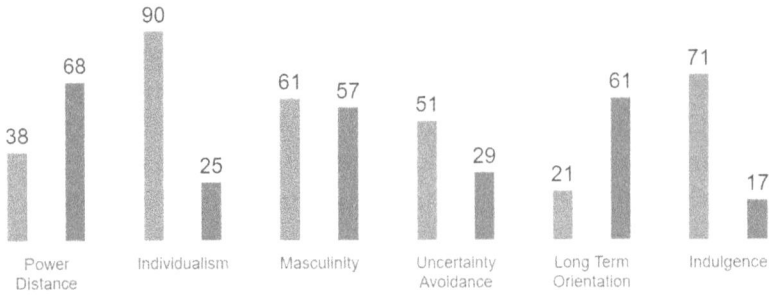

FIGURE 2.1 Six-dimension culture differences between Hong Kong (in purple) and Australia (in blue), derived from Hofstede-insights (2022).

and Carvalho (2003) compared number of countries with different contexts including Australia and Hong Kong as the benchmarking cases. They indicated these two nations in many ways as opposite cultures (Huang & Crotts, 2019). Likewise, Pan et al. (2008) compared the predictive impacts of the meaning of life and acculturative stressors on life satisfaction between Chinese students in Australia and Hong Kong. Students from Hong Kong experienced a significantly lower acculturative stressor than Australians. Lok and Crawford (2004) compared leadership styles and organisational culture on job satisfaction for managers in Australia and Hong Kong. In this study, Australian managers recorded higher scores in job satisfaction, organisational commitment, and innovative and supportive organisational cultures. National culture moderated the impact of respondents' age on satisfaction, which was more positive among managers from Hong Kong.

While Huang and Crotts' (2019) findings on cultural differences, and Lok and Crawford's (2004) results confirmed the substantial differences in organisational structure, the abovementioned three studies investigated the differences between Australian's and Hong Kongers' perceptions. To the best of our knowledge, no research has so far compared public perceptions of Australian Hong Kongers on AI for local government services. We aim to fill this gap in research.

3 RESEARCH DESIGN

This study adopted a case study approach to investigate public perceptions of AI in local government services. The study selected two case studies—i.e., Australia and Hong Kong. Following an ethics approval (#2000000257) granted by the University Human Research Ethics Committee, an online survey was developed to collect data from the public in Australia (from the most significant three cities: Sydney, Melbourne, Brisbane) and Hong Kong. The questionnaire is designed to capture data to address our research questions and validate or falsify our hypotheses. The study focused on addressing a knowledge gap and under-studied but emerging and important area of research, such as public perceptions on AI for local government services and particularly to capture the ease of use, attitude towards AI and usefulness of AI. The questions are developed considering literature about AI and public perceptions (Araujo et al., 2020; Neri & Cozman, 2020; Cui & Wu, 2021; Kassens-Noor et al., 2021; Selwyn & Gallo Cordoba, 2021; Yeh et al., 2021). The survey questionnaire is shown in Appendix 2.A.

The survey questionnaire was designed to capture and compare perceptions of the public on AI regarding local government services. There were sections of questions about current knowledge and experience with AI as well as views of AI use to support government services. Although most people in the developed world have some knowledge experience with AI through smartphone, smart home applications, and government services such as immigration ports at airports, it was not assumed or required for participants to know or have experience with AI at any level (Li et al., 2021). Their experience and knowledge were captured through questions and used in the analysis and conclusions.

Our findings and conclusions are not general but rather focused on Australia and Hong Kong. The same of very similar approach and analyses can be performed in other regions of the world to develop the corresponding insights and conclusions. For example, if Hong Kongers views and response behaviour are affected by unique factors such as their political and social interactions with neighbour China, our analysis and results should consider and be consistent with the corresponding data characteristics. That is, the responses of Hong Kongers are not expected to be representative of other groups that are not from the same place.

An online enterprise survey platform—i.e., Key Survey—was utilised to conduct the survey. The minimum number of participants (384 at confidence level 95% and margin of error 5%) was determined based on methods suggested in literature (Krejcie & Morgan, 1970; Rahman, 2023). Only adults (people over 18) were invited to the survey. The survey was open between November 2020 and March 2021. A professional survey panel company and social media channels were used to recruit participants. In total, 850 valid responses were received. The socioeconomic characteristics of the two samples are given in Table 2.2. Structural equation modelling (SEM)

TABLE 2.2 Socioeconomic characteristics of the sample

Characteristics	Categories	Total		Australia		Hong Kong	
		Sample	Proportion	Sample	Proportion	Sample	Proportion
	Total sample	850	100%	604	71.10%	246	28.90%
Age	18–24	207	24.30%	24	3.97%	140	56.91%
	25–34	162	19.00%	31	5.13%	46	18.70%
	35–44	155	18.30%	45	7.45%	30	12.20%
	45–54	133	15.60%	81	13.41%	18	7.32%
	55–64	90	10.60%	115	19.04%	9	3.66%
	65–74	48	5.60%	125	20.70%	3	1.22%
	75–84	31	3.60%	116	19.21%	0	0.00%
	85 and over	24	2.80%	67	11.09%	1	0.41%
Gender	Male	437	51.40%	298	49.34%	139	56.50%
	Female	408	48.10%	306	50.66%	107	43.50%
Education level	Primary school	53	6.20%	49	8.11%	4	1.63%
	Secondary School	26	3.10%	10	1.66%	16	6.50%
	Certificate	96	11.30%	83	13.74%	23	9.35%
	Diploma	106	12.50%	88	14.57%	8	3.25%
	Advanced diploma	98	11.50%	85	14.07%	13	5.28%
	Bachelor's degree	338	39.70%	192	31.79%	146	59.35%
	Postgraduate degree	133	15.70%	97	16.06%	36	14.63%
Industry	Accommodation, hospitality, and food services	33	3.90%	25	4.14%	8	3.25%
	Administration and support services	38	4.50%	27	4.47%	11	4.47%

Agriculture, forestry, and fishing	5	0.60%	2	0.33%	1	0.41%
Arts and recreation	24	2.80%	13	2.15%	11	4.47%
Construction	38	4.50%	29	4.80%	9	3.66%
Education and training	94	11.20%	56	9.27%	38	15.45%
Electricity, gas, water, and waste services	8	0.90%	6	0.99%	2	5.28%
Financial and insurance services	72	8.50%	26	4.30%	46	18.70%
Healthcare and social assistance	55	6.50%	45	7.45%	10	4.07%
Information, media, and telecommunications	45	5.30%	32	5.30%	13	5.28%
Manufacturing	28	3.30%	24	3.97%	4	1.63%
Mining	2	0.20%	2	0.33%	0	0.00%
Professional, scientific, and technical services	48	5.60%	36	5.96%	12	4.88%
Public administration and safety	36	4.20%	23	3.81%	13	5.28%
Rental, hiring, and real estate services	10	1.20%	6	0.99%	4	1.63%
Retail trade	59	6.90%	49	8.11%	10	4.07%
Transport, postal and warehousing	26	3.10%	23	3.81%	3	1.22%
Wholesale trade	13	1.50%	11	1.82%	2	0.81%
Other	216	25.40%	169	27.98%	47	19.11%

(Continued)

TABLE 2.2 (Continued)

Characteristics	Categories	Total		Australia		Hong Kong	
		Sample	Proportion	Sample	Proportion	Sample	Proportion
Income	No income	159	18.70%	69	11.42%	90	36.59%
	Low income	154	18.10%	95	15.73%	59	23.98%
	Medium-low income	185	21.70%	149	24.67%	36	14.63%
	Medium income	203	23.90%	177	29.30%	26	10.57%
	Medium-high income	94	11.00%	74	12.25%	20	8.13%
	High income	55	6.60%	40	6.62%	15	6.10%
Prior experience with AI	I have previously interacted with an AI technology	301	35.40%	187	30.96%	114	46.34%
	I may have previously interacted with an AI technology	323	38%	222	36.75%	101	41.06%
	I have never previously interacted with an AI technology	226	26.60%	195	32.28%	31	12.60%

was used for quantitative analysis of the survey data. Detailed information on this statistical approach is provided in the next section.

The survey team paid particular attention for not having selection bias and getting skewed results by only capturing feedback from a certain segment of the audience. Hence, the data is collected from numerous sources. Potential participants are approached through various social media channels and along with the efforts of a private survey panel company using its network. The hindside of this approach, however, is not possible to report the exact number of participants invited to the study. Our estimation is this figure being around 5,000 people in total from Australia and Hong Kong. This figure leads to an about 17% response rate. The survey team also paid particular attention for not having response bias. The survey questionnaire is designed to avoid bias and prevent leading questions that could result in measurement errors. The questions are designed to be straight forward, survey completion time is kept under 10 minutes to make it more convenient to complete and offered online for an easy access. To avoid question order bias, we have run the survey with a pilot group and revised the questions based on the received feedback. Lastly, to avoid survey sampling bias, the team randomised the answer option order, and limited the answer option list, but included a free response option to capture any choices the survey might have missed.

Furthermore, a minimum sample-size criteria was adopted for recruiting respondents. The minimum sample requirement for Australia was 385 (95% confidence level of 5% margin of error) and for Hong Kong was 175 (90% confidence level of 5% margin of error). Mann-Whitney U test (McKnight & Najab, 2010) was performed to evaluate if the samples from Australia and Hong Kong were significantly different in terms of their descriptive statistics. The results suggest that the two samples are significantly different in terms of age, education, and income.

4 ANALYSIS AND RESULTS

4.1 Exploratory Factor Analysis to Identify Underlying Latent Constructs

To determine latent constructs that influence perceived usefulness, indicators that significantly affect variability in data were identified. An exploratory factor analysis (EFA) was used to determine significant indicators and the underlying latent constructs in this study. EFA has been considered as one of the best tools to determine relationship between the measured indicators and their underlying latent constructs (Byrne, 2011).

To assess the adequacy of the sample and select important variables from the data for EFA, the Kaiser-Meyer-Olkin (KMO) adequacy measure was used (Kaiser, 1970). Based on the guidelines provided by Kaiser (1974), indicators with KMO greater than 0.6 were retained for EFA.

The overall measure of sampling adequacy (MSA) is 0.88 for the selected variables, with each variable's MSA above 0.7. In addition, Bartlett's Test of Sphericity (Tobias & Carlson, 1969) validate the EFA for the given dataset. Scree plot and parallel analysis suggest considering five factors for EFA (Çokluk et al., 2016).

Table 2.3 presents rotated factor loadings for identified three factors. Maskey et al. (2018) conducted a detailed review of EFA analysis used in maritime research and recommend considering factors with loadings above 0.5 as they represent practical significance. The analysis was conducted using the factanal() function of stats package in R. A cut-off value of 0.5 was imposed on factor loadings, which clearly indicated the presence of three constructs. The fourth and fifth factors were excluded from further analysis as their sum of squared (SS) loadings was less than one. To test potential correlation between the factors, an oblique 'promax' rotation was imposed on the factors. However, the correlation matrix indicated a considerably low correlation (≤0.5). Therefore, the factor analysis was conducted with an orthogonal rotation using 'varimax'.

The first factor is loaded by seven indicators, which capture the perceived usefulness of AI in improving local government services. The indicators include (a) AI's ability to improve objectivity in the delivery of local government services; (b) AI usefulness in enabling local governments to better monitor and respond to problems associated with urban infrastructure; (c) AI's ability to monitor and respond to environmental and climate crises; (d) AI usefulness to ensure safety and security of all residents; (e) AI usefulness in reducing public sector costs with savings used to reduce rates and other taxes; (f) AI in improving efficiency in delivering services; and (g) AI's ability to improve resource management.

The second factor captures respondents' attitude towards AI, measured by: (a) their opinion on how AI would impact society in future; (b) how they feel about AI; (c) their opinion on whether benefits outweigh the associated risks of AI; and (d) opinion on whether the respondents would be comfortable living and working in a fully autonomous environment was also included as an indicator for attitude.

TABLE 2.3 Rotated factor loadings from exploratory factor analysis

	Loadings			
Indicators	Factor 1	Factor 2	Factor 3	Factor 4
Age (AGE)				
Education level (EDU)				
Employment status (EMP)				
Do you feel like you understand the basic concepts of AI? (UND)			0.65	
Have you ever interacted with an AI technology? (EXP)			0.64	
How often would you estimate you interact with the previously identified applications? (FREQ)			0.66	
How long do you think it will be before AI has a noticeable impact on your daily life? (IMPACT)			0.66	
Do you think society will become better because of the future use of AI? (SOCIETY)		0.77		
How do you feel about AI in general? (FEEL)		0.77		
Do you think the benefits outweigh the risks when looking at the future use of AI? (BENVSRISK)		0.72		
Would you feel comfortable in a fully autonomous place to live or work? (COMFORT)		0.58		
Does the prospect of an AI future make you feel concerned? (CONCERFUT)				
Does the prospect of an AI future make you feel excited? (EXCITED)				
Does the prospect of an AI future make you feel optimistic? (OPTFUT)				

(Continued)

TABLE 2.3 (Continued)

Indicators	Loadings			
	Factor 1	**Factor 2**	**Factor 3**	**Factor 4**
Does the prospect of an AI future make you feel enthusiastic? (ENTHUFUT)				
Most promising aspect of future use of AI is to reduce error and mistakes (REDERR)				
The biggest disadvantage of AI is that it will be highly costly (COSTLY)				
Do you fear that the future use of AI could create economic inequality? (ECOINEQUALITY)				
AI can replicate and respond to human speech (SPEECH)				
AI will be most beneficial in automating data collection, management, and analysis (AUTOMATE)				
The most promising about the future use of AI is its ability to reduce crime and monitor illegal behaviours (REDCRIME)				
Most promising about the future use of AI is its ability to enhance productivity and innovation (ENHANCE)				
AI can help improve objectivity in the delivery of local government services (OBJECTIVITY)	0.77			
AI can help local governments monitor and respond to problems associated with urban infrastructure (INFRA)	0.80			
AI can help local governments monitor and respond to the environmental and climate crises (CLIMATE)	0.69			

(Continued)

TABLE 2.3   (Continued)

Indicators	Loadings			
	Factor 1	Factor 2	Factor 3	Factor 4
AI can be used to monitor urban areas and ensure safety and security of all residents (SECURITY)	0.81			
AI can be used to reduce public sector costs with savings used to reduce rates and other taxes (REDUCETAX)	0.82			
AI can be used to reduce public sector costs with savings to improve other local government services (REDUCECOST)				
AI can help with efficient delivery of local government services (DELIV)	0.82			
AI can free up resources so that local governments can spend more time focusing on resident needs and concerns (RES)	0.80			
Country of residence (AUS or HK)				0.98

The third factor explains the ease of using AI, which is captured by: (a) their ability to understand AI concepts; (b) their experience with AI; (c) usage frequency; and (d) their opinion on when AI will potentially have a noticeable impact on our daily life.

A detailed literature review revealed that the two countries have well-established cultural differences in terms of their power distance, individualism, orientation, and indulgence. A dummy variable, indicating the country of residence, was also included to investigate the effect of this cultural difference in the perceived usefulness of AI.

4.2 Analysis of Structural Relationships of Latent Constructs on Perception of AI Usefulness

Based on an EFA illustrated in the previous section and literature review, four underlying latent constructs regarding the usefulness of AI have been identified. Further analysis is required to determine the relationship between these constructs and assess their influence on the overall perceived usefulness of AI. SEM was adopted to test the direct/indirect relationships between the latent constructs, including attitude, ease of use of AI, and their overall influence on the perceived usefulness of AI in improving local government services. Considering the two research questions of the study (see Section 1), the following hypotheses were aimed to be tested in this study.

H1: Ease of using AI is significantly and positively influenced by attitude towards AI.

H2: The perceived usefulness of AI in improving local government services is significantly and positively influenced by the attitude towards AI technology along with cultural differences.

SEM can be divided into two fundamental components: (a) A measurement model component, which captures the effect of latent constructs using indicators given by Equation 2.1, where x is a vector of indicators describing a vector of latent variables ξ, Λ is a matrix of corresponding loading coefficients and ε is a vector of normally distributed random error terms; and (b) A structural model that represents the regression relationships between the considered endogenous and exogenous latent variables given by Equation 2.2, where y is a vector of endogenous latent variables, Γ is a matrix of the regression coefficients, and δ is a vector of random errors, also assumed to be normally distributed (Dastjerdi et al., 2019). A linear relationship is hypothesised between variables. The measurement models were defined using the indicators identified as significant and explanatory in the EFA. Figure 2.2 presents a conceptual model that includes hypotheses for estimating SEM.

$$X = \Lambda\xi + \varepsilon, \text{ where } \varepsilon \sim N\left(0, \psi_\varepsilon\right) \tag{2.1}$$

$$y = \Gamma\xi + \delta, \text{ where } \delta \sim N\left(0, \sigma^2\right) \tag{2.2}$$

4.3 Model Analysis and Results for the Entire Sample

Figure 2.3 presents the path diagram for the SEM model, which is based on the conceptual model. The path diagram presents the estimated loadings for the different components of the model along with the path of their association. The model results are discussed in the subsequent sections.

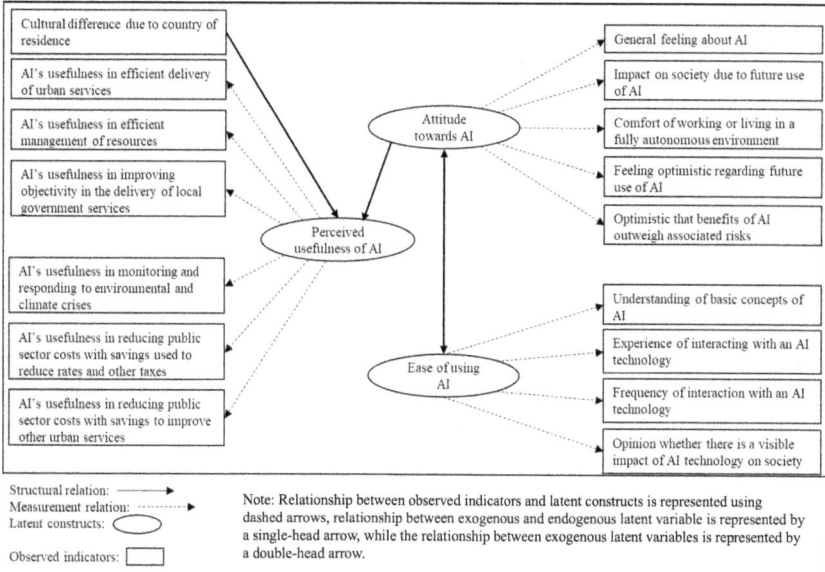

FIGURE 2.2 Conceptual model indicating relationships between the different latent constructs.

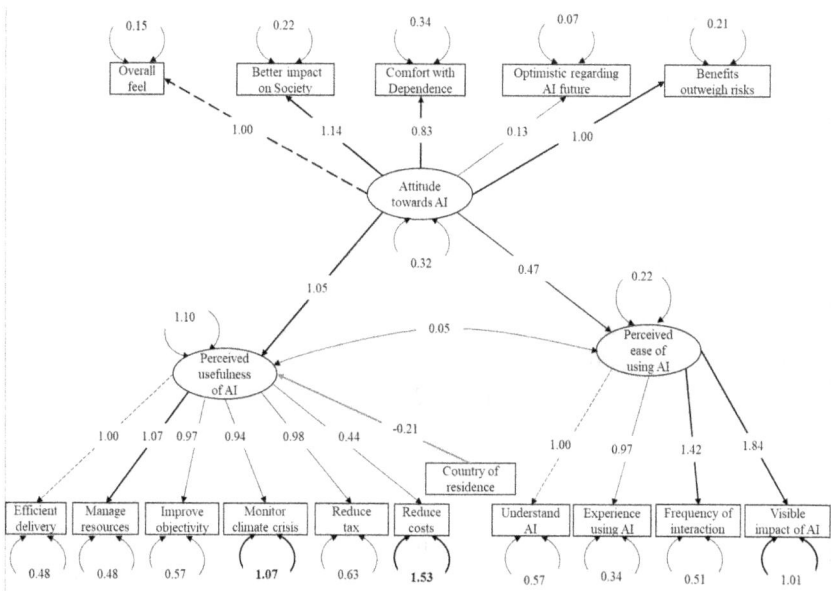

FIGURE 2.3 Path diagram from SEM based on the proposed hypotheses.

4.3.1 Measurement Models

The measurement models for the perceived usefulness of AI in improving local government services, ease of use of AI technology, and attitude towards AI were estimated based on the indicators that were identified in the EFA. Indicators which had statistical and behavioural significance in SEM were retained in the final models based on an extensive hypothesis testing. The results from the measurement model components are discussed in this section.

Table 2.4 presents the measurement model for the perceived usefulness of AI in improving local government services. The estimates suggest that the overall perceived usefulness of AI is indicated by its useability in improving efficiency in resource management and delivery of services, reducing the urban-service cost to reduce taxes, reducing overall services costs, improving residents' safety, and monitoring the environmental crisis. These indicators exhibit strong positive correlation with the perceived usefulness of AI. A high variance was found for AI's usefulness in overall reduction in service costs (1.53) and AI's usefulness in monitoring and responding to environmental and climate crises (1.08), indicating that their association with the perceived usefulness of AI considerably varied over the observed sample.

Table 2.5 presents the measurement model for ease of using AI technology. Among the various indicators for measuring the ease of using AI,

TABLE 2.4   Measurement model for perceived usefulness of AI in improving local government services

Parameter	Estimate	*t*–ratio
AI's usefulness in efficient delivery of local government services	1.00	
AI's usefulness in efficient management of resources	1.07	32.11***
AI's usefulness in improving objectivity in the delivery of local government services	0.97	30.10***
AI's usefulness in monitoring and responding to environmental and climate crises	0.94	21.76***
AI's usefulness in reducing public sector costs with savings used to reduce rates and other taxes	0.98	29.38***
AI's usefulness in reducing public sector costs with savings to improve other urban services	0.44	5.73***

Note: * = weakly significant ($p < 0.10, t > 1.645$), ** = significant ($p < 0.05, t > 1.96$), *** = strongly significant ($p < 0.01, t > 2.58$).

TABLE 2.5   Measurement model for ease of using AI technology

Parameter	Estimate	*t*-ratio
Understanding of basic concepts of AI	1.00	
Experience of interacting with an AI technology	0.97	13.51***
Visible impact of AI technology on society	1.84	13.16***
Frequency of interaction with an AI technology	1.42	12.25***

Note: * = weakly significant ($p < 0.10$, $t > 1.645$), ** = significant ($p < 0.05$, $t > 1.96$), *** = strongly significant ($p < 0.01$, $t > 2.58$).

the opinion regarding visible impact of AI on society, and frequency of interacting with AI technology was found to be the most significant. The estimates suggest that the ease of using AI potentially increases with an increase in the perception of a visible impact of AI on day-to-day lives and with the increase in the frequency of interaction. The other indicators include the ability to understand basic concepts of AI and experience with using an AI technology. Other indicators, including age and education levels, were also tested for an effect on ease of use of AI technology. While they slightly improved the model fit, the internal reliability measures for the latent variable significantly deteriorated. Therefore, the final model was estimated with only the four significant indicators.

Attitude towards AI is an important factor that can motivate AI usage, thereby influencing a positive perception regarding the usefulness of AI. Table 2.6 presents the measurement model for attitude towards AI technology. The opinion of residents regarding the positive impact of future use of AI on society was found to be the most significant indicator of attitude towards AI for the observed sample. The other indicators include opinions whether the benefits of AI outweigh the associated risks and comfort with working or living in a fully autonomous environment. Optimistic opinion regarding AI future was found to have the least loading on attitude towards AI.

4.3.2 Structural Models

The structural models indicate the results for the two hypotheses that were tested. Table 2.7 presents the structural model results.

The regression coefficient indicates that attitude has a significant positive effect on the ease of using AI technology. Users with a positive attitude

TABLE 2.6   Measurement model of attitude towards AI technology

Parameter	Estimate	t-ratio
General feeling about AI	1.00	
Better impact on society because of the future use of AI	1.14	29.59***
Feeling optimistic regarding the prospect of an AI future	0.13	7.33***
Comfort of working or living in a fully autonomous environment	0.84	20.36***
Optimistic whether benefits of AI outweigh associated risks	1.01	24.47***

Note: * = weakly significant ($p < 0.10$, $t > 1.645$), ** = significant ($p < 0.05$, $t > 1.96$), *** = strongly significant ($p < 0.01$, $t > 2.58$).

TABLE 2.7   Structural models to identify the factors influencing perceived usefulness of AI in improving local government services

Endogenous Latent Variables	Exogenous Latent Variables	Estimate	t-ratio
Ease of using AI technology	Attitude towards AI	0.48	9.78***
Perceived usefulness of AI in improving local government services	Attitude towards AI	1.00	11.02***
	Australian resident	−0.205	−2.385**
Perceived usefulness of AI in improving local government services			

Note: * = weakly significant ($p < 0.10$, $t > 1.645$), ** = significant ($p < 0.05$, $t > 1.96$), *** = strongly significant ($p < 0.01$, $t > 2.58$).

towards AI are more likely to use AI easily than those with a negative atti-tude. The coefficient estimates validate the rejection of the null hypothesis for H1 at a confidence level of 95%. The results also suggest that the per-ceived usefulness of AI in improving local government services is signifi-cantly and positively affected by users' attitudes towards AI technology.

In addition, the results indicate that the country of residence had a sig-nificant effect on the perceived usefulness of AI. The negative coefficient

TABLE 2.8   Goodness-of-fit measures for the structural equation model

Goodness-of-fit Measures	Estimates	Acceptable Criteria
Comparative Fit Index (CFI)	0.96	> 0.90
Root Mean Square Error of Approximation (RMSEA)	0.051	< 0.08
Standardised root mean square residual (SRMR)	0.06	< 0.08
Loglikelihood restricted (H0)	−14,947	
Loglikelihood unrestricted model (H1)	−14,785	
AIC	29,962	
BIC	30,124	

TABLE 2.9   Internal consistency and reliability measures for the structural equation model

Measures	Perceived Usefulness of AI in Improving Local Government Services	Ease of Using AI Technology	Attitude towards AI Technology
Internal consistency	0.89	0.74	0.80
Composite reliability	0.91	0.76	0.51

suggests that Australians were less likely to perceive AI's usefulness in improving local government services compared to the Hong Kong counterpart. These findings indicate that in a society, such as Australia, with low power distance, individualism, short-term orientation, and high indulgence, the perceived usefulness of AI is most likely to be lower than a society with high power distance, collectivism, long-term orientation, and restraint. The estimates are statistically significant, validating the rejection of the null hypothesis for H2 at a confidence level of 95%. The overall goodness-of-fit of the model is provided in Table 2.8. The fit measures comply with the criteria used in most SEM studies (Hassan et al., 2019) suggesting that the model is acceptable.

Moreover, reliability tests were also conducted to validate the consistency of model outcomes. Table 2.9 presents the test results. The internal consistency measure indicates whether the measured indicators were consistent with the underlying latent constructs. The associated internal consistency measure or the Cronbach alpha for the given specification for the perceived usefulness of AI in improving local government services and Ease of using AI technology is >0.7 which indicates excellent reliability for the given measures. However, a moderate reliability level was observed for Attitude towards AI technology.

4.4 Comparative Analysis of Perceived Usefulness of AI in Local Government Services

Out of the total respondents observed (n = 851), the Australian and the Hong Kong samples includes 604 and 246 observations, respectively. The Mann-Whitney U test results suggest that the two samples were statistically different in terms of their descriptive statistics. Therefore, a comparative analysis of perceived usefulness of AI was performed for the two samples. First separate EFA was performed on the two samples to evaluate if socio-demographic variables such as age, education, and income loaded significantly on the identified factors. The results indicated that the age, education, and income did not have any significant influence on the different factors identified that indicated perceived usefulness of AI. Then, separate SEMs were estimated for the two samples based on the conceptual model developed in Figure 2.3. Figures 2.4 and 2.5 present SEM path diagrams for Australia and Hong Kong samples, respectively. Analysis of results is presented in detail in subsequent sections.

4.4.1 Comparative Analysis of the Measurement Models

Table 2.10 presents the measurement models for the perceived usefulness of AI in improving local government services using the two samples. For both samples, efficient management of resources is an important indicator of their perception of AI usefulness. In addition, the parameters for the Australian sample suggest that their perception of AI's usefulness significantly influences their opinion regarding the use of AI in reducing public sector costs. However, in comparison for the Hong Kong sample, improving objectivity in the delivery of local government services is a significant indicator. In addition, a significant variance was found, indicating that the association between the perceived usefulness of AI in improving local government services and AI's usefulness in monitoring and responding to environmental and climate crises significantly varied over the observed Hong Kong sample.

Table 2.11 provides measurement models for the perceived ease of using AI in Australia and Hong Kong. The perception of a visible impact of AI on society was found to be a significant indicator of perceived ease of using

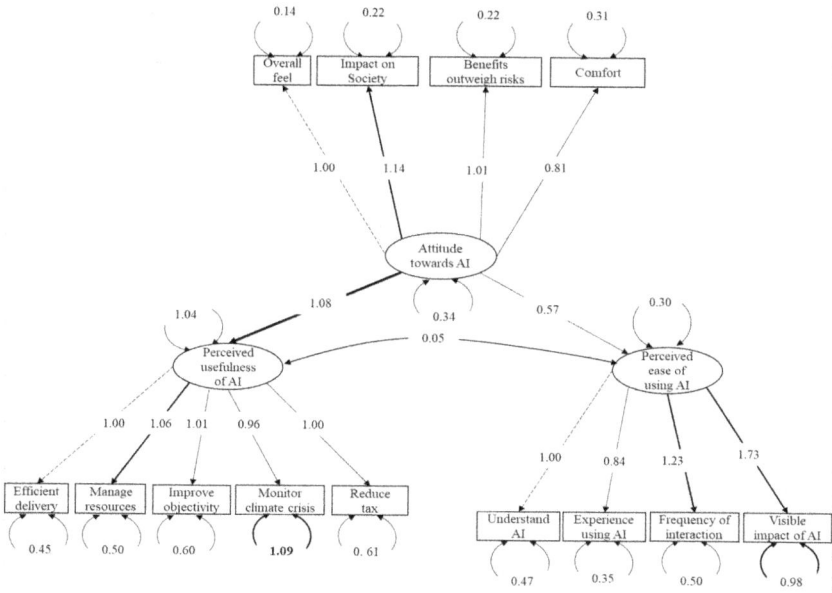

FIGURE 2.4   SEM path diagram based on the proposed hypotheses for Australian sample.

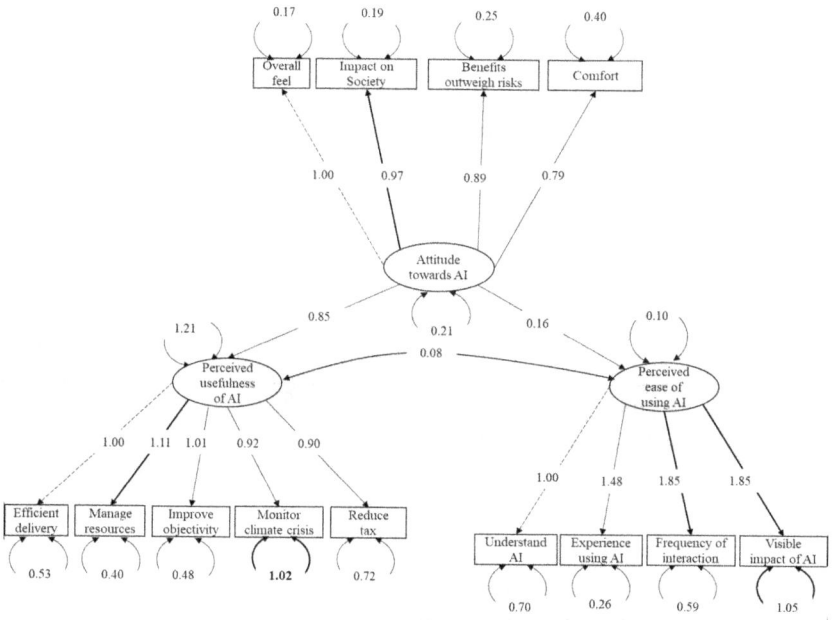

FIGURE 2.5   SEM path diagram based on the proposed hypotheses for Hong Kong sample.

TABLE 2.10   Measurement models for the perceived usefulness of AI in improving local government services: comparison between Australia and Hong Kong samples

Parameter	Australia Sample		Hong Kong Sample	
	Estimate	*t*-ratio	Estimate	*t*-ratio
AI's usefulness in the efficient delivery of local government services	1.00		1.00	
AI's usefulness in the efficient management of resources	1.06	28.8***	1.11	18.27***
AI's usefulness in improving objectivity in the delivery of local government services	0.96	26.26***	1.01	17.11***
AI's usefulness in monitoring and responding to environmental and climate crises	0.95	21.72***	0.92	13.21***
AI's usefulness in reducing public sector costs with the savings used to reduce rates and other taxes	1.00	26.87***	0.90	14.60***

Note: * = weakly significant ($p < 0.10$, $t > 1.645$), ** = significant ($p < 0.05$, $t > 1.96$), *** = strongly significant ($p < 0.01$, $t > 2.58$)

TABLE 2.11   Measurement models for perceived ease of using AI: comparison between Australia and Hong Kong samples

Parameter	Australia Sample		Hong Kong Sample	
	Estimate	*t*-ratio	Estimate	*t*-ratio
Understanding of basic concepts of AI	1.00		1.00	
Experience of interacting with an AI technology	0.84	13.59***	1.46	4.13***
Frequency of interaction with an AI technology	1.23	14.68***	1.85	4.11***
Visible impact on AI	1.75	14.70***	1.85	3.90***

Note: * = weakly significant ($p < 0.10$, $t > 1.645$), ** = significant ($p < 0.05$, $t > 1.96$), *** = strongly significant ($p < 0.01$, $t > 2.58$).

AI for the two samples. Similarly, the other indicators, including the frequency of interaction and experience with AI technology, loaded heavily on ease of AI use for both samples. Unlike Zhang et al.'s (2020) results, these results suggest that cultural differences (Hong Kong are of higher power distance; they are less individualistic, long-term oriented, and less indulgent) have no significant impact on the perceived ease of using AI.

TABLE 2.12 Measurement models for attitude towards AI technology: comparison between Australia and Hong Kong samples

Parameter	Australia Sample		Hong Kong Sample	
	Estimate	t-ratio	Estimate	t-ratio
The general feeling about AI	1.00		1.00	
Opinion that benefits of AI outweighs risks	1.01	21.11***	0.89	8.59***
Opinion regarding the societal impact due to the prospect of an AI future	1.14	22.12***	0.97	8.86***
Comfort to live and work in a fully autonomous environment	0.81	16.56***	0.79	6.68***

Note: * = weakly significant ($p < 0.10$, $t > 1.645$), ** = significant ($p < 0.05$, $t > 1.96$), *** = strongly significant ($p < 0.01$, $t > 2.58$).

Table 2.12 provides measurement models for attitude towards AI technology using the two samples. In both samples, the most significant factor that affected the attitude towards perceived usefulness of AI was the Opinion regarding the societal impact due to the prospect of an AI future. Similarly, no significant distinction was found in the effects of other indicators on the two samples. However, a significant variance in the effect of visible impact of AI was observed for the Australian sample. The variance suggests that while the opinion that there is already a substantial visible impact of AI on society suggests a positive influence on ease of using AI, the effect significantly varies across the observed Australian sample.

4.4.2 Comparative Analysis of the Structural Models

The structural models indicate the results for the two hypotheses tested for the two samples separately. Table 2.13 presents the structural model results. The results indicate that while the null hypothesis for H1 is rejected for the Australian sample at 95% confidence, it cannot be rejected for the Hong Kong sample. The observed Australians' attitude towards AI technology is more likely to be positive if they have found it easy to use an AI technology. However, for the observed Hong Kongers, there is no significant association between attitude and ease of use of AI. A strong correlation between ease of use and attitude towards AI was found in the Australian sample compared to the Hong Kong sample. The results indicate the null hypothesis for H2 can be rejected for both Australian and Hong Kong samples at 95% confidence. The correlation between the

TABLE 2.13   Structural models to identify the factors influencing perceived usefulness of AI in improving local government services: comparison between Australia and Hong Kong samples

		Australia Sample		Hong Kong Sample	
Endogenous Latent Variables	**Exogenous Latent Variables**	**Estimate**	**_t_-ratio**	**Estimate**	**_t_-ratio**
Ease of using AI technology	Attitude towards AI	0.55	10.006***	0.16	2.23
Perceived usefulness of AI in improving local government services	Attitude towards AI	1.08	11.96***	0.84	4.38***

Note: * = weakly significant ($p < 0.10$, $t > 1.645$), ** = significant ($p < 0.05$, $t > 1.96$), *** = strongly significant ($p < 0.01$, $t > 2.58$).

perceived usefulness of AI and attitude towards AI was found to be stronger for the Australian sample compared to Hong Kongers. In addition, a significant correlation was found between perceived usefulness and ease of use in the Australian sample. The estimates suggest that a positive perception regarding AI usefulness is likely to exist among residents if they find it easy to use AI-based applications.

Table 2.14 presents goodness-of-fit measures for the SEMs. The fit measures indicate that the two models are acceptable.

Table 2.15 presents the reliability measures for the two samples. While the measures indicate an excellent composite reliability for the two models, the internal consistency measures were found to indicate a moderate reliability for the ease of using AI variable for both samples. Further investigation is required to identify indicators which improve reliability for the defined latent construct.

5 DISCUSSION AND CONCLUSION

The rapidly advancing capabilities of AI have turned it into a highly alluring technology for many industry sectors to adopt (Cubric, 2020; Regona et al., 2022a, 2022b). The public sector, including local governments, is one of them (De Sousa et al., 2019; Nili et al., 2022; Yigitcanlar et al., 2022b, 2022c). Nonetheless, there is a limited understanding on public perceptions on AI in the context of local government services (Mikalef et al., 2019). Particularly, to the best of our knowledge, there are no cross-nation comparative studies available, to understand the similarities and differences

TABLE 2.14   Goodness of fit measures for SEM estimated for Australia and Hong Kong samples

Goodness-of-fit Measures	Estimates for Australian Sample	Estimates for Hong Kong Sample	Acceptable Criteria
Comparative Fit Index (CFI)	0.98	1.00	> 0.90
Root Mean Square Error of Approximation (RMSEA)	0.05	0.01	< 0.08
Standardised root mean square residual	0.03	0.04	< 0.08
Loglikelihood restricted (H0)	−9,529	−3,838	
Loglikelihood unrestricted model (H1)	−9,458	−3,806	
AIC	19,116	7,735	
BIC	19,244	7,836	

TABLE 2.15   Internal consistency and reliability measures for the structural equation model

Latent Constructs	Internal Consistency		Composite Reliability	
	Estimates for Australian Sample	Estimates for Hong Kong Sample	Estimates for Australian Sample	Estimates for Hong Kong Sample
Perceived usefulness of AI in improving local government services	0.92	0.91	0.92	0.91
Ease of using AI technology	0.67	0.58		
Attitude towards AI technology	0.74	0.74		

between country contexts concerning public views on the role of AI in local government services.

To help in the efforts to bridge this knowledge gap and also to provide empirical evidence and insights into public perceptions concerning the use of AI in local government services, this study conducted an online survey with 851 residents of Sydney, Melbourne, Brisbane, and Hong Kong. The statistical analysis of the data generated the following invaluable insights into public perceptions on AI in local government services.

First, the study disclosed that the ease of using AI is significantly and positively influenced by the attitude towards AI. This finding is relevant to both Australian and Hong Kong contexts. This is to say, people perceive AI positively as they get more familiar with AI and the technology becomes part of their daily lives. Other empirical studies also support this finding. For instance, Dennis et al. (2021, p. 93) shed light on the differences between people who have ridden a connected and autonomous vehicle (shuttle bus using AI technology) in downtown Las Vegas versus those who have not. The study found that "remarkably only few participants who had ridden the shuttle expressed negative sentiments toward connected and autonomous vehicles, suggesting that experiencing the connected and autonomous vehicle is largely related to positive feelings about this technology".

Second, the analysis found that perceived usefulness of AI in local government services is significantly and positively influenced by the attitude towards AI. This was the case in both case study country contexts. In other words, as advocated by Schepman and Rodway (2020, p. 1), "people's general attitudes towards AI play a large role in their acceptance of AI". Hence, the authorities are required to inform and engage citizens with AI adoption in local government services. On that very point, Neudert et al. (2020, p. 2) stated that

> putting AI to work for good governance will be a twofold challenge. Involving AI systems in public administration is going to require inclusive design, informed procurement, purposeful implementation, and persistent accountability. Additionally, it will require convincing citizens in many countries around the world that the benefits of using AI in public agencies outweigh the risks.

Next, the results also suggest that Australians were less likely to perceive AI's usefulness in improving local government services, compared to Hong Kongers (Table 2.13). The cultural differences between the two countries in aspects such as power distance, collectivism, short-term orientation, and indulgence could be the likely factors affecting such varied preferences for AI. This study displays similar findings to others. For example, high power distance and collectivism in Hong Kong negatively impact the perceived usefulness of technology (Mahomed et al., 2018). Communication can suffer from a fear of power. Avoidance behaviour is a result of emotional experiences like fear. Being silent seems to be a common avoidance strategy in this situation. Silence leads to reduced collaboration, information sharing, and subpar feedback (Dai et al., 2022). Likewise, previous research found that communication is more indirect in collectivist cultures (Campion & Wang, 2019). As tools' usefulness may require communication and knowledge sharing, high power distance places like Hong Kong may have lower perceived usefulness of AI in improving government services.

Fourth, the results indicate that, among the local government responsibility areas, AI is particularly seen useful in resource management and service delivery efficiency, urban-service cost reduction, public safety improvement, and environmental crisis monitoring effectiveness. This finding is also relevant to both contexts. While there is a lack of empirical studies investigating public view on which local government services would benefit AI adoption most, there is growing literature on how public sector benefits from AI technologies. For example, according to Valle-Cruz et al. (2019, p. 99), AI is "useful in the decision-making process, cost reduction, fighting corruption, detecting changes in the environment, disaster prevention and response, government-citizen interaction, personalisation of services, interoperability, and value creation".

Fifth, the analysis shed light on the cross-national differences in public perception of AI. To be precise, AI is more positively perceived by Australians in comparison to Hong Kongers, indicating the impact of contextual differences. A positive perception regarding AI usefulness is likely to exist between Australian users if they find it easy to use AI-based applications. While a study by Kostka et al. (2021, p. 686) found high-level acceptance of AI-based facial recognition technology in China, they also note that "China, the authoritarian political context might be reflected in the reported levels of social acceptance, as the dissent toward technologies officially endorsed by the government can be difficult".

Then, the findings of the study have important implications for actualising AI's affordances that contribute to the public value creation and quality service delivery in local governments. Understanding public perceptions and expectations for AI use of the local government services gives an opportunity for local authorities to take appropriate actions to improve local government service quality, effectiveness, and efficiency. These actions include (a) Understanding and addressing community technology adoption and acceptance challenges (Vu & Lim, 2021); (b) Increasing community awareness and readiness by informing local communities on the benefits of AI for urban services (Vitezić & Perić, 2021); (c) Establishing wider community and stakeholder engagement in decision-making, planning and deployment processes of local government AI technology (King & Cotterill, 2007; Lebcir et al., 2021); and (d) While making AI adoption decisions, factoring in cultural and contextual differences is critical as no one-size-fit all—even in the same country and city (Hardman, 2022).

Lastly, today AI technologies are rapidly becoming part of our societies and cities and bringing a transformational impact (Fatima et al., 2020). Many private organisations have made AI technologies an integral component of their business models, and public sector is also moving towards the same direction (Fatima et al., 2021). Furthermore, the planning of AI in local governments is starting to gain momentum in the recent year (Wang et al., 2020). Nevertheless, little is known as per how the public perceives AI for local government services, especially when cultural and contextual differences are concerned (Selwyn & Gallo-Cordoba, 2021). In this context, the findings of this chapter are critical in informing local government authorities—e.g., urban policymakers, managers, and planners—on their policy, planning, and implementation decisions concerning AI.

We also note that this study has the following limitations: (a) The cross-nation comparison only concerns two countries and only the residents of main cities of these countries were surveyed; (b) There might be some limitations on the representativeness of the samples. While this is a limitation, the sample sizes used in this study are like the ones used in most similar studies following a stated preference approach. However, future studies need to focus on recruiting representative samples based on different socioeconomic groups to enable detailed analysis for their perceived usefulness regarding AI technologies.

In addition, the results from the statistical analyses are sound and consistent; (c) The study only adopted a single lens to look at the AI in local government services issue that is the general public's point of view—rather than also capturing the views of local government policymakers, managers, and planners, or technology and consultancy companies that develop and deploy AI for local governments; and (d) Cultural differences could be perceived from vastly different fields and perspectives other than those mentioned in this chapter, like democracy and religious, as Sent and Annelie (2022) quoted Hodstede and Fink (2007)'s comments: "When you study cultures you have to be open to relevant information from various disciplines, from anthropology, from sociology, from social psychology, and even from individual psychology and from economics. All those disciplines play some role". It implies that there remains room for further research from vastly different aspects. For example, cultural differences could be perceived from perspectives other than those mentioned in this chapter, like democracy and religious. Yet, as the results of the World Values Surveys 2022 have not yet been released when the authors submitted the revised version, this research did not include these cultural differences. Our prospective research will focus on addressing these limitations.

ACKNOWLEDGEMENTS

This chapter, with permission from the copyright holder, is a reproduced version of the following journal article: Yigitcanlar, T., Li, R., Beeramoole, P., & Paz, A. (2023). Artificial intelligence in local government services: public perceptions from Australia and Hong Kong. *Government Information Quarterly*, 40(3), 101833.

APPENDIX 2.A. SURVEY QUESTIONNAIRE

PART A: YOUR BACKGROUND

What is your gender?

What is your age?

What is your highest education qualification obtained?

What is your current employment status?

Which industry are you employed in or have a business in?

What is your occupation/job title?

What is your gross weekly income?

What is your residential postcode?

PART B: YOUR PRIOR KNOWLEDGE ON AI

From which source did you mainly learn what AI is?

What is your first thoughts when you think of AI?

Do you feel like you understand the basic concepts of AI?

Which of the following best describes AI's abilities?

PART C: YOUR PRIOR EXPERIENCE WITH AI

Have you ever interacted with an AI technology?

Which of the following technologies have you used or encountered?

How often would you estimate you interact with the previously identified applications?

How long do you think it will be before AI has a noticeable impact on your daily life?

PART D: YOUR PERCEIVED BENEFITS OF AI

Which of the following is the most beneficial function of AI technology?

Which of the following is the most promising about the future use of AI?

PART E: YOUR PERCEIVED RISKS OF AI

Which of the following is the biggest disadvantage of AI technology?

Which of the following do you fear the most about the future use of AI?

PART F: YOUR VIEW ON THE AI FUTURE

Do you think society will become better or worse as a result of the future use of AI?

Do you think the benefits outweigh the risks when looking at the future use of AI?

PART G: YOUR COMFORT AND TRUST WITH AI

How do you feel about AI in general?

Which of the following would you be comfortable with AI operating autonomously?

Would you feel comfortable in a fully autonomous place to live or work?

How does the prospect of an AI future make you feel?

To what degree do the following scenarios concern you?

How much do you trust AI in your lifestyle?

PART H: YOUR VIEW ON AI IN LOCAL GOVERNMENT SERVICES

Which of the following local government services are most suited for the future application of AI technology?

When do you believe AI supported local government services will come into your city?

Where do you think AI is needed the most in local government services?

To what degree would AI be useful in the following?

Which of the following are the key challenges for local governments to adopt AI for local service delivery?

REFERENCES

Abdelhakim, A., Abou-Shouk, M., Ab Rahman, N., & Farooq, A. (2023). The fast-food employees' usage intention of robots: a cross-cultural study. *Tourism Management Perspectives*, 45, 101049.

Ahn, M., & Chen, Y. (2022). Digital transformation toward AI-augmented public administration: the perception of government employees and the willingness to use AI in government. *Government Information Quarterly*, 39, 101664.

Akour, I., Alshare, K., Miller, D., & Dwairi, M. (2006). An exploratory analysis of culture, perceived ease of use, perceived usefulness, and internet acceptance: the case of Jordan. *Journal of Internet Commerce*, 5, 83–108.

Allam, Z., & Dhunny, Z. (2019). On big data, artificial intelligence and smart cities. *Cities*, 89, 80–91.

Alshahrani, A., Dennehy, D., & Mäntymäki, M. (2021). An attention-based view of AI assimilation in public sector organizations: the case of Saudi Arabia. *Government Information Quarterly*. https://doi.org/10.1016/j.giq.2021.101617.

Amershi, B. (2019). Culture, the process of knowledge, perception of the world and emergence of AI. *AI & Society*, 35, 417–430.

Araujo, T., Helberger, N., Kruikemeier, S., & De Vreese, C. (2020). In AI we trust? Perceptions about automated decision-making by artificial intelligence. *AI & Society*, 35, 611–623.

Arteaga, C., Paz, A., & Park, J. (2020). Injury severity on traffic crashes: a text mining with an interpretable machine-learning approach. *Safety Science*, 132, 104988.

Assemi, B., Paz, A., & Baker, D. (2021). On-street parking occupancy inference based on payment transactions. *IEEE Transactions on Intelligent Transportation Systems*. https://doi.org/10.1109/TITS.2021.3095277.

Australian Bureau of Statistics (2022). Cultural diversity of Australia. https://www.abs.gov.au/articles/cultural-diversity-australia.

Australian Government (2022). Australia's digital economy. https://digitaleconomy.pmc.gov.au/fact-sheets/artificial-intelligence.

Bentafat, E., Rathore, M., & Bakiras, S. (2021). Towards real-time privacy-preserving video surveillance. *Computer Communications*, 180, 97–108.

Bogais, J., Garbuio, M., Groutsis, D., Peter, S., Riemer, K., Seno-Alday, S., Shields, J., Sutton-Brady, C., & Voola, R. (2022). Conquering 'the tyranny of distance': Australian–European economic and geopolitical relationships past, present and future. *European Management Journal*, 40, 310–319.

Byrne, B. (2011). *Structural equation modeling with mplus: basic concepts, applications, and programming.* London, Routledge.

Campion, L., & Wang, C. (2019). Collectivism and individualism: the differentiation of leadership. *Technology Trends*, 63, 353–356.

Chen, T., Guo, W., Gao, X., & Liang, Z. (2021). AI-based self-service technology in public service delivery: user experience and influencing factors. *Government Information Quarterly*, 38, 101520.

Chen, W., Zhao, L., Kang, Q., & Di, F. (2020). Systematizing heterogeneous expert knowledge, scenarios and goals via a goal-reasoning artificial intelligence agent for democratic urban land use planning. *Cities*, 101, 102703.

Chohan, S., Hu, G., Khan, A., Pasha, A., & Sheikh, M. (2020). Design and behavior science in government-to-citizens cognitive communication: a study towards an inclusive framework. *Transforming Government*, 15, 532–549.

Chopdar, P., Korfiatis, N., Sivakumar, V., & Lytras, M. (2018). Mobile shopping apps adoption and perceived risks: a cross-country perspective utilising the Unified Theory of Acceptance and Use of Technology. *Computers in Human Behavior*, 86, 109–128.

Çokluk, Ö., & Koçak, D. (2016). Using Horn's parallel analysis method in exploratory factor analysis for determining the number of factors. *Educational Sciences: Theory and Practice*, 16, 537–551.

Cubric, M. (2020). Drivers, barriers and social considerations for AI adoption in business and management: a tertiary study. *Technology in Society*, 62, 101257.

Cui, D., & Wu, F. (2021). The influence of media use on public perceptions of artificial intelligence in China: evidence from an online survey. *Information Development*, 37, 45–57.

Dai, Y., Li, H., Xie, W., & Deng, T. (2022). Power distance belief and workplace communication: the mediating role of fear of authority. *International Journal of Environmental Research and Public Health*, 19, 2932.

D'Amico, G., L'Abbate, P., Liao, W., Yigitcanlar, T., & Ioppolo, G. (2020). Understanding sensor cities: Insights from technology giant company driven smart urbanism practices. *Sensors*, 20(16), 4391.

Dasgupta, S., & Gupta, B. (2019). Espoused organizational culture values as antecedents of internet technology adoption in an emerging economy. *Information & Management*, 56, 103142.

Dastjerdi, A., Kaplan, S., Nielsen, O., & Pereira, F. (2019). Factors driving the adoption of mobility-management travel app: a Bayesian structural equation modelling analysis. Transportation Research Board 98th Annual Meeting (No. 19–02192). Washington, DC, USA.

De Sousa, W., de Melo, E., Bermejo, P., Farias, R., & Gomes, A. (2019). How and where is artificial intelligence in the public sector going? A literature review and research agenda. *Government Information Quarterly*, 36, 101392.

Dennis, S., Paz, A., & Yigitcanlar, T. (2021). Perceptions and attitudes towards the deployment of autonomous and connected vehicles: insights from Las Vegas, Nevada. *Journal of Urban Technology*, 28, 75–95.

Desouza, K., Dawson, G., & Chenok, D. (2020). Designing, developing, and deploying artificial intelligence systems: lessons from and for the public sector. *Business Horizons*, 63, 205–213.

Dong, J. (2011). User acceptance of information technology innovations in the Chinese cultural context. *Asian Journal of Technology Innovation*, 17, 129–149.

Dwivedi, Y., Hughes, L., Ismagilova, E., Aarts, G., Coombs, C., Crick, T., ... & Williams, M. (2021). Artificial intelligence (AI): multidisciplinary perspectives on emerging challenges, opportunities, and agenda for research, practice and policy. *International Journal of Information Management*, 57, 101994.

Fatima, S., Desouza, K., & Dawson, G. (2020). National strategic artificial intelligence plans: a multi-dimensional analysis. *Economic Analysis and Policy*, 67, 178–194.

Fatima, S., Desouza, K., Denford, J., & Dawson, G. (2021). What explains governments interest in artificial intelligence? A signalling theory approach. *Economic Analysis and Policy*, 71, 238–254.

Feldstein, S. (2019). *The global expansion of AI surveillance*. Washington, DC, Carnegie Endowment for International Peace.

Gursoy, D., Chi, O., Lu, L., & Nunkoo, R. (2019). Consumers acceptance of artificially intelligent (AI) device use in service delivery. *International Journal of Information Management*, 49, 157–169.

Hardman, L. (2022). Cultural influences on artificial intelligence: along the new Silk Road. In: Werthner, Hannes; Prem, Erich; Lee, Edward A.; Ghezzi, Carlo (eds.) *Perspectives on Digital Humanism* (pp. 233–239). Springer, Cham.

Hassan, H., Ferguson, M., Razavi, S., & Vrkljan, B. (2019). Factors that influence older Canadians' preferences for using autonomous vehicle technology: a structural equation analysis. *Transportation Research Record*, 2673, 469–480.

Hofstede, G., & Fink, G. (2007). Culture: Organisations, personalities and nations. Gerhard Fink interviews Geert Hofstede. *European Journal of International Management*, 1, 14–22.

Huang, S., & Crotts, J. (2019). Relationships between Hofstede's cultural dimensions and tourist satisfaction: a cross-country cross-sample examination. *Tourism Management*, 72, 232–241.

Huang, F., Teo, T., & Zhou, M. (2020). Chinese students' intentions to use the Internet-based technology for learning. *Educational Technology Research and Development*, 68, 575–591.

Izuagbe, R., Ifijeh, G., Izuagbe-Roland, E., Olawoyin, O., & Ogiamien, L. (2019). Determinants of perceived usefulness of social media in university libraries: subjective norm, image and voluntariness as indicators. *The Journal of Academic Librarianship*, 45, 394–405.

Kaiser, H. (1970). A second-generation little jiffy. *Psychometrika*, 35, 401–415.

Kaiser, H. (1974). An index of factorial simplicity. *Psychometrika*, 39, 31–36.

Kankanamge, N., Yigitcanlar, T., & Goonetilleke, A. (2021). Public perceptions on artificial intelligence driven disaster management: evidence from Sydney, Melbourne and Brisbane. *Telematics and Informatics*, 65, 101729.

Kassens-Noor, E., Wilson, M., Kotval-Karamchandani, Z., Cai, M., & Decaminada, T. (2021). Living with autonomy: public perceptions of an AI-mediated future. *Journal of Planning Education and Research*. https://doi.org/10.1177/0739456X20984529.

King, S., & Cotterill, S. (2007). Transformational government? The role of information technology in delivering citizen-centric local public services. *Local Government Studies*, 33, 333–354.

Kostka, G., Steinacker, L., & Meckel, M. (2021). Between security and convenience: facial recognition technology in the eyes of citizens in China, Germany, the United Kingdom, and the United States. *Public Understanding of Science*, 36, 671–690.

Krejcie, R., & Morgan, D. (1970). Determining sample size for research activities. *Educational and Psychological Measurement*, 30, 607–610.

Landman, T., & Carvalho, E. (2003). *Issues and methods in comparative politics: an introduction*. London, Routledge.

Lebcir, R., Hill, T., Atun, R., & Cubric, M. (2021). Stakeholders' views on the organizational factors affecting application of artificial intelligence in healthcare: a scoping review protocol. *BMJ Open*, 11, e044074.

Li, R., Tang, B., & Chau, K. (2019). Sustainable construction safety knowledge sharing: a partial least square-structural equation mdelling and a feedforward neural network approach. *Sustainability*, 11, 5831.

Li, W., Yigitcanlar, T., Erol, I., & Liu, A. (2021). Motivations, barriers and risks of smart home adoption: from systematic literature review to conceptual framework. *Energy Research & Social Science*, 80, 102211.

Lin, Y., Chen, M., & Flowerdew, J. (2022). Same, same but different': representations of Chinese mainland and Hong Kong people in the press in post-1997 Hong Kong. *Critical Discourse Studies*, 19, 364–373.

Liu, W., Xu, Y., Fan, D., Li, Y., Shao, X., & Zheng, J. (2021). Alleviating corporate environmental pollution threats toward public health and safety: the role of smart city and artificial intelligence. *Safety Science*, 143, 105433.

Lok, P., & Crawford, J. (2004). The effect of organizational culture and leadership style on job satisfaction and organizational commitment. *Journal of Management Development*, 23, 321–338.

Lopes, H. F. (2023). The impact of refugees in Neutral Hong Kong and Macau, 1937–1945. *The Historical Journal*, 66(1), 210–236.

Mahomed, A., Mcgrath, M., & Yuh, B. (2018). The impact of Hofstede's National Culture on usage of emails among Academician in Malaysian Public Universities. *International Journal of Engineering & Technology*, 7, 990–996.

Maskey, R., Fei, J., & Nguyen, H. (2018). Use of exploratory factor analysis in maritime research. *The Asian Journal of Shipping and Logistics*, 34, 91–111.

McKay, C. (2019). Predicting risk in criminal procedure: actuarial tools, algorithms, AI and judicial decision-making. *Current Issues in Criminal Justice*, 32, 22–39.

McKnight, P., & Najab, J. (2010). Mann-Whitney U test. In: George Stricker (ed.) *The Corsini Encyclopedia of Psychology* (pp. 1–10). Washington DC, John Wiley & Sons.

Mikalef, P., Fjørtoft, S., & Torvatn, H. (2019). Artificial Intelligence in the public sector: a study of challenges and opportunities for Norwegian municipalities. In: *Conference on e-Business, e-Services and e-Society* (pp. 267–277). Springer, Cham.

Neri, H., & Cozman, F. (2020). The role of experts in the public perception of risk of artificial intelligence. *AI & Society*, 35, 663–673.

Neudert, L., Knuutila, A., & Howard, P. (2020). *Global attitudes towards AI, machine learning & automated decision making.* Oxford: Oxford Commission on AI & Good Governance.

News.gov.hk. (2022). HK to focus on AI tech. Accessed on 15 April 2022 from https://www.news.gov.hk/eng.

Nili, A., Desouza, K., & Yigitcanlar, T. (2022). What can the public sector teach us about deploying artificial intelligence technologies? *IEEE Software*, 39(6), 58–63.

Pan, J., Wong, D., Joubert, L., & Chan, C. (2008). The protective function of meaning of life on life satisfaction among Chinese students in Australia and Hong Kong: a cross-cultural comparative study. *Journal of American College Health*, 57, 221–232.

Parboteeah, D., & Parboteeah, K. (2005). Perceived usefulness of information technology: a cross-national model. *Journal of Global Information Technology Management*, 8, 29–48.

Pencheva, I., Esteve, M., & Mikhaylov, S. (2020). Big data and AI: a transformational shift for government: so, what next for research? *Public Policy and Administration*, 35, 24–44.

Phau, I., & Kea, G. (2007). Attitudes of university students toward business ethics: a cross-national investigation of Australia, Singapore and Hong Kong. *Journal of Business Ethics*, 72, 61–75.

Polyakova, A., & Meserole, C. (2019). Exporting digital authoritarianism: the Russian and Chinese models. Policy Brief, Democracy and Disorder Series, Washington, DC: Brookings.

Rahman, M. (2023). Sample size determination for survey research and non-probability sampling techniques: a review and set of recommendations. *Journal of Entrepreneurship, Business and Economics*, 11(1), 42–62.

Regona, M., Yigitcanlar, T., Xia, B., & Li, R. (2022a). Opportunities and adoption challenges of AI in the construction industry: a PRISMA review. *Journal of Open Innovation*, 8, 45.

Regona, M., Yigitcanlar, T., Xia, B., & Li, R. (2022b). Artificial intelligent technologies for the construction industry: how are they perceived and utilized in Australia? *Journal of Open Innovation*, 8, 16.

Ritchie, K., Cartledge, C., Growns, B., Yan, A., Wang, Y., Guo, K., Kramer, R., Edmond, G., Martire, K., Roque, M., & White, D. (2021). Public attitudes towards the use of automatic facial recognition technology in criminal justice systems around the world. *PLoS One*, 16, e0258241.

Roski, J., Maier, E., Vigilante, K., Kane, E., & Matheny, M. (2021). Enhancing trust in AI through industry self-governance. *Journal of the American Medical Informatics Association*, 28, 1582–1590.

Schepman, A., & Rodway, P. (2020). Initial validation of the general attitudes towards qrtificial intelligence scale. *Computers in Human Behavior Reports*, 1, 100014.

Selwyn, N., & Gallo-Cordoba, B. (2021). Australian public understandings of artificial intelligence. *AI & Society*. https://doi.org/10.1007/s00146-021-01268-z.

Sent, E., & Kroese, A. (2022). Commemorating Geert Hofstede, a pioneer in the study of culture and institutions. *Journal of Institutional Economics*, 18, 15–27.

Shao, X., Li, Y., Suseno, Y., Li, R., Gouliamos, K., Yue, X.G., & Luo, Y. (2021). How does facial recognition as an urban safety technology affect firm performance? The moderating role of the home country's government subsidies. *Safety Science*, 143, 105434.

Sharma, G., Yadav, A., & Chopra, R. (2020). Artificial intelligence and effective governance: a review, critique and research agenda. *Sustainable Futures*, 2, 100004.

Son, T., Weedon, Z., Yigitcanlar, T., Sanchez, T., Corchado, J., & Mehmood, R. (2023). Algorithmic urban planning for smart and sustainable development: systematic review of the literature. *Sustainable Cities and Society*, 94, 104562.

The Hong Kong Special Administrative Region (2017). Application of artificial intelligence and opening up government data. https://www.info.gov.hk/gia/general/201712/13/P2017121300429.htm.

Tobias, S., & Carlson, J. (1969). Brief report: Bartlett's test of sphericity and chance findings in factor analysis. *Multivariate Behavioral Research*, 4, 375–377.

Ullah, F., Qayyum, S., Thaheem, M., Al-Turjman, F., & Sepasgozar, S. (2021). Risk management in sustainable smart cities governance: a TOE framework. *Technological Forecasting and Social Change*, 167, 120743.

Valle-Cruz, D., Alejandro Ruvalcaba-Gomez, E., Sandoval-Almazan, R., & Ignacio Criado, J. (2019). A review of artificial intelligence in government and its potential from a public policy perspective. In: *Proceedings of the 20th Annual International Conference on Digital Government Research* (pp. 91–99). Dubai, United Arab Emirates.

Vitezić, V., & Perić, M. (2021). Artificial intelligence acceptance in services: connecting with generation Z. *The Service Industries Journal*, 41, 926–946.

Vu, H., & Lim, J. (2021). Effects of country and individual factors on public acceptance of artificial intelligence and robotics technologies: a multilevel SEM analysis of 28-country survey data. *Behaviour & Information Technology*. https://doi.org/10.1080/0144929X.2021.1884288.

Wang, Y., Zhang, N., & Zhao, X. (2020). Understanding the determinants in the different government AI adoption stages: evidence of local government chatbots in China. *Social Science Computer Review*. https://doi.org/10.1177/0894439320980132.

Wilson, C., & Van Der Velden, M. (2022). Sustainable AI: an integrated model to guide public sector decision-making. *Technology in Society*, 68, 101926.

Wirtz, B., Weyerer, J., & Geyer, C. (2019). Artificial intelligence and the public sector: applications and challenges. *International Journal of Public Administration*, 42, 596–615.

Xiong, J., Hswen, Y., & Naslund, J. (2020). Digital surveillance for monitoring environmental health threats: a case study capturing public opinion from twitter about the 2019 Chennai water crisis. *International Journal of Environmental Research and Public Health*, 17, 5077.

Yeh, S., Wu, A., Yu, H., Wu, H., Kuo, Y., & Chen, P. (2021). Public perception of artificial intelligence and its connections to the sustainable development goals. *Sustainability*, 13, 9165.

Yigitcanlar, T. (2021). Greening the artificial intelligence for a sustainable planet: an editorial commentary. *Sustainability*, 13, 13508.

Yigitcanlar, T., & Cugurullo, F. (2020). The sustainability of artificial intelligence: an urbanistic viewpoint from the lens of smart and sustainable cities. *Sustainability*, 12, 8548.

Yigitcanlar, T., Butler, L., Windle, E., Desouza, K., Mehmood, R., & Corchado, J. (2020a). Can building 'artificially intelligent cities' protect humanity from natural disasters, pandemics and other catastrophes? an urban scholar's perspective. *Sensors*, 20, 2988.

Yigitcanlar, T., Desouza, K., Butler, L., & Roozkhosh, F. (2020b). Contributions and risks of artificial intelligence (AI) in building smarter cities: insights from a systematic review of the literature. *Energies*, 13, 1473.

Yigitcanlar, T., Kankanamge, N., Regona, M., Maldonado, M., Rowan, R., Ryu, A., Desouza, K., Corchado, J., Mehmood, R., & Li, R. (2020c). Artificial intelligence technologies and related urban planning and development concepts: how are they perceived and utilized in Australia? *Journal of Open Innovation*, 6, 187.

Yigitcanlar, T., Corchado, J., Mehmood, R., Li, R., Mossberger, K., & Desouza, K. (2021a). Responsible urban innovation with local government artificial intelligence (AI): a conceptual framework and research agenda. *Journal of Open Innovation*, 7, 71.

Yigitcanlar, T., Mehmood, R., & Corchado, J. (2021b). Green artificial intelligence: towards an efficient, sustainable and equitable technology for smart cities and futures. *Sustainability*, 13, 8952.

Yigitcanlar, T., Li, R., Inkinen, T., & Paz, A. (2022a). Public perceptions on application areas and adoption challenges of AI in urban services. *Emerging Sciences Journal*, 6(6), 1199–1236.

Yigitcanlar, T., Agdas, D., & Degirmenci, K. (2022b). Artificial intelligence in local governments: perceptions of city managers on prospects, constraints and choices. *AI & Society*. https://doi.org/10.1007/s00146-022-01450-x.

Yigitcanlar, T., Degirmenci, K., & Inkinen, T. (2022c). Drivers behind the public perception of artificial intelligence: insights from major Australian cities. *AI & Society*. https://doi.org/10.1007/s00146-022-01566-0.

Yoon, C. (2009). The effects of national culture values on consumer acceptance of e-commerce: online shoppers in China. *Information & Management*, 46, 294–301.

Young, A., Amara, D., Bhattacharya, A., & Wei, M. (2021). Patient and general public attitudes towards clinical artificial intelligence: a mixed methods systematic review. *The Lancet Digital Health*, 3, e599–e611.

Yun, J., Lee, D., Ahn, H., Park, K., & Yigitcanlar, T. (2016). Not deep learning but autonomous learning of open innovation for sustainable artificial intelligence. *Sustainability*, 8, 797.

Zhang, Z., Xia, E., & Huang, J. (2022). Impact of the moderating effect of national culture on adoption intention in wearable health care devices: meta-analysis. *JMIR Mhealth Uhealth*, 10, e30960.

Zuiderwijk, A., Chen, Y., & Salem, F. (2021). Implications of the use of artificial intelligence in public governance: a systematic literature review and a research agenda. *Government Information Quarterly*, 38, 101577.

CHAPTER 3

Perceptions on Artificial Intelligence in Urban Disaster Management

1 INTRODUCTION

During the last several decades, the frequency of disaster occurrence along with the magnitude of disaster impacts has been increasing across the globe at a considerable rate (Ludwig, 2017; Al Qundus et al., 2020; Kankanamge et al., 2020). For example, in Australia, between 1910 and 2015, 8,698 people lost their lives due to natural disasters, with 4,555 of these deaths (52.4%) attributed to extreme heatwaves related (Seeta, 2020). The frequency of heatwaves is on the rise as illustrated in Figure 3.1. According to King (2017), the number of hot days in Australia has doubled in the last 50 years. This is no surprise as Australia has warmed by over 1.4°C since 1910, and with the continuing climate crisis, the temperatures are further increasing (Mortoja & Yigitcanlar, 2020). What is worst, the magnitude of disasters has also increased during the past decades in Australia. For instance, the 2010–2011 Queensland Floods resulted in more than 78% of the State being designated as a disaster zone. This area is bigger than the total land extent of France and Germany combined (Seeta, 2002).

Accordingly, the emergency responses need to be prompter and more relevant to the crisis situations that are now occurring in higher frequencies and magnitudes. The lack of timely and adequate emergency

DOI: 10.1201/9781003521440-4

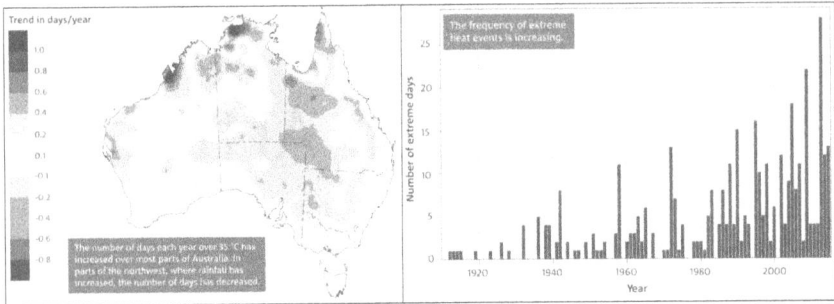

FIGURE 3.1   Increasing heatwave frequency and magnitude in Australia 1910–2015 (King, 2017).

response will increase the already high disaster risks further (Lee et al., 2021). As a solution to this issue, the most recent disaster management practices strive to incorporate disaster management theories and practices with novel digital technology solutions, where artificial intelligence (AI) applications are the most popular (Kemper & Kemper, 2020; Tan et al., 2021).

The use of AI is becoming a norm in disaster management efforts (Sun et al., 2020). In many cases, the public is also directly or indirectly engaged with AI in disaster management activities in various ways. Just to give some examples, these activities range from AI-driven analysis of public-generated social media information on a disaster (Alam et al., 2017), public engagement in AI-driven crowdsourced disaster information provision (Poblet et al., 2018) and AI-driven gamification applications, which generate public disaster awareness through serious games (Uskov & Sekar, 2015). While the public gets engaged with AI-aided disaster management activities, there are only limited empirical investigations and understanding of public perceptions concerning AI for disaster management (Neri & Cozman, 2019; Yigitcanlar et al., 2020b).

Against this backdrop, the chapter aims to expand our understanding on how the public perceives the benefits and risks in application of AI in disaster management. The study was based on an online survey undertaken in Australia. Following this introductory section, Section 2 provides a critical review of research literature. Section 3 introduces the study design. Section 4 presents the results. Section 5 provides a discussion on the findings. Section 6 provides a succinct summary of the key findings of the study.

2 LITERATURE BACKGROUND

2.1 Artificial Intelligence

With the emergence of the novel concepts of smart cities, knowledge cities, and innovation districts, AI has become the most disruptive innovation (Yigitcanlar et al., 2020a). There is no universal definition for AI, where it is described by many scholars in different ways. Mostly, AI is interpreted as "machines or computers that mimic cognitive functions that humans associate with the human mind, such as learning and problem solving" (Schalko, 1990). McCarthy (1988) defined AI as a technology that is used with methods of achieving goals in which the information has a certain complex character. However, in simple terms, AI is an attempt to reproduce human cognitive functions in computer systems (Wang, 2008).

Applications of AI range from banking, marketing, gaming, urban planning, construction, agriculture and health to defence, security, disaster management, and many more (Pannu et al., 2015; Saravi et al., 2019; Wirtz et al., 2019; Zhao et al., 2019). The use of AI in the context of urban planning and disaster resilience is rapidly evolving with the emergence of the smart city and sensor city concepts (D'Amico et al., 2020). AI has become a central topic for cities due to the large volume of data produced in this digital era and the increasing importance of, for example, disaster early warning and resilience issues (Sun et al., 2020). Consequently, AI technologies are in demand across many dimensions of urban planning and disaster resilience such as the economy, society, environment, and governance (Yigitcanlar et al., 2021a). AI applications and impacts of AI are clear in terms of the economy (e.g., computational investing, converting homo economics to machine economics, emergence of automated supply chains, presence and digital tourism, use of digital currency such as bitcoin) (Benyon et al., 2014; Böhme et al., 2015; Haseeb et al., 2019; Wagner, 2020); society (e.g., understanding different languages, virtual social networking mechanisms, virtual serious gaming or training, virtual site visits) (Makridakis, 2017; Floridi et al., 2018); governance (e.g., use of machine learning [ML] to forecast human abilities and responses to client needs via chatbots, preventing cyber-attacks on public digital data, image or face recognition for safety and security) (Van Oudheusden, 2014; Janowski, 2016; Linkov et al., 2018). In the context of urban planning and disaster resilience, AI also has many application areas in all these domains (Savari et al., 2019).

AI applications concerning the natural and built environments are also an increasing area of research. For example, AI is used to manage and monitor environmental conditions, weather patterns, reduce unnecessary pollution generation, and manage the supply of renewable energy through intelligent grid systems (Cortès et al., 2000; Haupt et al., 2008; Ortega-Fernández et al., 2020). However, use of AI in some of the aspects related to the environment can be challenging and often contested through numerous misconceptions (Kankanamge et al., 2020b). Therefore, more scientific investigations are needed to understand the application and significance of applying AI in various fields—that also includes the disaster management area.

2.2 Artificial Intelligence in Disaster Management

Disaster management is the process of organising and managing resources and responsibilities to reduce the negative impacts that people could face from a disaster situation. The main objectives of managing disasters are: (a) To minimise losses from disasters through effective response to the affected people; and (b) To achieve the optimum recovery (Lamsal & Kumar, 2020). Disasters are managed under the following three main phases: (a) Pre-disaster (actions related to prevention, mitigation, and preparedness); (b) During-disaster (actions required to meet the needs of the affected people and to alleviate and minimise the associated distress); and (c) Post-disaster (to achieve rapid and durable recovery) (Lamsal & Kumar, 2020; Sun et al., 2020).

Within the most contemporary disaster management practices, AI technology is widely used to enhance the efficiency in disaster management practices in pre-, during-, and post-disaster phases (Sun et al., 2020). Table 3.1 presents some of the AI applications and related technologies used in current disaster management practices.

As shown in Table 3.1, AI in disaster management is a trending research area as more and more studies have identified new applications and advantages of its usage. The AI methods and their related applications were cited simultaneously to reflect the relationships that exist between these methods and related applications. According to Nunavath and Goodwin (2019), ML and deep learning (DL) are the widely used AI technologies to manage disasters. Still, being a rapidly evolving area of research, this might change in the future. These tools also provide opportunity for community-based disaster management (Zhang et al., 2013). Nevertheless, the public perceptions on the use of AI for disaster management is an understudied area of research in this field.

TABLE 3.1 Applications of AI in disaster management

Disaster Management Phases	Applications of AI Technologies	AI Methods
Pre-disaster	• Crisis event detection (Cheng & Hoang, 2016; Gauthier et al., 2017; Yuan & Moayedi, 2020) • Understanding public reaction, increasing disaster awareness (Chang & Chien, 2007; Kankanange et al., 2020b) • Assess vulnerabilities (Song et al., 2014; Liu et al., 2016; Yuan & Moayedi, 2020) • Serious games/gamified applications (Lin et al., 2013; Kankanange et al., 2020c) • Disaster-related data mining and knowledge management, big data analytics from the web or social web (Lin et al., 2013; Ansari et al., 2015; Khare et al., 2019)	• Linear regression (Gauthier et al., 2017) • Deep learning (Robertson et al., 2019) • Non-linear regression, Logistic regression, Deep neural networks (Sankaranarayanan et al., 2019) • Object-level understanding of images/object-level annotations, support vector machine (Choubin et al., 2019) • Naïve bayes, decision tree, random forest (Kankanmage et al., 2020) • K-nearest neighbours, neural networks, Hierarchical clustering, K-means clustering (Liu et al., 2016) • Fuzzy clustering (Ansari et al., 2015)
During-disaster	• Situation recognition (Choubin et al., 2019; Sankaranarayanan et al., 2019; Kankanange et al., 2020) • Understanding public reaction, eyewitness identification, crisis communication (Cheng & Hoang, 2016; Robertson et al., 2019) • Launch disaster relief activities, i.e. drone-based disaster relief, human/life sign-detecting drones (Restas, 2015; Tariq et al., 2018) • Mental health chatbots (Cheng & Jiang, 2020) • Crowd evacuation (Muhammed et al., 2018; Khalilpourazari & Pasandideh, 2021)	• Principal component analysis (Cheng & Hoang, 2016) • Hidden Markov models (Song et al., 2014) • Neural networks—Convolutional neural networks, Recurrent neural networks, Recursive neural networks (Dekanová et al., 2018; Rauter & Winkler, 2018; Tinoco et al., 2019) • Multi-layer perception (Yuan & Moayedi, 2020) • Q-learning (Lin et al., 2013) • Policy gradient, Genetic algorithm (Chang & Chien, 2007) • Particle swarm optimisation (Romlay et al., 2016) • Simulated annealing (Hosseini et al., 2019) • Convolutional Neural Network (CNN) (Hartawan et al., 2019)
Post-disaster	• Damage recognition, detecting socioeconomic recovery (Eguchi et al., 1998; Deryugina, 2017; Frank et al., 2017; Ladds et al. 2017; Shibuya & Tanaka, 2019) • Understanding public reactions, damage assessments (Frank et al., 2017) • Human loss estimation (Aghamohammadi et al., 2013	• Artificial Neural Network (Aghamohammadi et al., 2013) • Natural Language Processing (Frank et al., 2017)

3 RESEARCH DESIGN

The following research procedure was employed in this study (Figure 3.2), and all the steps of this procedure are discussed in the following sections.

3.1 Case Study

This research utilises a case study approach to investigate public perceptions towards the use of AI technology in disaster management. The study selected three case study cities and their greater city regions from Australia—namely, Greater Sydney, Greater Melbourne, and Greater Brisbane (Figure 3.3). These are the largest and the most globalised and populated Australian cities (Pancholi et al., 2018; Yigitcanlar et al., 2021a). Although these cities are dominated by tourism, finance and insurance sectors and business and property services; there are also several clusters specialising in AI-intensive innovation industries such as aerospace, biotechnology, engineering, and environmental technologies (Esmaeilpoorarabi et al., 2020). However, similar to other global cities, over the years, jobs and innovative activities are mostly concentrated in their central business districts (CBDs) (Hu, 2012).

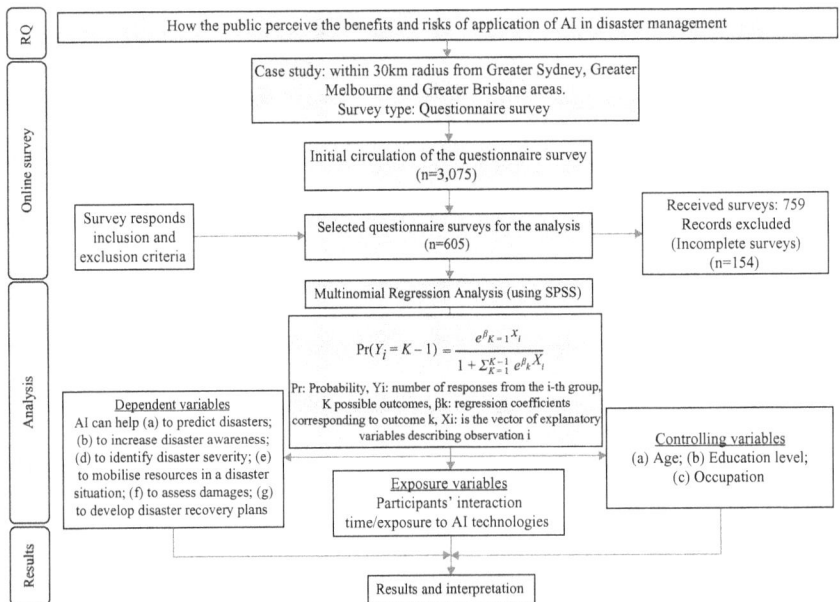

FIGURE 3.2 Research methodology.

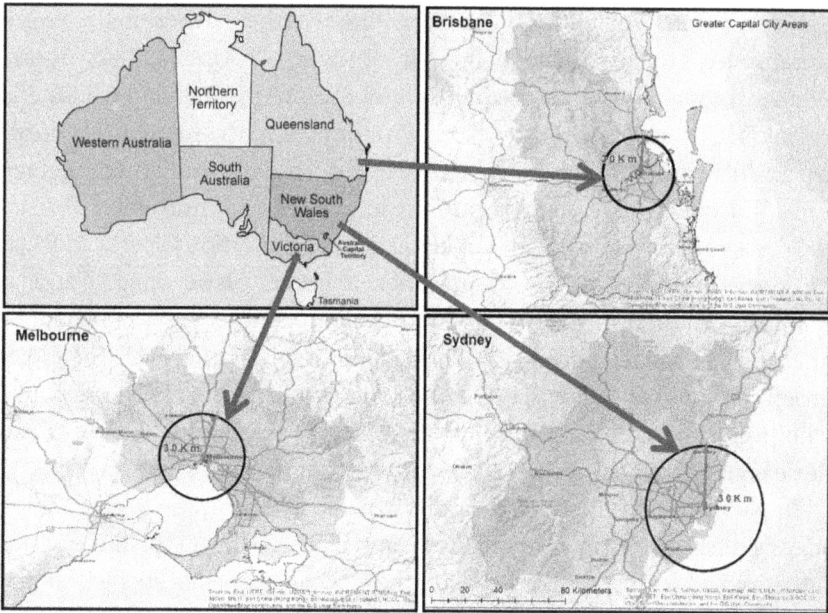

FIGURE 3.3   Locations of the case studies.

Consequently, a 30 km-distance was identified as the radius containing the highest density of AI technology-related industries, which are mostly located central to the Greater Sydney, Melbourne, and Brisbane areas. Folio3, Mode Games, Five2One, H2 ventures, BetaShares ETFs, Oovvuu, Metigy, Faethm, and Averoft are some of the leading AI development and service companies in the Greater Sydney Area. Beyond Analysis, Businessware Technologies, Trimantium GrowthOps (3wks), AOS, CX Central, and inGenious AI companies are located within the Greater Melbourne Area. Solentive, Maxwell Plus, SSW, Blackbook.ai, Relialytics, Nous Group, Quantium, Aginic, Valenta BPO, and BRIKS are the leading companies located within the Greater Brisbane Area. Most of these companies are located within the 30 km radius from their CBDs. The 30 km radius to a degree overlaps with the Greater City boundaries of Sydney, Melbourne, and Brisbane.

Furthermore, the state governments in Australia also provide incentives to expand AI-based developments in these areas. For instance, Queensland State Government has initiated a project called 'Queensland AI Hub' with the vision of "Queensland to be recognised nationally and internationally as a leader in AI-enabled social, economic and workplace positive transformation" (QAH, 2021). Under this initiative, they have created different community groups—Queensland AI, Brisbane AI User Group, Young Women Leaders in AI, Australian School of Entrepreneurship—Future is Female HQ, Gateway Industry Skills Program, with the intention of creating an AI aware community (QAH, 2021). Similarly, Australian Technology Park, Macquarie Park Innovation District, Westmead Innovation District in Sydney; Parkville National Employment and Innovation Cluster, Monash National Employment and Innovation Cluster, Dandenong National Employment and Innovation Cluster are located within the selected boundaries in Melbourne and Sydney (Esmaeilpoorarabia et al., 2018). The target group for data collection via an online survey included the general public who live in the postcode zones within 30 km of the selected CBDs, where these boundaries also accommodate most of the AI companies.

In particular, in the context of disaster management, the selected three states reflect differences and shared similarities in terms of the disaster intensity and frequency. For instance, the postcode zones located within the 30 km radius from the Sydney CBD were affected mostly by flood events. The number of flood events that occurred in Sydney and adjacent areas increased significantly from 38 in 2008–2009 to 150 in 2010–2011 (Sewell et al., 2016). Similarly, as a riverine city, flooding is the most frequent disaster experienced by the postcode zones located around the Brisbane CBD. The 2011 Brisbane flood was considered as one of the largest flood events experienced by Brisbane, with over 200,000 people affected and around 3,570 business premises inundated (Van den Honert & McAneney, 2011). Unlike Sydney and Brisbane, the disaster intensity in Melbourne is low for both, flooding and bushfires. For instance, the Multiple Disaster Index (MDI) for Melbourne CBD and its surrounding areas vary from 0.058 (Murrindini local government area (LGA)) to 1.056 (Wyndham LGA). In contrast, for Sydney and surrounding areas, MDI varies from 1.12 (Sydney) to 2.506 (Lithgow LGA). In Brisbane, MDI varies from 1.126 (Brisbane) to 2.157 (Somerset) (Nicholos & Evershed, 2020). The impact of disasters also motivates and promotes the adoption of new technologies such as AI to manage disasters (Sun et al., 2020). Therefore, selecting interviewees living in postcodes, with both disaster and technological experiences was a must.

3.2 Data Collection

An online survey was developed to collect data from the general public in Greater Sydney, Greater Melbourne, and Greater Brisbane. The questionnaire is presented in Table 3.2. The Key Survey tool—an enterprise survey

TABLE 3.2   Survey questions

No	Question
1	What is your gender?
2	What is your age?
3	What is your highest education qualification obtained?
4	Which industry are you employed in or have a business in?
5	What is your residential postcode?
6	Have you ever interacted with an AI technology?
7	How often would you estimate you interact with the previously identified AI applications?
8	How do you feel about AI in general?
9	Would you feel comfortable in a fully autonomous place to live or work?
10	How does the prospect of an AI future make you feel?
11	To what degree can AI help in predicting disasters, and providing early warnings?
12	To what degree can AI be used in gaming applications to increase community disaster awareness?
13	To what degree can AI help in determining disaster hotspots and severity?
14	To what degree can AI help in mobilise or allocate resources during a disaster?
15	To what degree can AI help in determining disaster damages and risky constructions and locations?
16	To what degree can AI help to develop disaster recovery plans?
17	Do you have any other relevant comments on AI in the context of disasters?

platform—was used to distribute the questionnaire. The survey questions were developed after considering state-of-the-art literature on AI, disaster management, and public perceptions (Zhou et al., 2003; Neri & Cozman, 2020; Sun et al., 2020; Yigitcanlar et al., 2020b; Cui, & Wu, 2021; Selwyn & Gallo Cordoba, 2021; Tan et al., 2021).

The number of participants was determined based on the methods proposed by Krejcie and Morgan (1970). Accordingly, the sample size for population over 100,000 is 384—confidence level 95%, margin of error 5%. People under 18 were excluded from the data collection. Services of a professional survey company were acquired to identify participants who live in the targeted postcode areas. The survey link was sent to individuals' email addresses in November 2020 by the selected professional survey company (n = 3,075 people). Participants were provided with a brief description of the project and definition of AI to reduce the risk of biased interpretations and misunderstanding. In total, 759 participants completed the survey. The participation rate was approximately 24.7%. The responses were checked for completeness and cases with missing data were excluded from

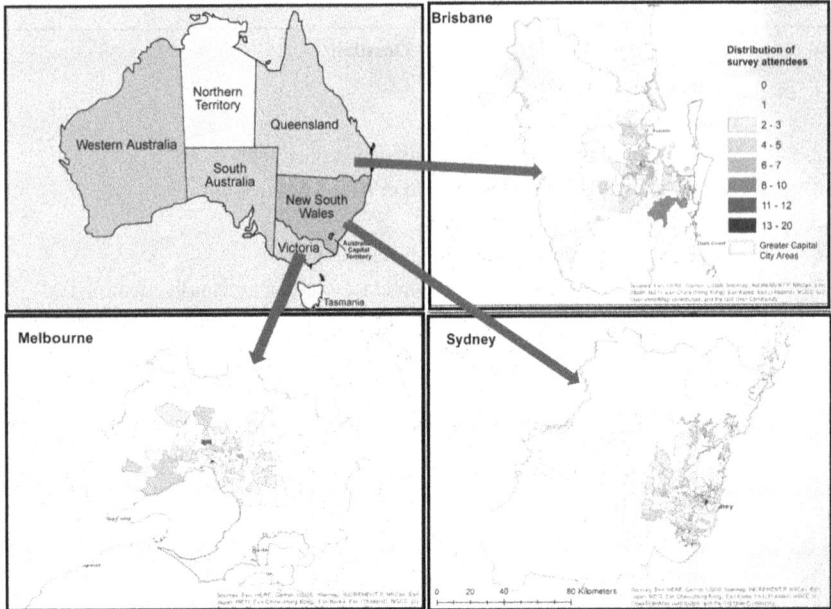

FIGURE 3.4   Distribution of the survey participants across the postcode areas.

further analysis. This resulted in 605 valid responses, which reduced the response rate to about 19.7%. The distribution of the survey participants across the postcode areas is given in Figure 3.4.

3.3 Data and Variables

3.3.1 Dependent Variables

Initially, the general perceptions of the community towards AI were interrogated using three questions. These were: (a) How do you feel about AI in general? (b) Would you feel comfortable in a fully autonomous place to live or work? (c) How does the prospect of an AI future make you feel? These questions were asked to gain an initial understanding about the use of AI among the people, and they were not considered for the multinomial regression analysis.

Second, public perception in using AI in disaster management was studied by requesting participants to rank their perceptions from strongly agree to strongly disagree range about the: (a) Use of AI in pre-disaster phase; (b) Use of AI in during-disaster phase; and (c) Use of AI in post-disaster phase. These three areas were shaped based on the three major phases of the disaster management cycle (Sun et al., 2020). This chapter used data on these three areas as the dependent variables. Particularly, the participants were asked to rank six statements based on their personal perceptions. The six statements were: (a) Pre-disaster phase—(i) AI can help to predict disasters; (ii) AI can help to increase disaster awareness; (b) During-disaster phase—(i) AI can help to identify disaster severity; (ii) AI can help to mobilise resources in a disaster situation, and; (c) Post-disaster phase—(i) AI can help in to assess damages that happened due to disasters; (ii) AI can help to develop disaster recovery plans. The participants were provided with a seven-point Likert scale to rank their perceptions, with number one representing the strongly disagree and number seven representing strongly agree.

3.3.2 Exposure Variable

Respondent's exposure to AI in disaster-related activities was measured by the participants' interaction time/exposure to AI technologies. Exposure time is a commonly used variable in most research studies (Thomas, 1988). Accordingly, the participants were asked to mention their interaction time with AI technologies. This enabled the study to evaluate the effect of the exposure on the three questions listed above. The idea here was that people who work closely with AI technologies should perceive the potential benefits of using AI in disaster management.

3.3.3 Controlling Factors

The research identified three socio-demographic variables which have a significant influence on the perception levels in relation to AI. These were: (a) Age; (b) Education level, and; (c) Occupation. Age was selected as a controlling factor, since the study expected to understand community perceptions across different age groups about using AI. As a newly emerging technology within the context of disaster management, it is important to see how the youth and older generation perceive such technology (ITU, 2019). It will help to decide to which degree and in what kind of AI-based technologies need to be used in the present and future disaster management approaches. Education-level can create a major impact on disaster management-related decision-making (SAMHSA, 2014). For instance, Fothergill and Peek (2004) identified education as a decisive factor for forming a community group—disaster preparedness, to face a disaster. Occupation is also an important factor in understanding the perceptions about using AI to manage disasters. Especially, the study expected to understand the perceptions of the respondents who work in areas directly related to managing disasters.

As a recently evolved technology, it is significant to understand the level of perception that exists among different age groups about the use of AI in disaster management. To derive a clearer understating, the study categorised age into seven groups as 18–24, 25–34, 35–44, 45–54, 55–64, 65–74, 75 and over. Education level was evaluated under six categories. These were: (a) Year 10 or below; (b) Year 11 or equivalent; (c) Certificate; (d) Advanced diploma; (e) Bachelor's degree; and (f) Postgraduate degree. Table 3.3 shows the descriptive statistics for the survey participants. As well as the education level, occupation is also an influential factor in relation to how a person perceives disaster management (Xu et al., 2020). As given in Table 3.3, the participants were given 20 categories to classify their occupations. The occupation categories having over 5% respondents were selected for the analysis. Accordingly, occupations in wholesale trade; Utilities—electricity, gas, water, and waste services; Arts and recreation; Transport, postal and warehousing; Agriculture, forestry, and fisheries; and Rental, hiring, and real-estate services were removed from further analysis.

TABLE 3.3 Salient characteristics of survey participants

Attribute	Category	Cumulative (%)	Sydney (%)	Melbourne (%)	Brisbane (%)
Age	18–24	10.9%	10.4	9.5	12.8
	25–34	19.3%	20.4	16.4	21.2
	35–44	20.8%	22.4	20.9	19.2
	45–54	19.0%	18.4	19.9	18.7
	55–64	13.4%	11.9	14.9	13.3
	65–74	7.4%	5.5	9.0	7.9
	75 and over	9.1	10.9	9.5	6.9
Gender	Female	50.2	52.7	50.7	47.3
	Male	49.4	46.8	48.8	52.7
	Other	0.3	0.5	0.5	0.0
Education	Year 10 or below	8.1	5.5	12.4	6.4
	Year 11 or equivalent	16.0	15.4	18.4	14.3
	Certificate	13.7	14.4	9.5	17.2
	Advanced diploma	14.0	12.4	13.9	15.8
	Bachelor degree	31.9	33.3	33.3	29.1
	Postgraduate degree	16.2	18.9	12.4	17.2
Occupation	Education and training	9.7	8.0	8.08	12.94
	Construction	5.3	6.5	3.03	6.47
	Information, media, and telecommunications	5.5	5.5	5.56	5.47

(Continued)

TABLE 3.3   Salient characteristics of survey participants

Attribute	Category	Cumulative (%)	Sydney (%)	Melbourne (%)	Brisbane (%)
	Wholesale trade	1.8	2.5	2.02	1.00
	Electricity, gas, water, and waste services	1.0	1.0	0.51	1.49
	Healthcare and social assistance	8.3	8.0	10.10	6.97
	Arts and recreation	2.2	2.5	3.54	0.50
	Retail trade	9.2	10.0	9.09	8.46
	Manufacturing	4.5	2.0	6.57	4.98
	Transport, postal and warehousing	3.8	2.0	6.57	2.99
	Homemaker	4.7	7.0	4.55	2.49
	Public administration and safety	6.2	7.0	3.54	7.96
	Financial and insurance services	4.3	6.0	4.55	2.49
	Accommodation, hospitality and food services	3.8	2.5	3.54	5.47
	Administration and support services	5.5	8.0	4.55	3.98
	Agriculture, forestry and fisheries	0.3	0.0	0.00	1.00
	Mining and natural resources	0.3	1.0	0.00	0.00
	Professional, scientific and technical services	7.5	8.0	4.55	9.95
	Rental, hiring, and real-estate services	1.0	1.5	1.01	0.50
	Other	15.0	11.4	18.69	14.93

3.4 Data Analysis

In addition to conducting the descriptive analysis, the study conducted ordinal regression modelling. This was undertaken to understand whether the participants who work closely with AI technologies perceive the applications of AI in disasters differently from those who do not use AI technologies often. Likewise, this analysis was used to understand how people from different age groups, education levels, and occupational categories perceive the application of AI in disaster management. Given that the research collected data for the three categories of applications, three multinomial regression models were derived. Multinomial Logistic Regression is one of the important methods for categorical data analysis. Particularly, the multinomial logistic regression is conducted when the dependent variable is nominal with more than two classes (El-Habil, 2012). For instance, the research problem—how the public perceive the benefits and risks of application of AI in disaster management—was addressed by studying community perceptions based on different age groups, education levels, and occupation categories. Variance Inflation Factor (VIF) was calculated to understand whether multicollinearity exists among the independent variables/predictors. It is a basic rule that to run a multinomial regression analysis, there should not be multicollinearity among dependent variables (Böhning, 1992). The mean VIF of the independent variables considered in this study is 1.2, which means that there is no high correlation among the independent variables considered.

Furthermore, the ordered nature of the 7-point Likert scale of the outcome variables justified the selection of multinomial regression logistic model. The log odds (logit) regression coefficients were estimated. Standard interpretation of the ordered logit coefficient (i.e., 0.5) is that for a one-unit increase in the predictor (i.e., interaction time: always), the response variable level is expected to change by 0.5 in the ordered log-odds scale while the other variables in the model are held constant. Except for the log odds ratio to interpret the relationships among the variables, ANOVA (one way), −2 Log Likelihood were used to measure the performance and to verify the significant differences between the three groups considered (gender, education, and occupation). In these models, in addition to the exposure variable, all other controlling factors were included. Nevertheless, only the statically significant ($p < 0.05$) factors were retained by stepwise exclusion of insignificant factors. All models were estimated using IBM SPSS Statistical Package.

4 ANALYSIS AND RESULTS

4.1 General Observations

From the 605 responses, 201 were from Sydney, 201 were from Melbourne and 203 were from Brisbane. Results of the questions that investigated people's general perception about using AI reflected their positive (72%) perception about using AI. In contrast, 5% had a neutral opinion and 23% had a negative opinion towards the use of AI. Further, 61% of people said 'Yes' to live in a fully autonomous home, and 27% were opposed. Further, 52% of respondents were either enthusiastic, excited, or optimistic about the future of AI whilst 36% were either concerned, unsure, or confused about the future of AI (Table 3.4).

TABLE 3.4   Community perceptions on AI

How Does the Prospect of an AI Future Make You Feel?

Perception	%
Concerned	12
Unsure	19
Confused	5
Neutral	12
Enthusiastic	23
Excited	14
Optimistic	15

TABLE 3.5   ANOVA test results (use of AI in disaster response phase)

Source of Variation	Sum of Squares	df	Mean Square	F	Sig
Between groups	99.745	2	52.876	6.567	0.137
Within groups	242.875	32	16.374		
Total	342.62	34			

The results to the questions, which investigated the use of AI in disaster management, revealed that the majority of the public—close to three-quarters (72.3%)—has a positive perception of using AI technology in disaster and emergency prediction and management to increase the efficiency and efficacy in urban services. Further, close to two-thirds (62.7%) believe that AI is one of the mostly needed technologies in local government and urban services.

4.2 Community Perceptions on Artificial Intelligence Use for Disaster Preparedness

As the study analysed AI, based on the three groups of gender, education, and occupation, an ANOVA statistical test was conducted. This was to further understand the significant differences between aforesaid groups. The null hypothesis of the ANOVA was "there is no difference between the community perceptions in using AI for disaster response among the three groups". The alternative hypothesis was "there is a difference between the community perceptions in using AI for disaster response among the three groups". In Table 3.5, the significance value is 0.137, which is above 0.05 and, therefore, there is no statistically significant difference in the community perceptions according to the gender, education, and occupation levels in using AI for disaster response.

The outcomes of the Multinomial Logistic Regression analysis further elaborate the findings given in Table 3.6. It indicated that age and the educational level have a strong impact on public perceptions in using AI technology in the disaster preparedness phase. Table 3.6 shows that people between age 24 and 54 trusted AI technologies in disaster awareness than using AI technologies for predicting disasters, with high significance. This group is considered as the prime working age group in Australia. Also, people with only secondary school education did not believe in using AI technology to predict disasters. Albeit people at all education levels believe in using AI technologies to increase disaster awareness. The people's occupation was also considered in this study. As depicted in Table 3.5, people who work in administration and support services (B = −1.647, sig. 0.984), technology professionals (B = −0.351, sig. 0.761), information, media, and telecommunications sectors (B = −0.031, sig. 0.969) did not trust in using AI to predict disasters and increase awareness. Nevertheless, people who were in professional, scientific and technical services (B = 0.17, sig. 0.879), healthcare and social assistance (B = 0.72, sig. 0.826), and education and training (B = 1.076, sig. 0.339) believed in using AI to predict disasters and to increase awareness. Figure 3.5 shows the multinomial logistics regression analysis of the probability of use of AI to increase awareness and predict disasters. The solid dots are the corresponding observed probabilities.

4.3 Community Perceptions on the Use of Artificial Intelligence for Disaster Response

The null hypothesis of the ANOVA was "there is no difference between the community perceptions in using AI for disaster response among the three groups". The alternative hypothesis was "there is a difference between the community perceptions in using AI for disaster response among the three groups". In Table 3.7, the significance value is 0.021, which is below 0.05 and, therefore, there is a statistically significant difference in the community perceptions according to the gender, education, and occupation levels in using AI for disaster response.

Table 3.8 further elaborates the findings of the ANOVA test for the community perceptions towards the use of AI for disaster response. Accordingly, there is a positive relationship between the use of AI technology and people's trust in using AI technology to determine disaster severity and mobilise resources during the disaster response phase. People from all age groups and all educational levels trust in using AI technology to mobilise resources. However, people above age 65 did not trust the AI technology's capacity to

TABLE 3.6 Multinomial logistics regression model showing community perceptions on AI use for disaster preparedness

	Regression Model: Possibility of Using AI to:									
	Predict Disasters					Increase Awareness				
				95% C.I for Exp (B)[4]					95% C.I for Exp (B)	
Explanatory Variables	B[1]	Sig[2]	Exp(B)[3]	Lower	Upper	B	Sig	Exp(B)	Lower	Upper
Interaction time with AI applications	0.211	0.166	1.235	0.916	1.664	0.142	.152	.878	0.101	.349
Age: 18–24 years										
25–34	−0.471	0.840	1.198	0.207	6.929	0.180	0.609	0.624	0.103	3.787
35–44	0.193	0.769	1.287	0.239	6.933	0.252	0.818	1.213	0.234	6.296
45–54	0.316	0.588	1.584	0.300	8.350	0.460	0.703	1.371	0.271	6.930
55–64	0.273	0.948	1.056	0.206	5.409	0.550	0.745	1.313	0.255	6.770
65–74	−0.275	0.708	0.727	0.138	3.842	−0.318	0.745	0.760	0.145	3.988
75+	−0.410	0.518	0.553	0.092	3.335	−0.593	0.633	0.664	0.123	3.577
Education: Advanced diploma	−0.123	0.357	0.426	0.069	2.618	−0.854	0.892	0.885	0.151	5.196
	0.282	0.581	1.326	0.487	3.607	0.025	0.572	0.002	0.241	0.965
Bachelor's degree	0.743	0.101	2.103	0.865	5.111	0.249	0.483	0.266	0.078	0.606
Certificate	−0.837	0.104	2.310	0.843	6.332	0.075	0.574	0.017	0.126	0.896
Postgraduate degree	0.338	0.531	1.402	0.487	4.033	0.231	0.579	0.159	0.101	0.690
Year 10 or below	−1.064	0.095	0.345	0.099	1.202	0.484	0.637	0.579	0.204	0.447
Year 11 or equivalent	−1.892	0.156	0.151	0.011	2.055	0.458	1.351	0.115	0.001	0.734
Occupation: Administration and support services	−1.647	0.984	0.799	0.799	0.799	−1.704	0.871	0.182	0.182	0.182
Professional, scientific, and technical services	0.170	0.879	1.186	0.133	10.598	2.781	0.212	0.062	0.001	4.878

(Continued)

TABLE 3.6 CONTINUED

Regression Model: Possibility of Using AI to:

Explanatory Variables	Predict Disasters					Increase Awareness				
	B[1]	Sig[2]	Exp(B)[3]	95% C.I for Exp (B)[4]		B	Sig	Exp(B)	95% C.I for Exp (B)	
				Lower	Upper				Lower	Upper
Healthcare and social assistance	0.720	0.826	2.054	0.003	79.226	1.108	0.837	0.330	0.000	92.474
Technology professional	−0.351	0.761	0.704	0.073	6.762	−2.970	0.218	0.051	0.000	5.801
Information, media, and telecommunications	−0.031	0.969	0.969	0.201	4.663	−2.773	0.027	0.062	0.005	0.724
Education and training	1.076	0.339	2.932	0.323	26.590	0.853	0.488	2.347	0.211	26.156
−2 Log Likelihood		165.286						1627.889		
Chi-Square		35.322						59.224		

[1] Estimated multinomial logistic regression coefficients for the models, [2] p-values of the coefficients, [3] odds ratios for the predictors, [4] Confidence Interval (CI) for an individual multinomial odds ratio

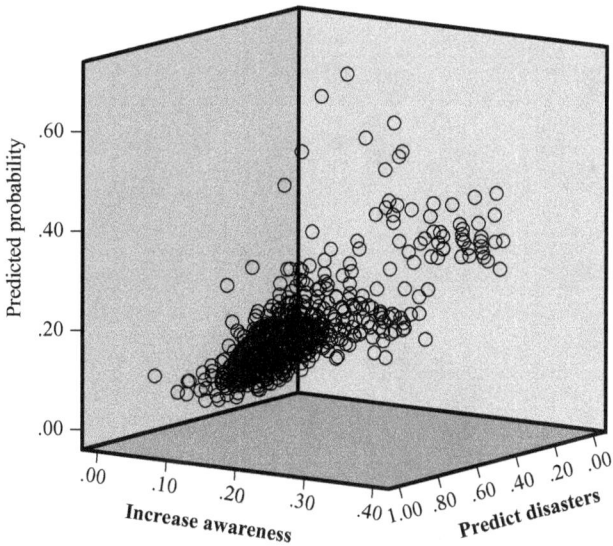

FIGURE 3.5   Results of multinomial logistic regression analysis of the probability of pre-disaster phase.

TABLE 3.7   ANOVA test results (use of AI in disaster response phase)

Source of Variation	Sum of Squares	df	Mean Square	F	Sig
Between groups	92.457	2	48.744	4.465	0.021
Within groups	278.500	28	12.347		
Total	370.957	30			

determine disaster severity. Also, people with only secondary school education did not trust in AI for determining disaster severity during the disaster response phase. People who are employed in administration and support services (B = −0.052; sig. −0.106), information, media and telecommunications (B = −0.478; sig. −1.322), and education and training (B = −0.334; sig. −0.39) did not perceive AI as a useful technology as a useful technology with lower significance levels to be used in determining disaster severity and to mobilise resources. Although, professionals in scientific and technical services (B = −0.385), and healthcare and social assistance (B = −5.384) did not perceive AI as a technology to be employed in determining disaster severity. However, they believed in AI's applicability in mobilising resources during a disaster. In contrast, technology professions (B = 0.477; sig. 0.762) believed in using AI for assessing disaster severity, but not in mobilising resources

TABLE 3.8 Multinomial logistics regression model showing the community perceptions on AI use for disaster response

	Regression Model: Possibility of Using AI to:									
	Determine Disaster Severity					Mobilise Resources				
				95% C.I for Exp (B)[4]					95% C.I for Exp (B)	
Explanatory Variables	B[1]	Sig[2]	Exp(B)[3]	Lower	Upper	B	Sig	Exp(B)	Lower	Upper
Interaction time with AI applications	0.398	0.080	0.671	0.430	1.049	0.422	0.007	1.525	1.125	2.067
Age: 18–24 years	0.687	0.616	1.988	0.135	29.271	0.203	0.800	0.816	0.170	3.929
25–34	0.863	0.503	2.370	0.190	29.522	0.147	0.833	0.863	0.219	3.397
35–44	0.282	0.827	1.326	0.106	16.670	0.330	0.636	1.390	0.356	5.435
45–54	0.146	0.908	1.158	0.097	13.874	0.291	0.670	1.338	0.351	5.093
55–64	0.201	0.876	1.223	0.098	15.283	0.335	0.629	1.398	0.359	5.444
65–74	−0.848	0.582	0.428	0.021	8.771	0.384	0.582	1.468	0.375	5.749
75+	−0.646	0.678	0.524	0.025	11.029	0.240	0.753	1.271	0.284	5.686
Education: Advanced diploma	0.290	0.696	0.749	0.175	3.196	0.006	0.991	1.006	0.348	2.911
Bachelor's degree	0.010	0.988	1.010	0.288	3.548	0.227	0.633	1.255	0.494	3.186
Certificate	0.331	0.655	1.393	0.326	5.952	0.229	0.704	0.795	0.244	2.590
Postgraduate degree	1.103	0.193	0.332	0.063	1.745	0.222	0.669	1.249	0.451	3.454
Year 10 or below	−1.258	0.302	0.284	0.026	3.098	0.722	0.227	2.058	0.639	6.629
Year 11 or above	−1.679	0.993	0.000	0.154	0.678	0.200	0.863	1.221	0.127	11.775
Occupation: Administration and support services	−0.052	1.000	94.73	0.000	43.71	−0.106	0.75	0.899	0.899	0.899

	β[1]	p[2]	OR[3]	CI lower[4]	CI upper	β[1]	p[2]	OR[3]	CI lower[4]	CI upper
Professional, scientific, and technical services	−0.385	0.814	1.470	0.059	36.337	0.677	0.465	1.969	0.320	12.127
Healthcare and social assistance	−5.384	0.065	217.805	0.720	90.177	0.966	0.745	2.628	0.008	884.66
Technology professional	0.477	0.762	1.612	0.073	35.530	−0.449	0.654	0.638	0.090	4.552
Information, media, and telecommunications	−0.478	0.691	1.612	0.153	16.978	−1.322	0.074	0.267	0.063	1.135
Education and training	−0.334	0.879	0.716	0.010	52.614	−0.390	0.726	0.677	0.076	6.006
−2 Log Likelihood			1586.775					1565.881		
Chi-Square			24.673					22.854		

[1] Estimated multinomial logistic regression coefficients for the models, [2] p-values of the coefficients, [3] odds ratios for the predictors, [4] Confidence Interval (CI) for an individual multinomial odds ratio

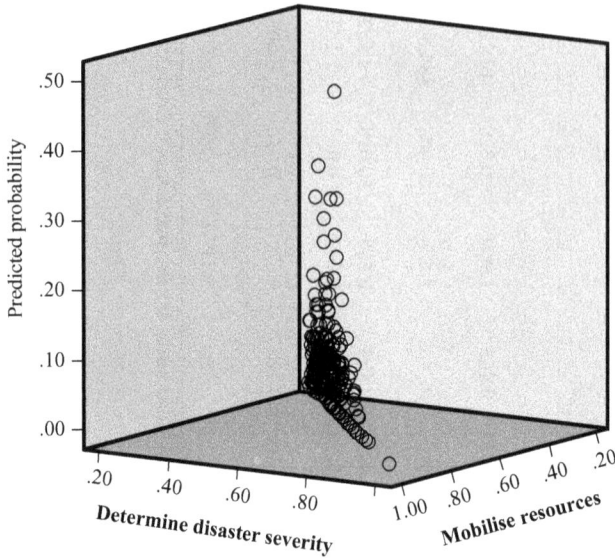

FIGURE 3.6   Results of MLR analysis of the probability of during disaster phase.

TABLE 3.9   ANOVA test results (Use of AI in disaster recovery phase)

Source of Variation	Sum of Squares	df	Mean Square	F	Sig
Between groups	89.365	2	52.741	6.535	0.027
Within groups	146.500	32	18.437		
Total	370.957	34			

(B = −0.449; sig. 0.654). Figure 3.6 shows the multinomial logistics regression analysis of the probability of use of AI to determine disaster severity and mobilise resources.

4.4 Community Perceptions on Artificial Intelligence Use for Disaster Recovery

The null hypothesis of the ANOVA was "there is no difference between the community perceptions in using AI for disaster recovery among the three groups". The alternative hypothesis was "there is a difference between the community perceptions in using AI for disaster recovery among the three groups". In Table 3.9, the significance value is 0.027, which is below 0.05 and, therefore, there is a statistically significant difference in the community perceptions according to the gender, education, and occupation levels in using AI for disaster recovery.

According to Table 3.10, people who interacted with AI technologies did not perceive its use to assess disaster damages and to develop disaster

TABLE 3.10 Multinomial logistics regression model showing community perceptions on AI use for disaster recovery

	Regression Model: Possibility of Using AI to:									
	Assess Damages					Develop Disaster Recovery Plans				
				95% C.I for Exp (B)[4]					95% C.I for Exp (B)	
Explanatory Variables	B[1]	Sig[2]	Exp(B)[3]	Lower	Upper	B	Sig	Exp(B)	Lower	Upper
Interaction time with AI applications	−0.577	0.012	0.561	0.357	0.882	−0.021	0.928	0.979	0.620	1.546
Age: 18–24 years	2.504	0.145	12.231	0.420	356.068	1.529	0.415	4.611	0.117	182.076
25–34	1.761	0.288	5.817	0.227	149.282	0.788	0.664	2.200	0.063	76.889
35–44	1.319	0.421	3.739	0.150	93.100	1.281	0.475	3.601	0.107	121.177
45–54	1.153	0.481	3.168	0.128	78.373	−0.652	0.712	1.919	0.060	61.460
55–64	−1.254	0.453	3.506	0.132	92.909	−1.328	0.460	3.772	0.112	127.318
65–74	−0.518	0.792	0.595	0.013	28.135	−1.165	0.538	3.206	0.078	131.037
75+	−0.218	0.914	0.804	0.015	42.044	−0.496	0.798	1.642	0.037	73.675
Education: Advanced diploma	−1.203	0.332	0.271	0.256	0.777	0.894	0.343	2.445	0.386	15.493
Bachelor's degree	−1.087	0.004	0.652	0.197	0.856	0.350	0.675	1.420	0.276	7.308
Certificate	0.231	0.342	0.786	0.638	2.322	0.556	0.583	1.743	0.240	12.637
Postgraduate degree	−0.765	0.553	1.340	0.152	0.483	1.666	0.058	5.293	0.946	29.604
Year 10 or below	1.976	0.000	1.212	0.904	0.982	−0.886	0.557	0.412	0.021	7.913
Year 11 or above	0.234	0.331	1.365	0.908	2.786	2.362	0.087	10.612	0.712	158.128
Occupation: Administration and support services	2.871	0.999	0.278	0.085	10.62	0.797	0.218	2.219	2.219	2.219
Professional, scientific and technical services	−0.469	0.658	0.626	0.079	4.991	−0.792	0.702	0.453	0.008	26.049
Healthcare and social assistance	1.238	0.672	3.447	0.011	56.71	2.651	0.227	14.161	0.192	1043.641

(Continued)

TABLE 3.10 Continued

	Regression Model: Possibility of Using AI to:									
	Assess Damages					Develop Disaster Recovery Plans				
				95% C.I for Exp (B)[4]					95% C.I for Exp (B)	
Explanatory Variables	B[1]	Sig[2]	Exp(B)[3]	Lower	Upper	B	Sig	Exp(B)	Lower	Upper
Technology professional	0.361	0.723	0.697	0.095	5.112	0.102	0.971	0.903	0.004	210.577
Information, media, and telecommunications	−0.664	0.364	0.515	0.123	2.159	−0.519	0.732	0.595	0.031	11.525
Education and training	−1.295	0.199	3.649	0.507	26.264	−2.997	0.024	20.035	1.488	269.827
−2 Log Likelihood			1625.525					1630.397		
P-value			0.039					0.045		

[1] Estimated multinomial logistic regression coefficients for the models, [2] p-values of the coefficients, [3] odds ratios for the predictors, [4] Confidence Interval (CI) for an individual multinomial odds ratio

recovery plans. Especially, people above 55 years of age did not trust the versatility of using AI technology to assess damages. Although people in the age group 45 to 54 thought that AI is a possible tool for assessing disaster damages (B = 1.153), people above age 45 did not see AI as a technology for use in developing disaster recovery plans. Most significantly, people with tertiary education did not accept AI as a robust method to be used in assessing disaster damages and developing disaster recovery plans. Employees in administration and support services (B = 2.871; sig. 0.797), healthcare and social assistance (B = 1.238; sig. 2.651), and technology professionals (B = 0.361; sig. 0.102) had a positive attitude to using AI for assessing damages and developing disaster recovery plans with higher statistical significance. The employees in professional, scientific and technical services (B = −0.469; sig. −0.792), information, media and telecommunications (B = −0.664; sig. −0.519), and education and training (B = −1.295; sig. −2.997) did not believe in using AI for assessing damages and developing disaster recovery plans. However, their results reflected a lower significance. Figure 3.7 shows the multinomial logistics regression analysis of the probability of use of AI to assess damages and develop disaster recovery plans.

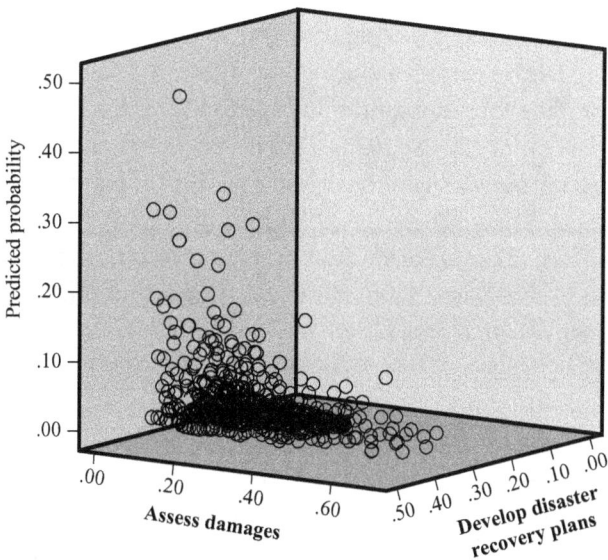

FIGURE 3.7   Results of MLR analysis of the probability of post-disaster phase.

5 FINDINGS AND DISCUSSION

Disaster management activities can be divided into three phases: pre-, during-, and post-disaster management (Sun et al., 2020). These phases do not require the same services nor the same technical knowledge from disaster management authorities (Kankanamge et al., 2020) including preparedness, rescue, and recovery activities. Therefore, application of AI in disaster management in the above phases takes different forms. Simultaneously, the way the community perceive the application of AI in disaster management has a significant impact when introducing AI induced disaster management activities to people as AI is identified as an advanced technology that is used by experts and professionals. Therefore, research is required to understand to what extent the general public trust AI as a useful tool to manage disasters, before introducing AI-driven applications in highly sensitive contexts such as disaster management. Accordingly, the study shows how significant is the age, education, and occupation in perceiving AI use in disaster management practices.

In considering the age factor, the study outcomes revealed, people between age 18 and 44 trust the possibility of using AI in disaster management for increasing disaster awareness, determining disaster severity, mobilising resources, assessing damages, and developing disaster recovery plans. The younger generation's positive attitude towards AI in disaster management is often evident through digital humanitarian networks (DHNs), which are active during disasters to help affected communities (Meier, 2015). DHN is a consortium of volunteer and technical communities, who interface with humanitarian organisations that seek their service during a crisis event (DHN, 2020). DHNs connect new digitally enabled young people's groups to help the disaster victims in many ways.

For example, during the Australian bushfires in 2009, 65% of tweets contained information deemed important for emergency response (Meier, 2011). Further, the young volunteers who started to call themselves as 'Mapsters', launched a campaign to manually monitor social media platforms such as Facebook, Twitter, YouTube, and Flicker to extract time-critical information related to the bushfires (Sinnappan et al., 2010; Kankanamge et al., 2020c). These examples illustrate how significant the role of young people is in AI-driven technologies to manage disasters. While appreciating the immense role played by rescue managers and fire fighters, this study emphasises the significance of the role played by young digital volunteers empowered by AI technologies.

Unlike the relationship that exists between AI and age categories, the study findings did not reflect a clear relationship between the education levels and the public perceptions towards applying AI in disaster management. In general, the analysis reflected the positive attitude the community had in using AI in disaster management activities considered in this research study. Albeit the people with an education to year 11 and below did not trust the use of AI in determining disaster severity and damage assessments. This is acceptable as the application of AI technologies in determining disaster severity and damage assessments is still at the experimental level, and such knowledge may not have penetrated to the general public yet (Cervone et al., 2016; Kankanamge et al., 2019; Kankanamge et al., 2020b).

In terms of the occupation factor, for occupations/professions considered, people who work in administration and support services only trusted the possibility of using AI in post-disaster phase, e.g., assessing damages and developing disaster recovery plans, which are directly related to administrative work. Similarly, technology professionals too trusted the use of AI in post-disaster phase. Nonetheless, the professional, scientific, and technical service providers only trusted the possibility of using AI in mobilising resources during disaster response (during-disaster), and pre-disaster phase activities, e.g., predicting disasters and increasing disaster awareness, but not in post-disaster-phase activities. Healthcare and social assistance workers trusted the use of AI for all the considered applications, except the use of AI in determining disaster severity.

Most significantly, the people who work in information, media, and telecommunication sectors did not trust the use of AI either in the pre-, during- nor post-disaster phase. Although the use of media for disaster management is a trending activity (Rajdev & Lee, 2015; Kankanamge et al., 2020), their applications are often challenged due to the spread of misinformation, fake news, and spam or phishing messages (Tran et al., 2019; Valenzuela et al., 2019; Flores-Saviaga & Savage, 2020). Education and training sector employees appear to only trust using AI in pre-disaster phase activities, such as increasing awareness and predicting disasters.

6 CONCLUSION

This study which adopted a quantitative approach investigated public perceptions in using AI for disaster management. The findings of the study contribute to deepening our understanding of community perceptions for using AI in the pre-, during- and post-disaster management phases. Understanding general public perceptions across different social strata is significantly important, especially in applying advanced technologies in highly sensitive domains such as disaster management (AlQahtany & Abubakar, 2020).

First, the study identified that the younger generation has a positive attitude towards accepting AI as a potential tool to be used in the various phases of the disaster management cycle—i.e., use of social media to raise awareness and crowd mapping. As an emerging technology, this can be considered as a positive sign for future applications. Therefore, the study emphasises the need for creating novel platforms and opportunities for younger people to be a part of AI-driven crowdsourcing platforms.

Second, the study revealed that occupations with a technical orientation trusted the use of AI for disaster management. Nevertheless, the lack of confidence in the use of AI for disaster management that exists among employees in professions such as healthcare and social assistance, information, media, and telecommunication needs to be taken into consideration. According to Eguchi et al. (2008), Boccardo and Tonolo (2015) Rawat et al. (2015), Adeel et al. (2019), and Novellino et al. (2019), AI-driven technologies generate large volumes of data including real-time and simulation data. Therefore, data storage requires attention and was not a concern prior to the 2000s. Moreover, such data is not only extremely important in disaster management for conducting analytics and operations but also to better engage the community in the process. However, as per the study findings, except for the age cohorts, people with different education levels and occupation/professional types did not trust enough certain applications or premises of AI in disaster management. In other words, community trust in AI was found to be not at a desired level at the present. Therefore, this study emphasises the need for further developing community trust in using/benefiting from AI in disaster management. Otherwise, the community will tend to question the disaster management decisions derived through such technologies. This may result in lesser or weaker community support in the disaster management process.

Third, the analysis undertaken to derive relationships between the level of education and community perceptions in using AI for disaster management, emphasise that greater knowledge is needed to be imparted to the general public about the use of AI for disaster management across all the phases of the disaster management cycle (Yigitcanlar et al., 2020c). This, in return, would increase community confidence and trust in accepting AI as a prospective tool to manage disasters. On the other hand, AI systems and applications should also be designed and deployed to be more responsible to avoid failures in the areas where it is applied (Yigitcanlar et al., 2021b). This also includes the disaster management field. Also, the use of AI to manage disasters needs stronger governance. According to Sohn et al. (2020), the users tend to underestimate the value of AI technology when people do not get to know that AI has been used. Therefore, adequate awareness and publicity should be given to increase community trust about using AI to manage disasters.

ACKNOWLEDGEMENTS

This chapter, with permission from the copyright holder, is a reproduced version of the following journal article: Kankanamge, N., Yigitcanlar, T., & Goonetilleke, A. (2021). Public perceptions on artificial intelligence driven disaster management: evidence from Sydney, Melbourne and Brisbane. *Telematics and Informatics*, 65, 101729.

REFERENCES

Adeel, A., Gogate, M., Farooq, S., Ieracitano, C., Dashtipour, K., Larijani, H., & Hussain, A. (2019). A survey on the role of wireless sensor networks and IoT in disaster management. In: *Proceedings of geological disaster monitoring based on sensor networks, Singapore*, pp. 57–66.

Aghamohammadi, H., Mesgari, M.S., Mansourian, A., & Molaei, D. (2013). Seismic human loss estimation for an earthquake disaster using neural network. *International Journal of Environmental Science and Technology*, 10(5), 931–939.

Alam, F., Imran, M., & Ofli, F. (2017). Image4act: Online social media image processing for disaster response. In: *Proceedings of the 2017 IEEE/ACM International Conference on Advances in Social Networks Analysis and Mining*, pp. 601–604. Sydney Australia.

AlQahtany, A., & Abubakar, I. (2020). Public perception and attitudes to disaster risks in a coastal metropolis of Saudi Arabia. *International Journal of Disaster Risk Reduction*, 44, 101422.

Al Qundus, J., Dabbour, K., Gupta, S., Meissonier, R., & Paschke, A. (2020). Wireless sensor network for AI-based flood disaster detection. *Annals of Operations Research*, https://doi.org/10.1007/s10479-020-03754-x.

Ansari, A., Firuzi, E., & Etemadsaeed, L. (2015). Delineation of seismic sources in probabilistic seismic-hazard analysis using fuzzy cluster analysis and Monte Carlo simulation. *Bulletin of the Seismological Society of America*, 105(4), 2174–2191.

Benyon, D., Quigley, A., O'keefe, B., & Riva, G. (2014). Presence and digital tourism. *AI & Society*, 29(4), 521–529.

Boccardo, P., & Tonolo, F. (2015). Remote sensing role in emergency mapping for disaster response. In: *Proceedings of engineering geology for society and territory*, Cham, pp.17–24.

Böhme, R., Christin, N., Edelman, B., & Moore, T. (2015). Bitcoin: economics, technology, and governance. *Journal of Economic Perspectives*, 29(2), 213–238.

Böhning, D. (1992). Multinomial logistic regression algorithm. *Annals of the institute of Statistical Mathematics*, 44(1), 197–200.

Cervone, G., Sava, E., Huang, Q., Schnebele, E., Harrison, J., & Waters, N. (2016). Using Twitter for tasking remote-sensing data collection and damage assessment: 2013 Boulder flood case study. *International Journal of Remote Sensing*, 37(1), 100–124.

Chang, T.C., & Chien, Y.H. (2007). The application of genetic algorithm in debris flows prediction. *Environmental Geology*, 53(2), 339–347.

Cheng, M.Y., & Hoang, N.D. (2016). Slope collapse prediction using Bayesian framework with k-nearest neighbor density estimation: case study in Taiwan. *Journal of Computing in Civil Engineering*, 30(1), 04014116.

Cheng, Y., & Jiang, H. (2020). AI-Powered mental health chatbots: Examining users' motivations, active communicative action and engagement after mass-shooting disasters. *Journal of Contingencies and Crisis Management*, 28(3), 339–354.

Choubin, B., Borji, M., Mosavi, A., Sajedi-Hosseini, F., Singh, V.P., & Shamshirband, S. (2019). Snow avalanche hazard prediction using machine learning methods. *Journal of Hydrology*, 577, 123929.

Cortès, U., Sànchez-Marrè, M., Ceccaroni, L., R-Roda, I., & Poch, M. (2000). Artificial intelligence and environmental decision support systems. *Applied Intelligence*, 13(1), 77–91.

Cui, D., & Wu, F. (2021). The influence of media use on public perceptions of artificial intelligence in China: evidence from an online survey. *Information Development*, 37, 45–57.

D'Amico, G., L'Abbate, P., Liao, W., Yigitcanlar, T., & Ioppolo, G. (2020). Understanding sensor cities: insights from technology giant company driven smart urbanism practices. *Sensors*, 20(16), 4391.

Dekanová, M., Duchoň, F., Dekan, M., Kyzek, F., & Biskupič, M. (2018, May). Avalanche forecasting using neural network. In *Proceedings of the Conference 2018 ELEKTRO*, pp. 1–5. Mikulov, Czech Republic.

Deryugina, T. (2017). The fiscal cost of hurricanes: disaster aid versus social insurance. *American Economic Journal: Economic Policy*, 9(3), 168–198.

DHN (2020) Digital humanitarian networks, Digital Humanitarian Networks. Accessed on 1 March 2020 from https://www.digitalhumanitarians.com.

Eguchi, R.T., Goltz, J.D., Taylor, C.E., Chang, S.E., Flores, P.J., Johnson, L.A., Seligson, H.A., & Blais, N.C. (1998). Direct economic losses in the Northridge earthquake: a three-year post-event perspective. *Earthquake Spectra*, 14(2), 245–264.

Eguchi, R., Huyck, C., Ghosh, S., & Adams, B. (2008). The application of remote sensing technologies for disaster management. In: *Proceedings of the 14th World Conference on Earthquake Engineering*. Beijing, China.

El-Habil, A.M. (2012). An application on multinomial logistic regression model. *Pakistan Journal of Statistics and Operation Research*, 8(2), 271–291.

Esmaeilpoorarabi, N., Yigitcanlar, T., & Guaralda, M. (2018). Place quality in innovation clusters: an empirical analysis of global best practices from Singapore, Helsinki, New York, and Sydney. *Cities*, 74, 156–168.

Esmaeilpoorarabi, N., Yigitcanlar, T., Kamruzzaman, M., & Guaralda, M. (2020). How can an enhanced community engagement with innovation districts be established? Evidence from Sydney, Melbourne and Brisbane. *Cities*, 96, 102430.

Flores-Saviaga, C., & Savage, S. (2020). Fighting disaster misinformation in Latin America: the# 19S Mexican earthquake case study. *Personal and Ubiquitous Computing*, https://doi.org/10.1007/s00779-020-01411-5.

Floridi, L., Cowls, J., Beltrametti, M., Chatila, R., Chazerand, P., Dignum, V., Luetge, C., Madelin, R., Pagallo, U., Rossi, F., & Schafer, B. (2018). AI4People—an ethical framework for a good AI society: opportunities, risks, principles, and recommendations. *Minds and Machines*, 28(4), 689–707.

Fothergill, A., & Peek, L.A. (2004). Poverty and disasters in the United States: a review of recent sociological findings. *Natural Hazards*, 32, 89–110.

Frank, J., Rebbapragada, U., Bialas, J., Oommen, T., & Havens, T. (2017). Effect of label noise on the machine-learned classification of earthquake damage. *Remote Sensing*, 9(8), 803.

Gauthier, F., Germain, D., & Hétu, B. (2017). Logistic models as a forecasting tool for snow avalanches in a cold maritime climate: northern Gaspésie, Québec, Canada. *Natural Hazards*, 89(1), 201–232.

Hartawan, D.R., Purboyo, T.W., & Setianingsih, C. (2019). Disaster victims detection system using convolutional neural network (CNN) method. In: *Proceedings of 2019 IEEE International Conference on Industry 4.0, Artificial Intelligence, and Communications Technology*, pp. 105–111. Bali, Indonesia.

Haseeb, M., Mihardjo, L.W., Gill, A., & Jermsittiparsert, K. (2019). Economic impact of artificial intelligence: new look for the macroeconomic assessment in Asia-Pacific Region. *International Journal of Computational Intelligence Systems*, 12(2), 1295–1310.

Haupt, S. E, Pasini, A., & Marzban, C. (Eds.). (2008). *Artificial intelligence methods in the environmental sciences*. Springer Science & Business Media.

ITU (2019) Digital skills insights, International Telecommunication Union (ITU). Accessed on 13 July 2021 from https://www.itu.int/hub/publication/d-phcb-cap_bld-03-2019/

Janowski, T. (2016). Implementing sustainable development goals with digital government–Aspiration-capacity gap. *Government Information Quarterly*, 33(4), 603–613.

Kankanamge, N., Yigitcanlar, T., Goonetilleke, A., & Kamruzzaman, M. (2019). Can volunteer crowdsourcing reduce disaster risk? A systematic review of the literature. *International Journal of Disaster Risk Reduction*, 35, 101097.

Kankanamge, N., Yigitcanlar, T., & Goonetilleke, A. (2020a). How engaging are disaster management related social media channels? The case of Australian state emergency organisations. *International Journal of Disaster Risk Reduction*, 48, 101571.

Kankanamge, N., Yigitcanlar, T., Goonetilleke, A., & Kamruzzaman, M. (2020b). Determining disaster severity through social media analysis: testing the methodology with South East Queensland Flood tweets. *International Journal of Disaster Risk Reduction*, 42, 101360.

Kankanamge, N., Yigitcanlar, T., Goonetilleke, A., & Kamruzzaman, M. (2020c). How can gamification be incorporated into disaster emergency planning? A systematic review of the literature. *International Journal of Disaster Resilience in the Built Environment*, 11(4), 481–506.

Kemper, H., & Kemper, G. (2020). Sensor fusion, GIS and AI technologies for disaster management. *The International Archives of Photogrammetry, Remote Sensing and Spatial Information Sciences*, 43, 1677–1683.

Khalilpourazari, S., & Pasandideh, S.H. (2021). Designing emergency flood evacuation plans using robust optimization and artificial intelligence. *Journal of Combinatorial Optimization*, 41(3), 640–677.

Khare, P., Burel, G., & Alani, H. (2019). Relevancy identification across languages and crisis types. *IEEE Intelligent Systems*, 34, 19–28.

King, A. (2017). Are heatwaves 'worsening' and have 'hot days' doubled in Australia in the last 50 years. Accessed on 10 March 2021 from https://theconversation.com/are-heatwaves-worsening-and-have-hot-days-doubled-in-australia-in-the-last-50-years-79337.

Krejcie, R., & Morgan, D. (1970). Determining sample size for research activities. *Educational and Psychological Measurement*, 30(3), 607–610.

Lamsal, R., & Kumar, T.V.V. (2020). Artificial intelligence and early warning systems. In: Kumar, T.V.V., & Sud, K. (Eds.) *AI and Robotics in Disaster Studies. Disaster Research and Management Series on the Global South.* (pp. 13–32). Palgrave Macmillan, Singapore. https://doi.org/10.1007/978-981-15-4291-6_2

Lee, M., Mesicek, L., Bae, K., & Ko, H. (2021). AI advisor platform for disaster response based on big data. *Concurrency and Computation: Practice and Experience*, https://doi.org/10.1002/cpe.6215.

Lin, S.Y., Chao, K.M., Lo, C.C., & Godwin, N. (2013). Distributed dynamic data driven prediction based on reinforcement learning approach. In *Proceedings of the 28th Annual ACM Symposium on Applied Computing*, pp. 779–784.

Linkov, I., Trump, B., Poinsatte-Jones, K., & Florin, M. (2018). Governance strategies for a sustainable digital world. *Sustainability*, 10(2), 440.

Liu, K., Li, Z., Yao, C., Chen, J., Zhang, K., & Saifullah, M. (2016). Coupling the k-nearest neighbor procedure with the Kalman filter for real-time updating of the hydraulic model in flood forecasting. *International Journal of Sediment Research*, 31(2), 149–158.

Ludwig, T., Kotthaus, C., Reuter, C., Van Dongen, S., & Pipek, V. (2017). Situated crowdsourcing during disasters: managing the tasks of spontaneous volunteers through public displays. *International Journal of Human-Computer Studies*, 102, 103–121.

Makridakis, S. (2017). The forthcoming artificial intelligence (AI) revolution: its impact on society and firms. *Futures*, 90, 46–60.

McCarthy, J. (1988). Mathematical logic in artificial intelligence. *Dædalus*, 117, 297–311.

Meier, P. (2011). How to use technology to counter rumors during crises. Accessed on 28 February 2020 from https://irevolutions.org/2011/03/26/technology-to-counter-rumors/.

Meier, P. (2015). *Digital humanitarians: how big data is changing the face of humanitarian response*. New York, NY: CRC Press.

Mortoja, M., & Yigitcanlar, T. (2020). Local drivers of anthropogenic climate change: quantifying the impact through a remote sensing approach in Brisbane. *Remote Sensing*, 12(14), 2270.

Muhammad, K., Ahmad, J., & Baik, S.W. (2018). Early fire detection using convolutional neural networks during surveillance for effective disaster management. *Neurocomputing*, 288, 30–42.

Neri, H., & Cozman, F. (2020). The role of experts in the public perception of risk of artificial intelligence. *AI & Society*, 35(3), 663–673.

Nicholos, J., & Evershed, N. (2020). Interactive map: which areas of Australia were hit by multiple disasters in 2020? *The Guardian*. Accessed on 8 July 2021 from https://www.theguardian.com/news/datablog/2020/dec/22/interactive-map-which-areas-of-australia-were-hit-by-multiple-disasters-in-2020.

Novellino, A., Jordan, C., Ager, G., Bateson, L., Fleming, C., & Confuorto, P. (2019). Remote sensing for natural or man-made disasters and environmental changes. In: *Proceedings of Geological Disaster Monitoring Based on Sensor Networks,* Singapore, pp. 21–23.

Nunavath, V., & Goodwin, M. (2019). The use of artificial intelligence in disaster Management-A systematic literature review. In: *Proceedings of 2019 International Conference on Information and Communication Technologies for Disaster Management*, pp. 1–8. Paris, France.

Ortega-Fernández, A., Martín-Rojas, R., & García-Morales, V. (2020). Artificial intelligence in the urban environment: smart cities as models for developing innovation and sustainability. *Sustainability*, 12(19), 7860.

Pancholi, S., Yigitcanlar, T., & Guaralda, M. (2018). Societal integration that matters: place making experience of Macquarie Park Innovation District, Sydney. *City, Culture and Society*, 13, 13–21.

Pannu, A. (2015). Artificial intelligence and its application in different areas. *Artificial Intelligence*, 4(10), 79–84.

Poblet, M., García-Cuesta, E., & Casanovas, P. (2018). Crowdsourcing roles, methods and tools for data-intensive disaster management. *Information Systems Frontiers*, 20(6), 1363–1379.

QAH (2021) Queensland Artificial Intelligence Hub. Accessed on 2 March 2021 from https://www.qldaihub.com.

Rajdev, M., & Lee, K. (2015). Fake and spam messages: detecting misinformation during natural disasters on social media. In: *Proceedings of 2015 IEEE/WIC/ACM International Conference on Web Intelligence and Intelligent Agent Technology*, IEEE, pp. 17–20. Singapore.

Rauter, M., & Winkler, D. (2018). Predicting natural hazards with neuronal networks. Accessed on 13 July 2021 from https://arxiv.org/pdf/1802.07257.pdf.

Rawat, P., Haddad, M., & Altman, E. (2015). Towards efficient disaster management: 5G and Device to Device communication. In: *Proceedings of 2015 2nd International Conference on Information and Communication Technologies for Disaster Management*, IEEE, pp. 79–87. Paris, France.

Restas, A. (2015). Drone applications for supporting disaster management. *World Journal of Engineering and Technology*, 3(3), 316.

Robertson, B.W., Johnson, M., Murthy, D., Smith, W.R., & Stephens, K.K. (2019). Using a combination of human insights and 'deep learning' for real-time disaster communication. *Progress in Disaster Science*, 2, 100030.

SAMHSA. (2017). Disaster Technical Assistance Center Supplemental Research Bulletin Greater Impact: How Disasters Affect People of Low Socioeconomic Status, Substance Abuse and Mental Health Services Administration (SAMHSA). Accessed on 4 July 2021 from https://www.samhsa.gov/sites/default/files/dtac/srb-low-ses_2.pdf.

Sankaranarayanan, S., Prabhakar, M., Satish, S., Jain, P., Ramprasad, A., & Krishnan, A. (2020). Flood prediction based on weather parameters using deep learning. *Journal of Water and Climate Change*, 11(4), 1766–1783.

Saravi, S., Kalawsky, R., Joannou, D., Rivas Casado, M., Fu, G., & Meng, F. (2019). Use of artificial intelligence to improve resilience and preparedness against adverse flood events. *Water*, 11(5), 973.

Schalko, R. (1990). *Artificial intelligence: an engineering approach*. McGraw-Hill: New York.

Seeta, T. (2020). Natural disasters in Australia. Accessed on 12 March 2021 from https://www.canstar.com.au/home-insurance/natural-disasters-australia.

Selwyn, N., & Gallo Cordoba, B. (2021). Australian public understandings of artificial intelligence. *AI & Society*, https://doi.org/10.1007/s00146-021-01268-z.

Sewell, T., Stephens, R.E., Dominey-Howes, D., Bruce, E., & Perkins-Kirkpatrick, S. (2016). Disaster declarations associated with bushfires, floods and storms in New South Wales, Australia between 2004 and 2014. *Scientific Reports*, 6(1), 1–11.

Shibuya, Y., & Tanaka, H. (2019). Using social media to detect socio-economic disaster recovery. *IEEE Intelligent Systems*, 34(3), 29–37.

Sinnappan, S., Farrell, C., & Stewart, E. (2010). Swinburne priceless Tweets! A study on Twitter messages posted during crisis: Black Saturday. Accessed on 28 February 2020 from https://www.slideshare.net/slideshow/priceless-tweets/7634333

Sohn, K., Sung, C.E., Koo, G., & Kwon, O. (2020). Artificial intelligence in the fashion industry: consumer responses to generative adversarial network technology. *International Journal of Retail & Distribution Management*, 49(1), 61–80.

Song, X., Zhang, Q., Sekimoto, Y., & Shibasaki, R. (2014, August). Prediction of human emergency behavior and their mobility following large-scale disaster. In *Proceedings of the 20th ACM SIGKDD International Conference on Knowledge Discovery and Data Mining*, pp. 5–14. New York, NY, USA.

Sun, W., Bocchini, P., & Davison, B. (2020). Applications of artificial intelligence for disaster management. *Natural Hazards*, 103, 2631–2689.

Tan, L., Guo, J., Mohanarajah, S., & Zhou, K. (2021). Can we detect trends in natural disaster management with artificial intelligence? A review of modeling practices. *Natural Hazards*, 107(3), 2389–2417.

Tariq, R., Rahim, M., Aslam, N., Bawany, N., & Faseeha, U. (2018). Dronaid: A smart human detection drone for rescue. In: *Proceedings of 2018 15th International Conference on Smart Cities: Improving Quality of Life Using ICT & IoT*, pp. 33–37. Golden, Colorado USA.

Thomas, D. (1988). Models for exposure-time-response relationships with applications to cancer epidemiology. Annual Review of Public Health, 9(1), 451–482.

Tinoco, J., Correia, A.G., Cortez, P., & Toll, D. (2019, September). Combining artificial neural networks and genetic algorithms for rock cuttings slopes stability condition identification. *In Proceedings of International Conference on Information Technology in Geo-Engineering*, pp. 196–209.

Tran, T., Valecha, R., Rad, P., & Rao, H. (2019). An investigation of misinformation harms related to social media during humanitarian crises. In: *Proceedings of International Conference on Secure Knowledge Management in Artificial Intelligence Era*, Singapore, pp. 167–181.

Uskov, A., & Sekar, B. (2015). Smart gamification and smart serious games. In: Dharmendra Sharma, Margarita Favorskaya, Lakhmi C. Jain, Robert J. Howlett (Eds.), *Fusion of smart, multimedia and computer gaming technologies* (pp. 7–36). Springer, Cham.

Valenzuela, S., Halpern, D., Katz, J., & Miranda, J. (2019). The paradox of participation versus misinformation: social media, political engagement, and the spread of misinformation. *Digital Journalism*, 7(6), 802–823.

Van den Honert, R.C., & McAneney, J. (2011). The 2011 Brisbane floods: causes, impacts and implications. *Water*, 3(4), 1149–1173.

Van Oudheusden, M. (2014). Where are the politics in responsible innovation? European governance, technology assessments, and beyond. *Journal of Responsible Innovation*, 1(1), 67–86.

Wagner, D. (2020). Economic patterns in a world with artificial intelligence. *Evolutionary and Institutional Economics Review*, 17(1), 111–131.

Wang, P. (2008). What do you mean by AI? In: *Proceedings of AGI Conference*, pp. 362–373. Memphis, Tennessee, USA.

Wirtz, B., Weyerer, J., & Geyer, C. (2019). Artificial intelligence and the public sector: applications and challenges. *International Journal of Public Administration*, 42(7), 596–615.

Xu, D., Tsang, I., Chew, E., Siclari, C., & Kaul, V. (2019). A data-analytics approach for enterprise resilience. *IEEE Intelligent Systems*, 34(3), 6–18.

Xu, D., Zhuang, L., Deng, X., Qing, C., & Yong, Z. (2020). Media exposure, disaster experience, and risk perception of rural households in earthquake-stricken areas: evidence from rural China. *International Journal of Environmental Research and Public Health*, 17(9), 3246.

Yigitcanlar, T., & Cugurullo, F. (2020). The sustainability of artificial intelligence: an urbanistic viewpoint from the lens of smart and sustainable cities. *Sustainability*, 12(20), 8548.

Yigitcanlar, T., Desouza, K., Butler, L., & Roozkhosh, F. (2020a). Contributions and risks of artificial intelligence (AI) in building smarter cities: insights from a systematic review of the literature. *Energies*, 13, 1473.

Yigitcanlar, T., Butler, L., Windle, E., Desouza, K., Mehmood, R., & Corchado, J. (2020b). Can building "artificially intelligent cities" safeguard humanity from natural disasters, pandemics, and other catastrophes? An urban scholar's perspective. *Sensors*, 20(10), 2988.

Yigitcanlar, T., Kankanamge, N., Regona, M., Maldonado, A., Rowan, B., Ryu, A., Desouza, K., Corchado, J., Mehmood, R. and Li, R. (2020c). Artificial intelligence technologies and related urban planning and development concepts: how are they perceived and utilized in Australia? *Journal of Open Innovation: Technology, Market, and Complexity*, 6(4), 187.

Yigitcanlar, T., Kankanamge, N., & Vella, K. (2021a). How are smart city concepts and technologies perceived and utilized? A systematic geo-twitter analysis of smart cities in Australia. *Journal of Urban Technology*, 28(1), 135–154.

Yigitcanlar, T., Corchado, J., Mehmood, R., Li, R, Mossberger, K., & Desouza, K. (2021b). Responsible urban innovation with local government artificial intelligence (AI): a conceptual framework and research agenda. *Journal of Open Innovation*, 7(1), 71.

Yuan, C., & Moayedi, H. (2020). Evaluation and comparison of the advanced metaheuristic and conventional machine learning methods for the prediction of landslide occurrence. *Engineering with Computers*, 36(4), 1801–1811.

Zhang, X., Yi, L., & Zhao, D. (2013). Community-based disaster management: a review of progress in China. *Natural Hazards*, 65(3), 2215–2239.

Zhao, Y., Li, T., Zhang, X., & Zhang, C. (2019). Artificial intelligence-based fault detection and diagnosis methods for building energy systems: advantages, challenges and the future. *Renewable and Sustainable Energy Reviews*, 109, 85–101.

Zhou, Q., Li, J.Y., & Zhao, J.B. (2003). Study on index system of assessment of public disaster perception in the western China. *Chinese Geographical Science*, 13(3), 284–288.

CHAPTER 4

Perceptions on Artificial Intelligence in the Construction Industry

1 INTRODUCTION AND BACKGROUND

Artificial intelligence (AI) technologies have been widely adopted in many industry sectors [1–3]. Among those, the adoption level is significantly lower in the Australian construction industry. The Australian construction industry generates approximately 360 billion in revenue, accounting for 9% of the country's gross domestic product (GDP), and it is expected to grow to 11.5% of the total GDP in the next five years [4]. Nonetheless, its productivity has only increased by 1% over the past two decades. Thus, there are growing concerns regarding efficiency in the industry [5]. The slow growth is a direct result of the fundamental rules and characteristics of the construction market. The cyclical demand is further compounded, leading to low capital investment and limited standardisation [6].

In response to the slow growth, the need for investment and research into AI technologies is being explored to streamline the processes and increase productivity [1–3]. The benefits that AI can bring to the construction industry include preventing cost overruns, improving site safety, and managing projects efficiently [7–11]. There has already been substantial growth in the following AI areas of big data and analytics, robotics, automation, data integration, and wearable technology [12,13].

DOI: 10.1201/9781003521440-5

Implementing AI technologies and realising the benefits they may bring is difficult. Most algorithms require accurate data for training, and collecting data is costly and time-consuming at the beginning [2,14,15]. The implementation of AI in construction remains in the initial stages, even though some larger construction companies have already begun to enjoy the benefits of these technologies. This has resulted in an increased debate on the future of the construction workforce and how AI will impact jobs [16].

Despite the increasing importance of AI for the construction industry, there are only limited studies investigated the AI adoption prospects and constraints in the construction industry [17–19]. Although the industry transforms slowly to digitalise the construction process, firms have an increasing interest when they realise the benefits of AI-powered algorithms and analytics [20,21]. Nonetheless, there are growing but still limited studies reported in the international literature [22–25], and only a few of them look at this issue in the context of Australia [26,27].

A knowledge gap remains regarding how the public perceives the implementation of AI technologies. In addition, how they feel about the extensive application of automated technologies producing sustainable outcomes and making traditional jobs obsolete [7,28–30]. A good understanding of the public perception of AI technologies will inform governing bodies how to respond adequately to public demands and figure out the most efficient ways to implement AI without disrupting traditional work processes [31]. Therefore, it is necessary to study how AI directly interacts with individuals and how different AI technologies can positively or negatively impact individuals or companies in the construction industry.

This study, hence, focuses on the public perception of AI technologies and discusses the prospects and constraints that AI technologies may bring in Australia. We use the social media analytics method and conduct an opinion and content analysis of location-based Twitter messages from Australia. Following this introduction section, Section 2 introduces the methodological approach of the study. Section 3 presents the results of the analysis and observes the data that were collected. Section 4 discusses the study findings, general insights, research limitations, and future research recommendations. Lastly, Section 5 states the final remarks of the chapter.

2 RESEARCH DESIGN

We investigated the public perception of AI as it becomes more frequently used on construction sites, and project success in the future will be highly dependent on the efficient use of these technologies. The reasons behind this selection include: (a) some of Australia's major construction firms have begun to realise the benefits of AI. These firms successfully adopt AI for their projects to save costs and time; (b) Australia is developing a national AI strategy and roadmap, meaning that AI uptake in cities and industries is planned to avoid solely organic occurrence; (c) the use of social media in Australia is very popular, making it a source of information that can provide a generalised perception of AI in the construction industry. The number of internet users who use social media daily continues to grow in 2021; at present, 79.9% of the Australian population uses social media. This saw an increase of 8.8% in social media usage from 2020. Around 56% of people go online more than 10 times a day, and 26% of people go online more than 20 times a day. Among 79.9% of internet users that use social media, 20% of them have accessed Twitter, and one-third of them tweet daily [32]; (d) Although social media produce an abundance of data regarding AI in Australia, there are not any, to our knowledge, studies investigate the public perception of AI in the construction industry.

Instead of using a traditional data collection method, we employ the social media analysis for this research. As more people use social media to communicate and express their opinions, it has become a source of qualitative data [33]. This data collection method has been used in a wide range of research. Social media allows researchers to engage with a large group of people in an unbiased setting [34]. In addition, researchers can engage with people from a broad geographic area according to user locations, which are tagged in their posts [35].

Twitter was the only social media platform used to obtain data as it's a micro-blogging site. Among the four types of social media services, micro-blogging sites, specifically Twitter, collect information for sentiment analysis [36]. Twitter is one of the ten most visited websites that enable users to post and interact with short messages. The platform allows for opinion and provides very valuable information to scholars. Conducting a sentiment analysis on other social media platforms is not favourable as data are not readily available, unstructured, and often used in short form. This makes the data harder to be analysed.

A geo-Twitter analysis has proven to be a successful data collection method. The research method is efficient in analysing public opinions [37]. It offers an insight into new AI technologies that are developing and/or currently being used on construction sites through real-time information

[38]. For instance, social media analytics safeguards Australian cities and their residents from the coronavirus outbreak (COVID-19) in 2020 [39].

Initially, sentiment and content analysis were computed for the total number of location-based Twitter messages. To do this, the original dataset was obtained from the QUT Digital Observatory on 5 April 2021 (https://www.qut.edu.au/research/why-qut/infrastructure/digital-observatory). By using five data filtering processes—i.e., frequency analysis, location, date, bot, and relevance filters—11365 tweets were filtered down to 7906 tweets.

We selected a two-year period from 1 July 2019 to 1 July 2021 for analysis and removed all tweets outside Australia. The reason behind a two-year period was that a one-year period could not provide enough data for analysis or derive objective quantitative results from texts. In addition, a three-year period would not be able to capture the latest trends, as AI in construction is developing fast and public perception changes rapidly. Thus, a two-year period reflected more accurately people's sentiment towards AI in construction. The bot filter is employed to remove tweet repetition.

Second, to identify the main themes of tweets about AI applications in the construction industry, NVivo was also used to undertake the content analysis and to analyse word frequency, concepts, and technologies. Next, we conduct a word co-occurrence analysis on tweets that discuss AI technologies and construction-related ideas (or AI application areas).

Fourth, we conducted a spatial analysis to complement the content analysis. The tweets are classified according to themes, concepts, and technologies based on locations. This allows us to know more about Australia's most popular themes, concepts, and technologies in each state/territory. We used ArcGIS Pro software to visualise the spatial information. The relevance criteria were used to identify tweets related to or discussed AI technologies in construction-related concepts, noting key sentiment words. The scale adopted was as follows: 1 = very positive sentiment, 2 = positive sentiment, 3 = neutral sentiment, 4 = negative sentiment, 5 = very negative sentiment. We then processed these words via Weka software, which created a dataset for further analysis. We showcased the sensitivity of these specific words via Random Tree and Random Forest functions.

Finally, a network analysis was created to present the relationship between AI themes, concepts, and the relationships between AI technologies. In this analysis, we used Gephi software to understand the nodes and edges relationship found in the tweets. Figure 4.1 shows the research process that was used as the research model.

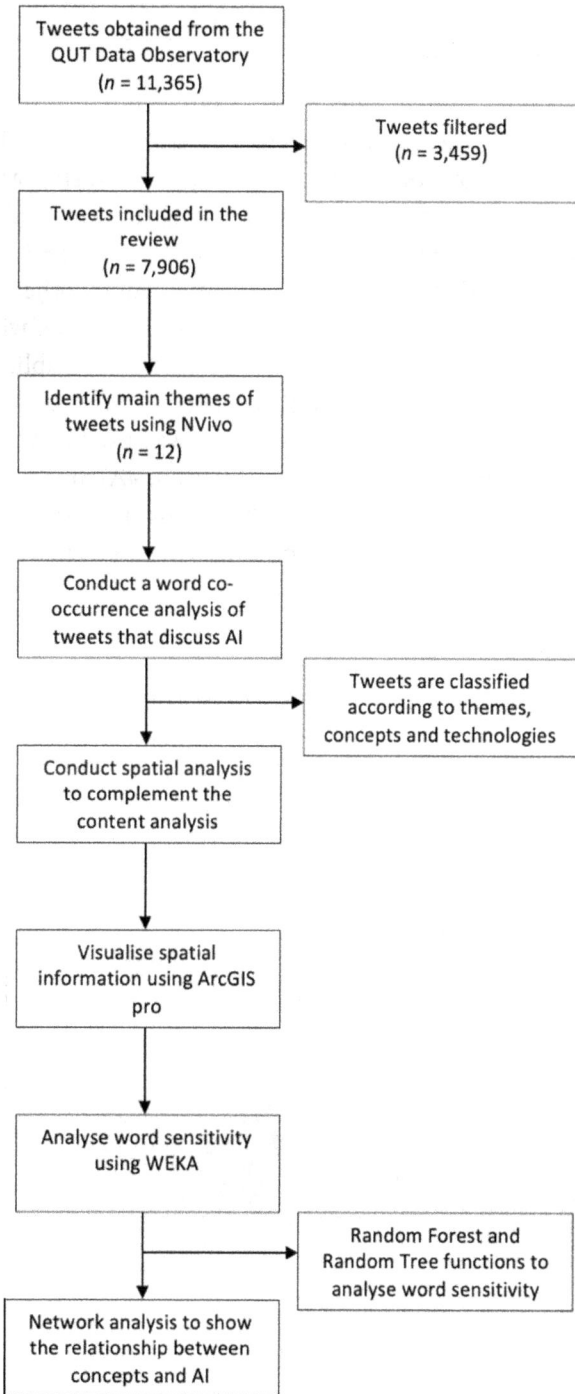

FIGURE 4.1 Process of conducting a sentiment analysis.

3 ANALYSIS AND RESULTS

3.1 General Observations

Among 7906 tweets, 39% (n = 2997) were from New South Wales (NSW), followed by 28% (n = 2214) from Victoria (VIC), 19% (n = 1540) from Queensland (QLD), 5% (n = 426) from Western Australia (WA), 5% (n = 364) from South Australia (SA), 3% (n = 258) from Australian Capital Territory (ACT), 1% (n = 55) from Tasmania (TAS), and 1% (n = 52) from Northern Territory (NT) (Figure 4.2). Compared to other states and territories, ACT and TAS only recorded 55 and 52 tweets, which represented a negligible percentage of 1%. This reveals a low public interest in ACT and TAS community regarding AI-related technologies in the construction industry. A wide range of hashtags was used in the circulated tweets. Among them, tags such as #Industry4.0, #AIconstruction, #IoT, #Predictiveanalytics, #Robotics, #MachineLearning, #Bigdata, #BIM, #Fourththindustryrevolution, #Datamining, and #Automation were the most popular keywords.

3.2 Community Sentiments

Out of the 7907 tweets, 49% (n = 3,396) of them were positive about AI technologies and applications within the context of construction.

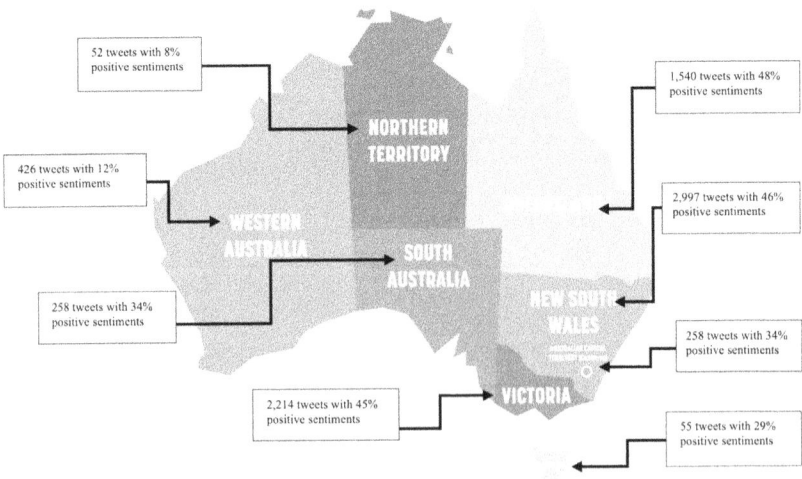

FIGURE 4.2 Tweet numbers and positive sentiment percentages by states and territories.

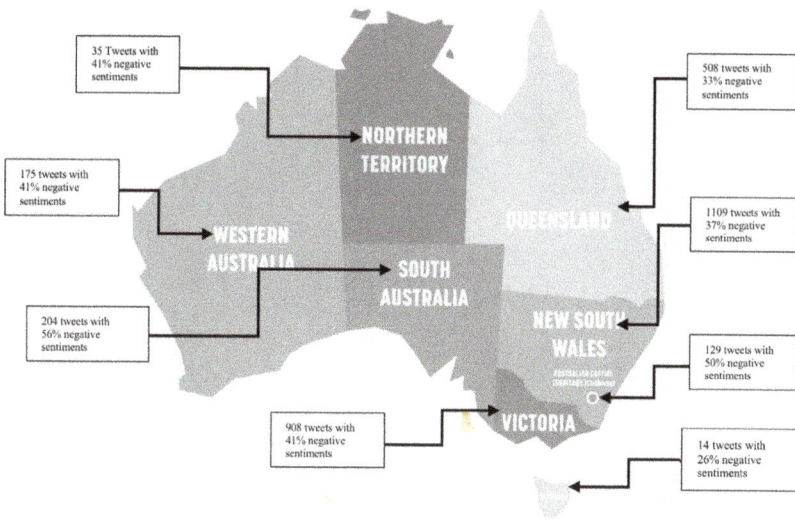

FIGURE 4.3   Tweet numbers and negative sentiment percentages by states and territories.

An analysis of positive sentiment in each state and territory is shown in Figure 4.2.

Around 37% (*n* = 3,085) were negative towards AI in construction. An analysis of negative sentiment in each state and territory is shown in Figure 4.3.

Furthermore, around 14% (*n* = 1,425) of tweets were neutral, where such tweets only used a set of hashtags to express their ideas rather than comments with elaboration. An analysis of neutral sentiment in each state and territory is shown in Figure 4.4. In addition, Table 4.1 is an overview of the sentiment analysis of each state and territory, from very positive sentiment to very negative sentiment.

The tweets from NSW (*n* = 2,997) and QLD (*n* = 1,540) recorded positive sentiment of 46% and 48%, respectively. Both states were the most positive towards AI. VIC had the second-highest number of tweets (*n* = 2,214) with 45% (*n* = 996) positive and 41% (*n* = 908) negative, respectively. Out of the 55 tweets from TAS, 45% (*n* = 25) were neutral. The remaining states perceived AI in construction as negative. From the tweets originating from WA (*n* = 426) and ACT (*n* = 258), 41% and 50% were negative, respectively. Out of the 364 tweets from SA, 35% (*n* = 127) were positive, and 204 (56%) were negative. NT had the lowest

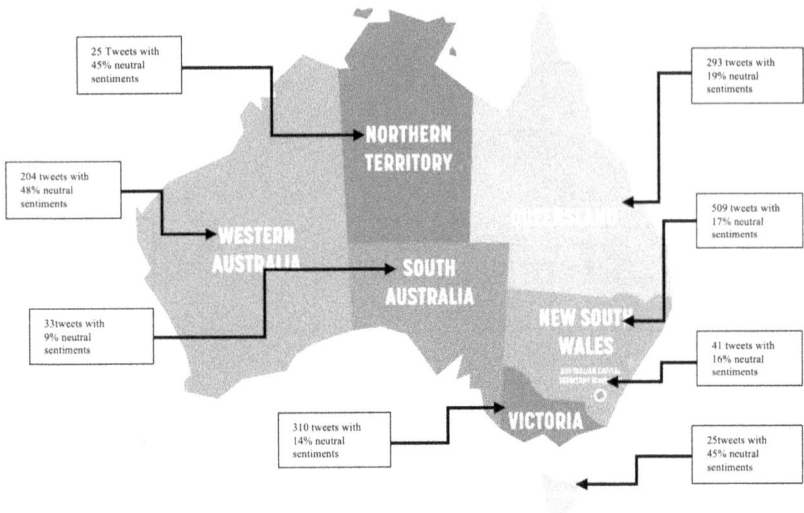

FIGURE 4.4   Tweet numbers and neutral sentiment percentages by states and territories.

TABLE 4.1   Tweet sentiment in percentages per state/territory

	QLD (%)	TAS (%)	NSW (%)	SA (%)	ACT (%)	VIC (%)	WA (%)	NT (%)	Australia (%)
Very positive sentiments	12	8	7	7	6	6	3	0	6
Positive sentiments	36	21	39	28	28	39	8	8	43
Neutral sentiments	19	45	17	9	16	14	48	18	14
Negative sentiments	23	18	25	44	43	32	29	56	28
Very negative sentiments	10	8	12	12	7	9	12	18	95
Total	100	100	100	100	100	100	100	100	N/A

number of tweets relating to AI in construction. Among them, 8% ($n = 4$) were positive, and 74% ($n = 38$) were negative. NT was the state with the highest percentage of negative sentiment as it was viewed as disruptive to the industry. Example tweets for each sentiment category are given in Table 4.2.

TABLE 4.2   Example tweets from each sentiment category

Date and Time	State/ Territory	Tweet	Sentiment
25 April 2020 12:46	NSW	Time is of the essence when dealing with project correspondence and how you can mitigate the risks using platform technology. #AI, #Construction, #Technology	Very positive
22 June 2021 23:19	ACT	Organisations are rethinking their industry framework, improving their manufacturing processes. Looking at a much more holistic supply chain and connected network to drive agility, efficiency, innovation, and sustainability. # Construction #Automation, #Innovation	Positive
12 November 2019 20:32	WA	Automation is standard practice in construction and has been for generations. If you support private enterprise, you also support automation. It will continue, and AI will delete more and more jobs as time goes on. #Future, #MachineLearning	Negative
28 July 2021 13:14	QLD	Do people realise that the Industrial Revolution will see massive job losses in construction, let's think about where the income stream will come from? #AI	Very negative
08 August 2021 08:30	VIC	The safety of workers and the profitability of construction projects are paramount concerns for any company. Bosses who put off investing in technology place businesses workers at a significant disadvantage. #Construction, #Digital, #Technology	Neutral

3.3 Artificial Intelligent Technologies in Construction

By counting word frequency, we identified 12 key AI-related construction technologies (Figure 4.5 and Table 4.3), including 'artificial intelligence' ($n = 341$), 'automation' ($n = 475$), 'big data' ($n = 457$), 'blockchain' ($n = 147$), 'deep learning' ($n = 406$), 'digital twin' ($n = 44$), 'IoT' ($n = 562$), 'machine learning' ($n = 522$), 'robotics' ($n = 931$), 'risk predictive modelling' ($n = 55$), 'simulation' ($n = 13$), and 'virtual learning' ($n = 75$).

The popularity of each construction technology is different in each state and territory. For instance, there were more tweets in NSW about 'big

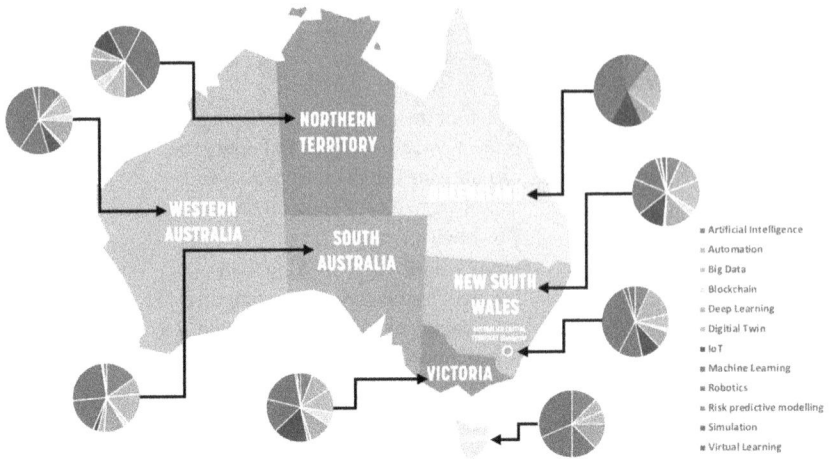

FIGURE 4.5 Distribution of tweets about AI technologies in construction per state/territory.

data' ($n = 169$) than QLD ($n = 123$). Conversely, 'robotics' was around three times more popular ($n = 431$) in QLD than in VIC ($n = 198$). Furthermore, different states had different popular topics of AI technologies. For example, 'machine learning' ($n = 194$) was the most popular tweet related to AI construction technology in NSW. While 'robotics' ($n = 198$) was the most tweeted AI construction technology in VIC, 'robotics' was the most tweeted technology in WA ($n = 67$), SA ($n = 31$), ACT ($n = 31$), and NT ($n = 6$). 'Virtual learning' ($n = 6$) was comparatively high among the tweets circulated in TAS. Table 4.4 provides examples of tweets that were related to each technology.

3.4 Prospects of Artificial Intelligence Technologies in Construction

Based on word frequency, 12 prospects that AI technologies will bring to a construction site were identified from AI related tweets (Figure 4.6 and Table 4.5). These include 'accountability' ($n = 51$), 'accuracy' ($n = 66$), 'consistency' ($n = 37$), 'cost reduction' ($n = 138$), 'digitalisation' ($n = 767$), 'efficiency' ($n = 109$), 'innovation' ($n = 691$), 'productivity' ($n = 232$), 'quality' ($n = 73$), 'reliability' ($n = 35$), 'safety' ($n = 84$), and 'time saving' ($n = 294$).

Digitalisation ($n = 767$) was the most discussed construction prospect derived from AI-related tweets, but its usability differed from one state/territory to another. While digitalisation was the most popular tweets technology concept in NSW ($n = 317$), QLD ($n = 156$), VIC ($n = 246$), and

TABLE 4.3 Distribution of tweets by AI technologies in construction per state/territory

	Artificial Intelligence	Automation	Big Data	Blockchain	Deep Learning	Digital Twin	IoT	Machine Learning	Robotics	Risk Prediction Modelling	Simulation	Virtual learning
NSW	80	147	169	42	159	10	154	194	163	26	4	30
QLD	149	192	123	25	86	9	205	104	431	16	6	12
VIC	64	109	120	68	116	18	178	162	198	10	1	18
WA	20	3	16	7	21	2	12	25	67	0	0	5
SA	20	10	20	3	13	4	3	22	31	1	0	2
TAS	2	1	1	0	2	0	0	2	3	0	0	5
ACT	6	13	6	1	7	0	8	10	32	0	2	3
NT	0	0	2	1	2	1	2	3	6	2	0	0
Total	341	475	457	147	406	44	562	522	931	55	13	75
NSW	23.46%	30.95%	36.98%	28.57%	39.16%	22.73%	27.40%	37.16%	17.51%	47.27%	30.77%	40.00%
QLD	43.70%	40.42%	26.91%	17.01%	21.18%	20.45%	36.48%	19.92%	46.29%	29.09%	46.15%	16.00%
VIC	18.77%	22.95%	26.26%	46.26%	28.57%	40.91%	31.67%	31.03%	21.27%	18.18%	7.69%	24.00%
WA	5.87%	0.63%	3.50%	4.76%	5.17%	4.55%	2.14%	4.79%	7.20%	0.00%	0.00%	6.67%
SA	5.87%	2.11%	4.38%	2.04%	3.20%	9.09%	0.53%	4.21%	3.33%	1.82%	0.00%	2.67%
TAS	0.59%	0.21%	0.22%	0.00%	0.49%	0.00%	0.00%	0.38%	0.32%	0.00%	0.00%	6.67%
ACT	1.76%	2.74%	1.31%	0.68%	1.72%	0.00%	1.42%	1.92%	3.44%	0.00%	15.38%	4.00%
NT	0.00%	0.00%	0.44%	0.68%	0.49%	2.27%	0.36%	0.57%	0.64%	3.64%	0.00%	0.00%
Total	100%	100%	100%	100%	100%	100%	100%	100%	100%	100%	100%	100%

TABLE 4.4 Example tweets for AI technologies in construction

Technology	Date and Time	State/Territory	Tweet	Sentiment
Artificial intelligence	08 July 2019 11:07	NSW	Artificial intelligence can play a transformative role in improving the efficiency and safety of construction sites by giving developers a transparent overview of their projects.	Positive
Automation	18 January 2021 12:37	QLD	The fourth revolution in the construction industry is characterised by connectivity, advanced analytics, automation, and advanced engineering that made a greater impact after Covid. #Analytics, #Connectivity, #Indsitry4.0, #Technology, #automation	Neutral
Big data	20 May 2020 21:58	QLD	The convergence of automation and intelligence is known as Hyper Automation. Hyper Automation is at the forefront of the industrial revolution as emerging technologies such as natural language processing and big data analytics are now being combined with automation. # Construction #AI	Neutral
Blockchain	25 November 2020 04:02	VIC	We live in a world that's not only powered by technology but also shaped by it. Blockchain and IoT can help improve construction efficiency. #Blockchain, #Technology, #IoT, #AI, #digitaltechnology	Neutral
Deep learning	10 January 2020 08:09	QLD	Aerial imagery company Nearmap has acquired deep learning and analytics technology that extracts data from D models to provide roof geometry for a variety of sectors, including construction.	Neutral
Digital twin	25 May 2019 05:40	VIC	Collaboration is needed to drive construction innovation and industry growth. The technologies include IoT, advanced automation, robotics, 4D printing, machine-to-machine communication, digital twins, and sensor technology.	Positive
IoT	10 January 2020 06:00	NSW	Artificial intelligence and IoT will connect construction sites of the future that work faster and more flexibly with minimal downtime. #Iot, #Automation, #AI	Positive
Machine learning	23 February 2019 09:31	VIC	Emerging tech like machine learning and automation is driving massive social change but are these the changes we want? Agree with David Thodey it's time for an informed national conversation on the social implications of the construction revolution.	Negative

Risk prediction modelling	15 March 2019 09:37	NSW	To thrive in the industrial revolution, construction companies are rapidly adopting agile practices. People also need to effectively manage risk.	Neutral
Robotics	19 February 2020 04:46	WA	True focus on founders who bring technology to market that eliminates repetitive manual labour and multiplies human productivity by automating routine tasks. #Construction, #Automation, #Artificialintelligence	Positive
Simulation	12 February 2019 20:42	ACT	Simulations teach humans how to manipulate the arms of the machine rather than having humans teach the machine how to dig for itself. #AI, #Construction	Neutral
Virtual learning	16 April 2021 14:22 p.m.	NSW	The impact of virtual learning technology is showing a major impact on the transformation of design and planning throughout construction stages. #Virtuallearning, #modelling, #design, #Bim, #Construction, #Technology	Neutral

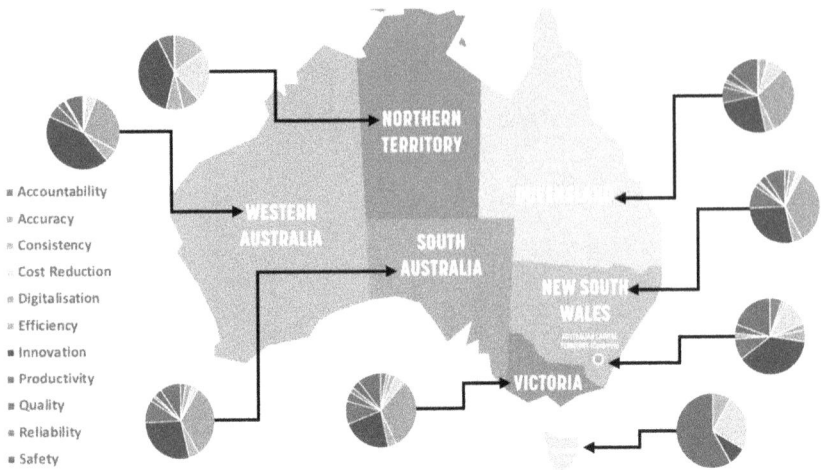

Key:
- Accountability
- Accuracy
- Consistency
- Cost Reduction
- Digitalisation
- Efficiency
- Innovation
- Productivity
- Quality
- Reliability
- Safety

FIGURE 4.6 Distribution of tweets by the prospects of AI technologies in construction per state/territory.

SA (n = 37), TAS's most tweeted technology prospect was 'timesaving' (n = 7). 'Innovation' was the most tweeted concept in ACT (n = 15) and WA (n = 62). Although there were fewer tweets in NT, most were related to the prospect of 'innovation' (n = 5). Table 4.6 provides examples of tweets related to AI technologies and prospects.

3.5 Constraints of Artificial Intelligence Technologies in Construction

A word frequency analysis was also conducted for the 12 constraints that derived from AI-related tweets (Figure 4.7 and Table 4.7). These included 'complexity' (n = 96), 'disruptiveness' (n = 90), 'higher initial costs' (n = 48), 'higher variability' (n = 53), 'implementation' (n = 76), 'lack of capabilities' (n = 110), 'lack of cohesion' (n = 33, 'project risk' (n = 93), 'resistance' (n = 73), 'security of data' (n = 156), and 'unstructured environment' (n = 95).

Security of data was the most discussed constraint in AI-related tweets, but its usability differed from one state/territory to another. While 'security of data' (n = 156) was the most tweeted AI constraint in NSW (n = 32), VIC (n = 21), and QLD (n = 27), 'unstructured environment' was the most popular constraint concept in WA (n = 9) and ACT (n = 5). Tweets from SA had more discussions related to 'disruptiveness' (n = 5), while 'higher initial costs' were predominately discussed in TAS (n = 3) and NT (n = 9). Table 4.8 provides examples of tweets related to AI technologies and prospects.

TABLE 4.5 Distribution of tweets by the prospects of AI technologies in construction per state/territory

	Accountability	Accuracy	Consistency	Cost Reduction	Digitalisation	Efficiency	Innovation	Productivity	Quality	Reliability	Safety	Timesaving
NSW	21	31	9	19	317	43	266	94	19	9	32	90
QLD	4	19	2	42	156	20	132	36	11	6	19	75
VIC	21	12	21	37	246	30	176	81	34	17	30	82
WA	1	0	1	7	38	8	62	8	5	1	0	12
SA	0	3	0	18	37	3	20	6	2	1	0	13
TAS	0	0	1	3	0	0	1	0	0	0	0	7
ACT	4	1	1	9	2	4	29	7	1	1	3	15
NT	0	0	2	3	1	1	5	0	1	0	0	0
Total	51	66	37	138	797	109	691	232	73	35	84	294
NSW	41.18%	46.97%	24.32%	13.77%	39.77%	39.45%	38.49%	40.52%	26.03%	25.71%	38.10%	30.61%
QLD	7.84%	28.79%	5.41%	30.43%	19.57%	18.35%	19.10%	15.52%	15.07%	17.14%	22.62%	25.51%
VIC	41.18%	18.18%	56.76%	26.81%	30.87%	27.52%	25.47%	34.91%	46.58%	48.57%	35.71%	27.89%
WA	1.96%	0.00%	2.70%	5.07%	4.77%	7.34%	8.97%	3.45%	6.85%	2.86%	0.00%	4.08%
SA	0.00%	4.55%	0.00%	13.04%	4.64%	2.75%	2.89%	2.59%	2.74%	2.86%	0.00%	4.42%
TAS	0.00%	0.00%	2.70%	2.17%	0.00%	0.00%	0.14%	0.00%	0.00%	0.00%	0.00%	2.38%
ACT	7.84%	1.52%	2.70%	6.52%	0.25%	3.67%	4.20%	3.02%	1.37%	2.86%	3.57%	5.10%
NT	0.00%	0.00%	5.41%	2.17%	0.13%	0.92%	0.72%	0.00%	1.37%	0.00%	0.00%	0.00%
Total	100%	100%	100%	100%	100%	100%	100%	100%	100%	100%	100%	100%

TABLE 4.6 Example tweets of prospects of AI technologies in construction per state/territory

Prospects of AI Adoption	Date and Time	State/Territory	Tweet
Accountability	30 December 2019 12:42	VIC	We need accountability and a good balance between automation and IoT. #Accountability, #Automation
Accuracy	19 May 2021 08:53	VIC	It's no surprise that there is interest in automation and technologies to lash all aspects of production together. Customers are after more productivity, higher accuracy, and more process driven.
Consistency	10 November 2019 09:15	ACT	The Fourth Industrial Revolution could spell more jobs, not fewer. But people will need different types of skills and need to unlearn and relearn consistency.
Cost reduction	25 December 2020 07:41	NSW	We arrived at the fourth industrial revolution, and it will lead to increased efficiency, new revenue opportunities, and overall cost reductions.
Digitalisation	18 August 2019 23:15	NSW	Robots are supposed to destroy most jobs occupied by females, and the immediate social impact on labour is Job Loss of Women. It's safe to say that Australia is perplexed by digitalisation.
Efficiency	29 September 2020 16:24	WA	With the help of telematics and fleet management technology, it can help benefit businesses by keeping track of your cranes, helping with compliance, efficiency, and boosting profits. #Construction, #Technology
Innovation	14 February 2019 07:04	QLD	The basic construction of a monetary system must not be confused with innovations in payment technology.
Productivity	23 March 2019 23:00	NSW	Construction has lagged behind other industries in harnessing the benefits of digitalisation. But it is now looking to catch up with new technology having enormous potential to make construction greener, safer, and smarter while boosting productivity. #MachineLearning, #AI
Quality	23 January 2020 20:29	NT	The next-generation construction sealant is a high-quality construction sealant based on hybrid technology. It cures under the influence of humidity to form a durable elastic rubber. # Construction
Reliability	24 April 2021 05:39	NSW	Construction is underway on the Victorian NSW Interconnector upgrade. It will allow cheaper generations to be transferred between the states. We're using a new technology called Smart Wires which will improve reliability and avoid the cost of upgrading existing infrastructure.
Safety	13 June 2021 12:24	NSW	Monash University files for privileges of new technology, which identifies safety features of construction machinery on building sites. #Construction
Timesaving	20 January 2020 17:01	QLD	An arm of construction giant BMD has already deployed Octant, and the company expects to reap savings of up to 30% in its turnover in urban development projects.

TABLE 4.7 Distribution of tweets by the constraints of AI technologies in construction per state/territory

	Complexity	Disruptiveness	Higher Initial Cost	Higher Variability	Implementation	Interpretation	Lack of Capabilities	Lack of Cohesion	Project risk	Resistance	Security of Data	Unstructured Environment
NSW	27	19	6	22	25	13	42	13	42	32	56	32
QLD	18	25	12	17	12	19	23	9	28	15	39	21
VIC	36	36	8	12	34	4	34	7	19	22	45	27
WA	8	2	2	0	4	0	3	1	2	0	4	9
SA	1	5	3	1	0	0	0	2	0	2	3	1
TAS	2	1	3	1	1	0	0	0	1	1	1	0
ACT	2	2	5	0	0	1	3	1	0	1	4	5
NT	2	0	9	0	0	1	5	0	1	0	4	0
Total	96	90	48	53	76	38	110	33	93	73	156	95
NSW	28.13%	21.11%	12.50%	41.51%	32.89%	34.21%	38.18%	39.39%	45.16%	43.84%	35.90%	33.68%
QLD	18.75%	27.78%	25.00%	32.08%	15.79%	50.00%	20.91%	27.27%	30.11%	20.55%	25.00%	22.11%
VIC	37.50%	40.00%	16.67%	22.64%	44.74%	10.53%	30.91%	21.21%	20.43%	30.14%	28.85%	28.42%
WA	8.33%	2.22%	4.17%	0.00%	5.26%	0.00%	2.73%	3.03%	2.15%	0.00%	2.56%	9.47%
SA	1.04%	5.56%	6.25%	1.89%	0.00%	0.00%	0.00%	6.06%	0.00%	2.74%	1.92%	1.05%
TAS	2.08%	1.11%	6.25%	1.89%	1.32%	0.00%	0.00%	0.00%	1.08%	1.37%	0.64%	0.00%
ACT	2.08%	2.22%	10.42%	0.00%	0.00%	2.63%	2.73%	3.03%	0.00%	1.37%	2.56%	5.26%
NT	2.08%	0.00%	18.75%	0.00%	0.00%	2.63%	4.55%	0.00%	1.08%	0.00%	2.56%	0.00%
Total	100%	100%	100%	100%	100%	100%	100%	100%	100%	100%	100%	100%

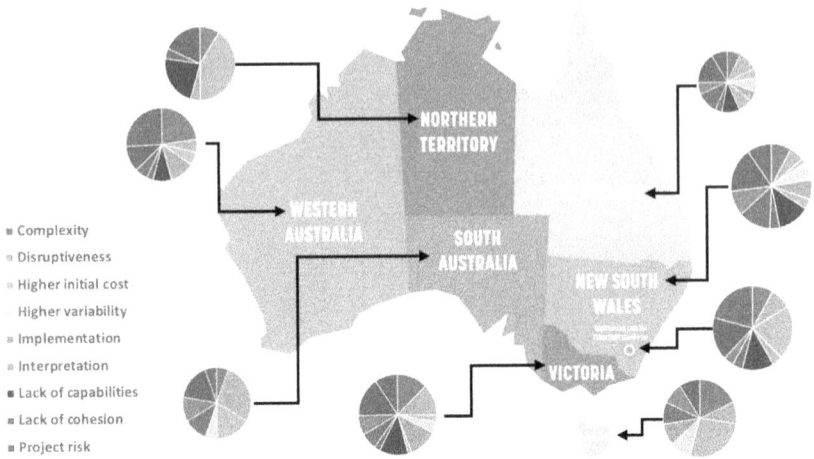

FIGURE 4.7   Distribution of tweets by the constraints of AI technologies in construction per state/territory.

3.6 Prospects and Constraints of AI Technologies in Australian States/Territories

Understanding the public perception of prospects and constraints that AI technologies may bring onto a construction site was at the forefront of this study. A word co-occurrence analysis was conducted, which identified the number of tweets that mentioned an AI technology and prospect or constraint.

Figures 4.8 and 4.9 represent the network topography developed based on word co-occurrence analysis. This network typology was initially generated by using Gephi software. Nonetheless, due to the crowdedness of the original figure, a less complex version was recreated by only showing the stronger relationships that occurred between AI technologies and prospect or constraints concepts. As the number of total tweets from prospects was 1319 and constraints was 609, we made two separate connection measurements. We identified a connection between 20 and 29 as more as semi-strong, from 30 to 39 as strong, and 40 or more as very strong. Furthermore, for constraints, we identified between 10 and 15 as semi-strong, from 16 to 20 as strong, and 20 or more as very strong.

3.6.1 Prospects in Relation to AI Technologies

As shown in Table 4.9, 'robotics' (*n* = 325) was the AI technology that will have the most positive influence over a construction site. This technology has a close relationship with the following prospects: 'time saving'

TABLE 4.8 Example tweets of constraints of AI technologies in construction per state/territory

Technology	Date and Rime	State/Territory	Tweet
Complexity	10 January 2020 23:47	VIC	The Fourth Industrial Revolution is complex as it is unprecedented technologies and characterised by a fusion of advanced robotics, blurring the line between the physical and digital world. #Construction, #Robotics
Disruptiveness	28 April 2021 12.35	NSW	The powerful combination of Artificial intelligence and The Internet of Things will transform entire industries and enable new disruptive services. #AI, #IoT, #Industry 4.0
Higher initial costs	14 February 2020 16.58	NSW	While industry trends have been building momentum for some time, McKinsey said that many are now at a point where their greater reliability and higher initial cost are starting to make sense for industrial applications.
High variability	13 July 2020 12.11	WA	Technology is continuing to drive change in the architecture and construction industries, and it is highly variable.
Implementation	06 April 2020 18.21	QLD	Construction technology implementation is only half of the battle. What is just as important is making that technology investment work in a way that extracts maximum value and keeps projects moving.
Interpretation	15 December 2019 18.39	NSW	Virtual reconstruction of a project that cannot be open to interpretation. #Bigdata, #Automation
Lack of capabilities	28 March 2019 01.30	SA	Artificial intelligence is being adopted so fast that its technical capabilities have outpaced the construction of an ethical governance framework. #Artificial Intelligence
Lack of cohesion	25 February 2021 11.44	NSW	Australia's construction industry is committed to a National Strategy to accommodate industry needs that promote technology cohesion.

(Continued)

TABLE 4.8 Continued

Technology	Date and Rime	State/Territory	Tweet
Project risk	2 February 2021 16.57	TAS	At the advent of the industrial revolution, we must simultaneously be aware of the considerable project risks that are likely to emerge as transformative technologies are assimilated across processes and functions of the industry, government, and broader society. #Construction, #AIconstruction
Resistance	11 May 2020 15.44	VIC	A great article on the Industrial Revolution by the Australian Treasurer. Resistance is futile; embrace the new technologies that will be implemented in projects. Employment will not improve. #Automation, #Bigdata, #Construction
Security of data	18 May 2020 19.52	QLD	Preparing for the industrial revolution requires leadership. Australia needs a construction framework for societal transformation, which includes addressing skills standards and security of data.
Unstructured environment	06 June 2020 03.16	VIC	The construction industry is replacing the traditional working algorithm by providing more intelligent manufacturing equipment and environments #Innovation, #Industry 4.0, #Construction

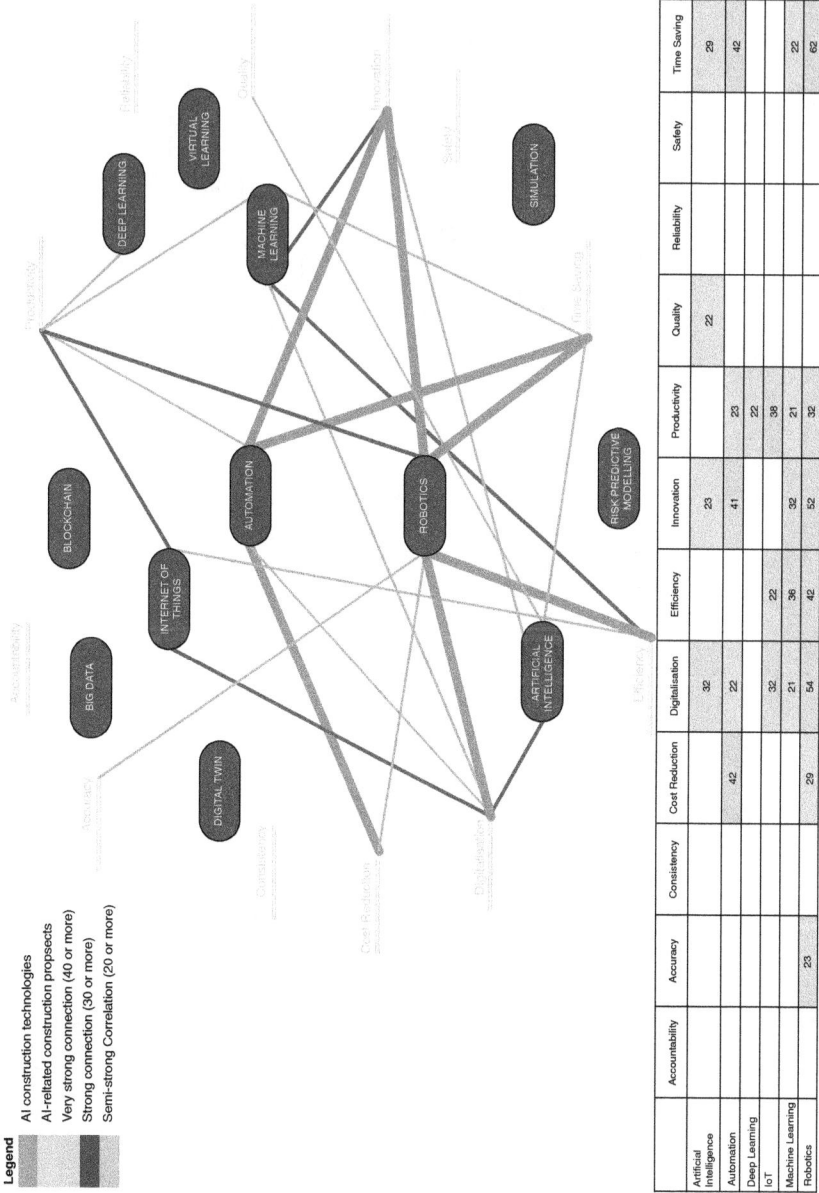

	Accountability	Accuracy	Consistency	Cost Reduction	Digitalisation	Efficiency	Innovation	Productivity	Quality	Reliability	Safety	Time Saving
Artificial Intelligence					32		23		22			29
Automation				42	22		41	23				42
Deep Learning						22		22				
IoT					32	36	32	38				
Machine Learning					21			21				22
Robotics		23		29	54	42	52	32				62

FIGURE 4.8 Relationship between AI technologies and their prospects.

Legend

AI construction technologies
AI-related construction prospects
Very strong connection (40 or more)
Strong connection (30 or more)
Semi-strong Correlation (20 or more)

Legend

- AI construction technologies
- AI-related construction constraints
- Very strong connection (20 or more)
- Strong connection (15 or more)
- Semi-strong Correlation (10 or more)

	Complexity	Disruptiveness	Higher initial cost	Higher variability	Implementation	Interpretation	Lack of capabilities	Lack of cohesion	Project risk	Resistance	Security of data	Unstructured environment
Artificial Intelligence	12	11					22		15	12		
Automation		12							12			
Big Data											12	
Blockchain											12	
Deep Learning							12				15	
IoT											14	
Machine Learning												
Robotics	19						15		11	18		
Risk predictive modelling									15			

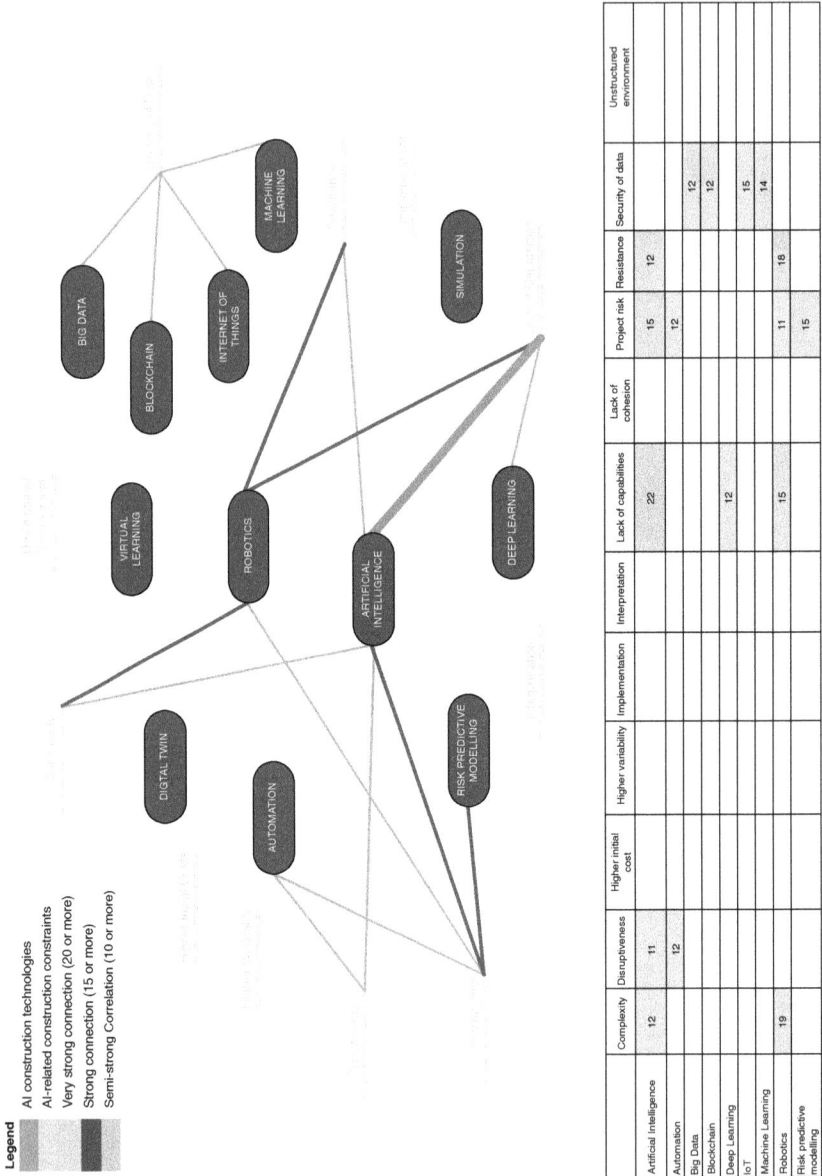

FIGURE 4.9 Relationship between AI technologies and their constraints.

TABLE 4.9   Distribution of tweets by the prospects of AI technologies

	Accountability	Accuracy	Consistency	Cost Reduction	Digitalisation	Efficiency	Innovation	Productivity	Quality	Reliability	Safety	Timesaving	Total
Artificial intelligence	8	5	2	12	32	12	23	2	22	3	4	29	154
Automation	7	5	3	42	22	19	41	23	4	2	12	42	222
Big data	3	2	5	3	12	14	14	16	3	4	0	19	95
Blockchain	2	2	2	3	8	12	12	2	0	1	0	4	48
Deep learning	3	1	1	1	8	18	9	22	2	5	2	8	80
Digital twin	1	2	0	2	2	3	0	0	0	0	0	2	12
IoT	4	8	8	3	32	22	17	38	7	2	1	18	160
Machine learning	11	4	2	2	21	36	32	21	3	6	2	22	162
Robotics	4	23	3	29	54	42	52	32	4	2	18	62	325
Risk prediction modelling	2	1	0	5	4	8	5	3	0	0	1	3	32
Simulation	0	1	2	0	3	1	0	1	0	0	0	1	9
Virtual learning	1	1	0	3	8	0	2	0	0	1	0	4	20
Total	46	55	28	105	206	187	207	160	45	26	40	214	1319

(n = 62), 'digitalisation' (n = 54), 'innovation' (n = 52), and 'efficiency' (n = 42). Second, the AI technology that was discussed most was 'automation' (n = 222), which had a close relationship with 'digitalisation' (n = 32), 'time saving' (n = 29), innovation (n = 23), and 'quality' (n = 22). The third popular technology was 'machine learning' (n = 162) as it will have a positive impact on construction by increased 'efficiency' (n = 36), 'innovation' (n = 32), 'timesaving' (n = 22), 'digitalisation' (n = 21), and 'productivity' (n = 21).

The relationship between the AI-related technologies and prospects—such as 'reliability' (n = 26), 'consistency' (n = 28), 'safety' (n = 40), 'quality' (n = 45), and 'accountability' (n = 46)—was frequently identified in relation to positive attributes that AI technologies can bring to the urban built environment. The tweets related to 'IoT' (n = 160) and the prospects of 'productivity' (n = 38), 'digitalisation' (n = 32), 'efficiency' (n = 22), 'time saving' (n = 18), and 'innovation' (n = 17) highlight the positive attributes that IoT will bring to the construction industry in public and private sectors. Nevertheless, AI technology tweets related to the prospects of 'simulation' (n = 9), 'digital twin' (n = 12), 'virtual learning' (n = 20), 'risk predictive modelling' (n = 32), and 'blockchain' (n = 48) were comparatively low.

3.6.2 Constraints in Relations to AI Technologies

As shown in Table 4.10, 'robotics' (n = 112) was the AI technology that will have the most negative influence over a construction site. The technology has a close relationship with 'complexity' (n = 19), 'resistance' (n = 18), 'lack of capabilities' (n = 15), and 'project risk' (n = 11). Second, 'artificial intelligence' (n = 98) had a close relationship with 'lack of capabilities' (n = 22), 'project risk' (n = 12), 'resistance' (n = 12), 'complexity' (n = 12), and 'disruptiveness' (n = 11). The third popular relationship was 'automation' (n = 71) as it was perceived to have a negative influence over a construction site by being highly 'disruptive' (n = 12), 'project risk' (n = 12), and will be difficult to 'implement' (n = 9).

Although there were some connections between AI-related technologies and constraints such as 'interpretation' (n = 22), less emphasis was placed on other negative attributes like 'lack of cohesion' (n = 26), 'unstructured environments' (n = 30), and 'higher initial costs' (n = 34) when discussing the challenges AI technologies may introduce to the urban built environment. Discussions related to AI technologies were 69% less than the prospect that technologies can bring to a construction site. The tweets related to 'machine learning' (n = 62) and the constraints of 'security of data' (n = 14),

TABLE 4.10 Distribution of tweets by the constraints of AI technologies

	Complexity	Disruptiveness	Higher Initial cost	Higher Variability	Implementation	Interpretation	Lack of Capabilities	Lack of Cohesion	Project Risk	Resistance	Security of Data	Unstructured Environment	Total
Artificial intelligence	12	11	6	2	3	1	22	2	15	12	6	6	98
Automation	4	12	2	3	9	6	2	4	12	8	5	4	71
Big data	2	4	8	2	9	5	2	2	5	2	12	2	55
Blockchain	2	2	0	4	2	0	9	0	9	2	12	1	43
Deep learning	8	8	2	6	1	1	12	7	3	1	2	5	56
Digital twin	2	4	2	1	1	5	2	1	0	0	3	2	23
IoT	7	8	3	6	2	0	4	5	5	3	15	0	58
Machine learning	8	8	4	6	6	2	8	1	3	1	14	1	62
Robotics	19	9	7	7	9	2	15	2	11	18	5	8	112
Risk prediction modelling	1	2	0	1	0	0	0	0	15	0	1	1	21
Simulation	1	0	0	0	0	0	0	1	1	0	0	0	3
Virtual learning	0	1	0	2	0	0	0	1	2	0	1	0	7
Total	66	69	34	40	42	22	76	26	81	47	76	30	609

'lack of capabilities' ($n = 8$), 'complexity' ($n = 8$), and 'disruptiveness' ($n = 8$) highlight the negative attributes that machine learning will bring to the construction industry in the public and private sectors. Nevertheless, AI technology tweets related to 'simulation' ($n = 3$), 'virtual learning' ($n = 7$), 'risk predictive modelling' ($n = 21$), and 'digital twin' ($n = 23$) were comparatively low. These four constraints were the same as the four prospects, which shows that they were not discussed frequently on Twitter and can be perceived as a neutral impact that they will bring to the construction industry.

4 FINDINGS AND DISCUSSION

The last five years have seen major advancements in AI, and it is beginning to gain traction in the construction industry from planning to construction. The potential of AI in the planning and design stages is an increase in the accuracy of cost estimates, accurate milestones, and reduction of onsite risk by using constructive alternative analysis. Furthermore, the benefits of AI in the construction stage are increasing productivity, improving work processes, and reducing the probability of onsite accidents.

Construction firms analyse vast amounts of internal and external unstructured data to provide insights from previous projects. This will allow businesses to generate more accurate estimates and reduce budgets and timeline deviation by an estimated 10%–20% and engineering hours by 10%–30% [40,41]. AI's potential in construction is to provide real-time insight that will help project managers ensure efficient use of resources, anticipate potential risk, and increase safety. Potential savings from data analytics and related technologies can reduce 10%–15% of total construction costs [21,42].

The sentiment towards AI in Australia is becoming more positive, as evident in the findings. The public's opinion has been highly influenced by the Australian government as they have committed $125 million through an 'AI Action Plan' to operate the digital frontier. Furthermore, through this plan, the government has attempted to address the issues identified in this study by investing in the R&D of AI [43]. Additionally, the AI Roadmap also outlines Australia's direction in implementing AI in construction by stating the direction for utilising AI in Australia to improve the built environment by capturing social, economic, and environmental benefits. This includes improving design, planning, construction, operational, and maintenance of infrastructure and buildings [44]. High construction costs and unplanned cost overruns will be fundamental in AI development, limiting the ability to improve Australian cities and infrastructure.

4.1 Sentiment Analysis

The sentiment analysis found that AI in construction is a growing ecosystem of hardware and software. It has recently gained popularity, and the public perception regarding the use of technology is an under-studied area of research [9,10]. AI is a powerful tool that has the power to reshape and disrupt the construction industry. Today, there is limited understanding of the trending construction technologies and their application areas. This is evidenced in the low number of tweets ($n = 7907$). In addition, there is limited knowledge on the public perception of AI technologies, their application area, and the AI-related policies that businesses need to follow when we incorporate AI. Hence, the study aimed to understand the relationship between AI technologies, their key prospects, and constraints in the construction industry.

Overall, the location-based Twitter analysis identifies that 'robotics' ($n = 931$), 'IoT' ($n = 562$), 'machine learning' ($n = 522$), and 'big data' ($n = 467$) are the most discussed topics of AI technologies for the construction industry across the entire Australia, despite their popularity differs by state and territory. The analysis also revealed that the public perception from the three largest states, NSW, QLD, and VIC, were primarily favourable towards AI being implemented in a construction project. While tweets in WA and TA were neutral, and SA, ACT, and NT were mostly negative.

4.1.1 Positive Sentiments

The overall Australian public was positive regarding AI in the construction industry (43% positive sentiment), which is evident in the presented findings. From the identified AI technologies, a prospect analysis was conducted and found that 'timesaving' ($n = 214$), 'innovation' ($n = 207$), 'digitalisation' ($n = 206$), and 'efficiency' ($n = 187$) are the most discussed benefits of AI in construction.

This positive sentiment is driven mainly by the three larger states by population (QLD, VIC, NSW), as they have already invested in the research and development (R&D) of potential AI technologies. In addition, there is a common agreeance between larger construction companies that operate in these states that inadequate project selection is a major challenge. To limit the risk that this challenge may impose, AI technology will need to be implemented into their projects as it will increase efficiency substantially. This has been the key driver that has influenced the construction landscape and increased positive opinion.

4.1.2 Negative Sentiments

Meanwhile, the public also raised concerns about the use of AI in construction, such as data security and a lack of capabilities to incorporate AI technologies. While many AI technologies remain in the R&D phase, they may impose a project risk that has cost implications. Reducing the impact of these three constraints will be necessary for the continuing development of AI in construction. Moreover, a constraint analysis was conducted and found that 'project risk' ($n = 81$), 'security of data' ($n = 76$), 'lack of capabilities' ($n = 76$), and 'disruptiveness ($n = 69$) were the most discussed disadvantages of AI in the construction industry. It is also noted that both the perceptions on the prospects and constraints were differed by states and territories.

The negative sentiment was driven mainly by the smaller states by population (NT, WA, SA, and ACT), as AI technology is still in the initial phase. These states are highly fragmented with smaller construction companies, and there is limited knowledge on the potential technologies may bring. AI is predominately seen as a disruption, as smaller companies cannot compete with larger companies to obtain data to train models. There is a strong focus on the disadvantages that technologies may bring and how they will directly impact the workforce negatively.

4.2 Research Limitations

The study has the following limitations:

- The scope of the research constrains the chapter in itself.
- AI in construction is still a broad concept; the relationship between technologies, constraints, and prospects is constantly changing. There is a lack of resemblance between companies.
- The study did not conduct strengths, weaknesses, opportunities, and threats (SWOT).
- The study was only able to analyse 7,906 tweets due to data availability limitations.
- The Random Forest and Random Tree software has difficulty detecting positive or negative words when looked at in isolation. For example, it struggles if the general user is sarcastic, ironic, or hyperbolic.

Our prospective research, nevertheless, will focus on addressing these constraints.

4.3 Research Directions

In the light of the analysis conducted in this study, the directions for future research are related to the barriers are identified as follows:

- Expanding on the current findings of the research and developing a better understanding of the relationship between AI technologies, prospects, and constraints.

- Using other social media big data. The sentiment analysis only gathered Twitter data from the Australian public. Future studies could obtain Twitter data from various countries to expand on the scope of the chapter.

- Expanding on the search parameters and including data obtained from other social media platforms.

- Supplementing the study with mixed methods. Future studies could conduct interviews with construction professionals and gather qualitative data to expand on current literature findings.

- Expanding on the current empirical studies and analysis is needed to further understand public perception towards AI in construction.

- Extending on the current research and assimilating the practical aspect of the technologies to enable guidelines to be produced within the industry for the construction community.

5 CONCLUSION

There is limited knowledge on the public perception of AI technologies, their application area, and the AI-related policies that businesses need to follow when we incorporate AI [45]. Hence, the study aimed to understand the relationship between AI technologies, their key prospects, and constraints in the construction industry.

Among all states and territories in Australia, QLD (46%), NSW (46%), and VIC (45%) had the highest degree of satisfaction regarding AI in the construction industry. In contrast, given that most states and territories had a positive sentiment, NT (74%), SA (56%), and ACT (50%) had the lowest degree of satisfaction. The states and territories that had the lowest interest in sharing their views on social media channels (i.e., Twitter) showed primarily neutral or negative sentiments. Furthermore, AI 'prospects' ($n = 1,319$) were mentioned twice the amount of 'constraints'

(n = 609). We also justified the close relationship between AI technologies and prospects in several analysis procedures, that is, sentiment and content analysis, frequency analysis, content analysis, co-occurrence analysis, and spatial analysis.

This study also highlighted AI as a powerful technology that has the power to reshape and disrupt the construction industry. Today, there is limited understanding of the trending construction technologies and their application areas. This is evidenced in the low number of tweets (n = 7,907). AI technologies received less attention on Twitter; additional empirical studies and analysis are needed to further understand public perception towards AI in construction. This will allow construction bodies to ease the transition from traditional management methods to management that incorporates machine and deep learning components to automate various construction stages. We believe the findings of this study inform the construction industry on public perceptions and prospects and constraints of AI adoption and advocate the search for finding the most efficient means to utilise AI technologies. This study captured the general public's perceptions of AI technologies in the construction industry, while our prospective research will concentrate on expanding and consolidating the understanding and relationship between AI technologies and the key actors of the construction industry.

ACKNOWLEDGEMENTS

This chapter, with permission from the copyright holder, is a reproduced version of the following journal article: Regona, M., Yigitcanlar, T., Xia, B., & Li, R. (2022). Artificial intelligent technologies for the construction industry: how are they perceived and utilised in Australia? *Journal of Open Innovation: Technology, Market, and Complexity*, 8(1), 16.

REFERENCES

1. Zhou, Z.; Goh, Y.; Shen, L. Overview and analysis of ontology studies supporting development of the construction industry. *J. Comput. Civil Eng.* 2016, 30, 04016026.
2. Yigitcanlar, T.; Butler, L.; Windle, E.; Desouza, K.; Mehmood, R.; Corchado, J. Can building "artificially intelligent cities" safeguard humanity from natural disasters, pandemics, and other catastrophes? An urban scholar's perspective. *Sensors.* 2020, 20, 2988.
3. Yun, J.; Lee, D.; Ahn, H.; Park, K.; Yigitcanlar, T. Not deep learning but autonomous learning of open innovation for sustainable artificial intelligence. *Sustainability.* 2016, 8, 797.

4. Hong, Y.; Hammad, A.; Zhong, X.; Wang, B.; Akbarnezhad, A. Comparative modeling approach to capture the differences in BIM adoption decision-making process in Australia and China. *J. Constr. Eng. Manag.* 2020, 146, 04019099.

5. McKinsey & Company. The Next Normal in Construction. 2020. Available online: https://www.mckinsey.com/~/media/McKinsey/Industries/Capital%20Projects%20and%20Infrastructure/Our%20Insights/The%20next%20normal%20in%20construction/The-next-normal-in-construction.pdf (accessed on 5 May 2021).

6. Adwan, E.; Al-Soufi, A. A review of ICT applications in construction. *Inter. J. Infom. Visual.* 2018, 2, 1–12.

7. Koskela, L.; Ballard, G.; Howell, G. Achieving change in construction. *In Proceedings of the International Group of Lean Construction 11th Annual Conference,* Virginia, USA, 5 June 2003.

8. Irani, Z.; Kamal, M.M. Intelligent systems research in the construction industry. *Expert Syst. Appl.* 2014, 41, 934–950.

9. McKinsey State of Machine Learning and AI. Available online: https://www.Forbes.com/sites/louiscolumbus/2017/07/09/mckinseys-state-of-machine-learning-and-ai-2017 (accessed on 21 June 2021). McKinsey & Company.

10. Lv, Z.; Chen, D.; Lou, R.; Alazab, A. Artificial intelligence for securing industrial-based cyber–physical systems. *Future Gener. Comput. Syst.* 2021, 117, 291–298.

11. Yigitcanlar, T.; Kankanamge, N.; Regona, M.; Ruiz Maldonado, A.; Rowan, B.; Ryu, A.; Li, R. Artificial intelligence technologies and related urban planning and development concepts: How are they perceived and utilized in Australia? *J. Open Innova. Tech., Market, Complex.* 2020, 6, 187.

12. Akinosho, T.D.; Oyedele, L.O.; Bilal, M.D.; Ajayi, A.O.; Delgado, M.D.; Akinade, O.O.; Ahmed, A.A. Deep learning in the construction industry: a review of present status and future innovations. *J. Build. Eng.* 2020, 5, 101827.

13. Darko, A.; Chan, A.P.; Adabre, M.A.; Edwards, D.J.; Hosseini, M.R.; Ameyaw, E.E. Artificial intelligence in the AEC industry: scientometrics analysis and visualisation of research activities. *Autom. Constr.* 2020, 112, 103081.

14. Blanco, J.; Fuchs, S.; Parsons, M.; Ribeirinho, M.J. Artificial intelligence: construction technology's next frontier. *Build. Econ.* 2018, 9, 7–13.

15. Yigitcanlar, T. Greening the artificial intelligence for a sustainable planet: an editorial commentary. *Sustainability.* 2021, 13, 13508.

16. Howard, J. Artificial intelligence: implications for the future of work. *Am. J. Ind. Med.* 2019, 62, 917–926.

17. Baker, H.; Hallowell, M.R.; Tixier, A.J. AI-based prediction of independent construction safety outcomes from universal attributes. *Autom. Constr.* 2020, 118, 103146.

18. Yaseen, Z.M.; Ali, Z.H.; Salih, S.Q.; Al-Ansari, N. Prediction of risk delay in construction projects using a hybrid artificial intelligence model. *Sustainability.* 2020, 12, 1514.

19. Choi, S.J.; Choi, S.W.; Kim, J.H.; Lee, E.B. AI and text-mining applications for analyzing contractor's risk in invitation to bid (ITB) and contracts for engineering procurement and construction (EPC) projects. *Energies*. 2021, 14, 4632.

20. Grover, P.; Kar, A.K.; Dwivedi, Y.K. Understanding artificial intelligence adoption in operations management: insights from the review of academic literature and social media discussions. *Ann. Oper. Res.* 2020, 308, 177–213.

21. Abdirad, H.; Mathur, P. Artificial intelligence for BIM content management and delivery: case study of association rule mining for construction detailing. *Adv. Eng. Informatics*. 2021, 50, 101414.

22. Mahbub, R. An investigation into the barriers to the implementation of automation and robotics technologies in the construction industry. Ph.D. Thesis, Queensland University of Technology, Queensland, Australia, 2008.

23. Oprach, S.; Bolduan, T.; Steuer, D.; Vössing, M.; Haghsheno, S. Building the future of the construction industry through artificial intelligence and platform thinking. *Digit Welt*. 2019, 3, 40–44.

24. Abioye, S.O.; Oyedele, L.O.; Akanbi, L.; Ajayi, A.; Delgado, J.M.; Bilal, M.; Ahmed, A. Artificial intelligence in the construction industry: a review of present status, opportunities, and future challenges. *J. Build. Eng.* 2021, 44, 103299.

25. Mohamed, M.A.; Ahmad, A.B.; Mohamad, D. The implementation of artificial intelligence (AI) in the Malaysia construction industry. *AIP Conf. Proc.* 2021, 2339, 020136, https://doi.org/10.1063/5.0044597.

26. Atuahene, B.T.; Kanjanabootra, S.; Gajendra, T. Benefits of big data application experienced in the construction industry: a case of an Australian construction company. *In Proceedings of the 36th Annual Association of Researchers in Construction Management Conference*, London, UK, 7 September 2020.

27. Palaniappan, K.; Kok, C.L.; Kato, K. Artificial Intelligence (AI) coupled with the Internet of Things (IoT) for the enhancement of occupational health and safety in the construction industry. In *Advances in Artificial Intelligence, Software and Systems Engineering: Proceedings of the AHFE 2021 Virtual Conferences on Human Factors in Software and Systems Engineering, Artificial Intelligence and Social Computing, and Energy*, 2021, 31–38, https://doi.org/10.1007/978-3-030-80624-8_4.

28. Kankanamge, N.; Yigitcanlar, T.; Goonetilleke, A.; Kamruzzaman, M. Can volunteer crowdsourcing reduce disaster risk? A systematic review of the literature. *Int. J. Disaster Risk Reduct.* 2019, 35, 101097.

29. Yigitcanlar, T.; Cugurullo, F. The sustainability of artificial intelligence: An urbanistic viewpoint from the lens of smart and sustainable cities. *Sustainability*. 2020, 12, 8548.

30. Yigitcanlar, T.; Mehmood, R.; Corchado, J.M. Green artificial intelligence: towards an efficient, sustainable and equitable technology for smart cities and futures. *Sustainability*. 2021, 13, 8952.

31. Eber, W. Potentials of artificial intelligence in construction management. *Organ. Technol. Manag. Constr. Int. J.* 2020, 12, 2053–2063.

32. ConsumerStatistics.Availableonline:https://www.yellow.com.au/wp-content/uploads/sites/2/2020/07/Yellow-Social-Media-Report-2020-Consumer-Statistics.pdf (accessed 8 September 2021).

33. Kankanamge, N.; Yigitcanlar, T.; Goonetilleke, A.; Kamruzzaman, M. Determining disaster severity through social media analysis: testing the methodology with Southeast Queensland Flood tweets. *Int. J. Disaster Risk Reduct.* 2020, 42, 101360.

34. Kankanamge, N.; Yigitcanlar, T.; Goonetilleke, A. How engaging are disaster management-related social media channels? The case of Australian state emergency organisations. *Int. J. Disaster Risk Reduct.* 2020, 48, 101571.

35. Kankanamge, N.; Yigitcanlar, T.; Goonetilleke, A. Public perceptions on artificial intelligence driven disaster management: evidence from Sydney, Melbourne and Brisbane. *Tele. Informatics.* 2021, 65, 101729.

36. Drus, Z.; Khalid, H. Sentiment analysis in social media and its application: systematic literature review. *Procedia Comput. Sci.* 2019, 161, 707–714.

37. Alomari, E.; Katib, I.; Mehmood, R. Iktishaf: A big data road-traffic event detection tool using Twitter and spark machine learning. *Mob. Networks Appl.* 2023, 28(2), 603–618.

38. Alomari, E.; Katib, I.; Albeshri, A.; Yigitcanlar, T.; Mehmood, R. A big data tool with automatic labeling for road traffic social sensing and event detection using distributed machine learning. *Sensors.* 2021, 21, 2993.

39. Yigitcanlar, T.; Kankanamge, N.; Preston, A.; Gill, P. S.; Rezayee, M.; Ostadnia, M.; Ioppolo, G. How can social media analytics assist authorities in pandemic-related policy decisions? Insights from Australian states and territories. *Health Inf. Sci. Syst.* 2020, 8, 1–21.

40. Artificial Intelligence in the Construction Industry. Available online: https://www.rolandberger.com/en/Insights/Publications/Artificial-intelligence-in-the-construction-industry.html (accessed 19 September 2021).

41. You, Z.; Feng, L. Integration of industry 4.0 related technologies in construction industry: a framework of cyber-physical system. *IEEE Access.* 2020, 8, 122908–122922.

42. Pillai, V.S.; Matus, K.J. Towards a responsible integration of artificial intelligence technology in the construction sector. *Sci. Public Policy.* 2020, 47, 689–704.

43. Data61. Artificial Intelligence Roadmap. Available online: https://data61.csiro.au/en/Our-Research/Our-Work/AI-Roadmap (accessed 2 October 2021).

44. Australian Government. Australia's Artificial Intelligence Action Plan. Available online: https://www.industry.gov.au/data-and-publications/australias-artificial-intelligence-action-plan (accessed 2 October 2021).

45. Kassens-Noor, E.; Wilson, M.; Kotval-Karamchandani, Z.; Cai, M.; Decaminada, T. Living with autonomy: public perceptions of an AI-mediated future. *J. Plan. Educ. Res.* 2021, https://doi.org/10.1177/0739456x20984529.

PART 2

Responsible Urban Artificial Intelligence

This part of the book is dedicated to offering a thorough understanding of responsible urban artificial intelligence. It covers a range of crucial areas, starting with the implementation of AI by local governments in a responsible manner. It also delves into the conceptual framework of responsible urban artificial intelligence, providing a structured approach to ethical AI deployment. Additionally, this book discusses an assessment framework for evaluating the responsibility of AI applications in urban settings. Furthermore, it examines the relationship between artificial intelligence and sustainable development goals, highlighting how responsible AI can contribute to sustainable urban development.

DOI: 10.1201/9781003521440-6

Responsible Local Government Artificial Intelligence

1 INTRODUCTION

Over the last 50 years, the pace of technological development has increased significantly. We owe this remarkable progress to the efforts of the stakeholders of the global innovation ecosystem that activated two ground-breaking digital revolutions [1–3]. The First Digital Revolution occurred in the 1980s and 1990s—some scholars even date it back to the 1970s, when the development of the personal computer commenced [4]. These technological developments resulted in mass digitisation, an increasing number of products and services being encoded in the cyberspace, and the diffusion of the internet on a pervasive scale [5]. Today, the world is on the verge of the Second Digital Revolution—where an increasing number of computing- and internet-enabled objects and devices allow for ubiquitous computing and open innovation opportunities in our everyday lives [6–8].

Moreover, Makridakis [9] estimates that the next digital revolution will take place within the next couple of decades, and calls it the 'artificial intelligence (AI) revolution'. He further predicts that it will have a greater impact than both the first and the second digital revolutions combined. However, we are already on track towards the AI revolution. For instance, the Internet-of-Things (IoT) links objects wirelessly

DOI: 10.1201/9781003521440-7

to a network that enables data sharing, and within this network, AI is simultaneously analysing IoT data and making decisions autonomously [10–12]. The smart home can be offered as an example of the popular application areas for this technology [13,14]. While highly innovative technologies—e.g., artificially intelligent IOT (AI-IoT) [15]—are disrupting the industrial processes—i.e., Industry 4.0 [16]—, they are disrupting our cities and societies as well—i.e., smart city and smart community [17–19].

Nonetheless, this disruption is not necessarily solely generating positive externalities and delivering the desired outcomes or the desired outcomes for all [20]. For instance, on the one hand, autonomous vehicles—in the form of autonomous shuttle buses—could increase public transport coverage and patronage, and hence decrease the carbon emissions associated with transport [21,22]. On the other hand, autonomous vehicles—in the form of private autonomous cars—could increase mobility and urban sprawl, and thus increase transport carbon emissions [23]. Issues similar to these bring up the need for technological innovation in the context of cities, or in other words urban innovation, to become responsible for maximising the desired outcomes and positive impacts for all and minimising the unwanted ones [24–26].

Responsible innovation is vital in order to tackle the challenges our cities face, irrespective of whether they are related to natural resource degradation, climate change, economic progress, or social welfare [27]. According to Von Schomberg [28, p. 51],

> responsible innovation is a transparent, interactive process by which societal actors and innovators become mutually responsive to each other with a view to the ethical acceptability, sustainability and societal desirability of the innovation process and its marketable products in order to allow for the proper embedding of scientific and technological advances in our society.

Responsible urban innovation can be defined as "a collective commitment of care for the urban futures through responsive stewardship of science, technology and innovation in the present" [29, p. 27]. That is to say, responsible urban innovation challenges us not only to generate science, technology, and innovation which can have a positive impact on our cities and societies today, but also makes us think about and act upon our responsibility to build the desired urban futures for all [30].

This perspective chapter is written with the purpose of contributing to the existing responsible urban innovation discourse—that is an under-studied and a relatively under-advocated area. With a specific focus on technology for responsible urban innovation, the chapter concentrates on AI and its use as part of local government systems. The rationale behind this selection is as follows: (a) AI, a technology with an increasing number of applications in the urban context, is referred to as one of the most powerful technologies of our time with both positive and negative externalities for cities [31,32]; (b) AI is an integral part of a smart city structure that provides the required efficiencies and automation ability in the delivery of local infrastructures, services and amenities [33,34], and; (c) there is a trend among local government agencies to adopt AI for managing routine, complex and complicated urban issues, where the knowledge and the experience of the staff in the area of responsible innovation, in general, are fairly limited [35–37].

As for the methodological approach, this perspective chapter undertakes a review of the literature, research, developments, trends, and applications concerning responsible urban innovation with local government AI systems, and develops a conceptual framework. In the light of the findings, the chapter advocates the need for balancing the costs, benefits, risks and impacts of developing, adopting, deploying, and managing local government AI systems targeting responsible urban innovation.

Following this introduction, Section 2 provides an overview of the notion of responsible urban innovation. Subsequently, Section 3 focuses on local government AI systems including their common application areas. Next, Section 4 presents the concept of responsible local government AI and its necessity for obtaining the desired urban outcomes as well as for showcasing responsible urban innovation practices. Section 5 introduces a conceptual framework of responsible urban innovation with local government AI. Lastly, Section 6 closes the chapter with some concluding remarks and prospective lines of research.

2 RESPONSIBLE URBAN INNOVATION

Cities continue to experience significant challenges—e.g., resource demands, governance complexity, socioeconomic inequality, and environmental threats—where innovation is seen as an important means of addressing these problems [38–40]. In other words, innovation is considered necessary for tackling urbanisation problems and ensuring smart, sustainable, and inclusive growth [41–43]. While local governments conducted urban experiments to trial combatting urban problems in a novel way with technological innovation [44], this also created a lucrative business opportunity for high-tech companies—such as Cisco, IBM, Siemens, Huawei, and Sidewalk Labs—which merged technology solutions with urban planning and development under the popular 'smart city' brand [45,46].

Some initiatives are renowned for their success in the use of advanced technologies, such as AI to guide urban planners in making improvements in the city. The following are just some of the success cases: "(a) Massachusetts Institute of Technology (MIT) Media Lab's agent-based simulation to explore possible designs for busy public spaces, including a regenerated Champs-Élysées in Paris; (b) the AI application of Topos, a New York based startup, including image recognition and natural language processing, to help understand how the layout of a city affects those living in it, and to identify how different areas of New York were used by the residents; (c) University of Melbourne's AI utilization for future urban design decisions through the use of generative adversarial networks (GANs) to reproduce Google Street View images in the style of Melbourne's neighborhoods with public health characteristics" [47, p. 4]; (d) the Array of Things (AoT) project of the University of Chicago that comprise "a network of interactive, modular devices, or nodes, that are installed around Chicago to collect real-time data on the city's environment, infrastructure, and activity. These measurements are also shared as open data for research and public use" [48, p. 1]—also see Hawthorne [49] for how citizen devices are being used for tracking Chicago's pollution hot spots, and; (e) additional examples can be given with the cities' living lab experiments that utilise quadruple helix as a form of local innovation system [50], deploying innovative technology to encourage citizen participation in urban decisions [51], and social innovation initiatives concerning urban problems, such as sustainable development and climate change [52].

Nonetheless, the short-term profit-at-any-cost mindset of many disruptive technology companies has been generating innovation with more negative externalities—e.g., increased energy demand, pollution, damage to physical and mental health, and waste of taxpayers' money—than positive externalities [53,54]. This is to say, innovation without responsibility creates more problems than it solves—e.g., technology push, negligence of fundamental ethical principles, policy pull, and lack of precautionary measures and technology foresight [55,56].

Some of the common examples of the negative consequences of urban innovation involving advanced technologies, such as AI, include, but are not exclusively limited to: (a) the failure of algorithmic decision-making and predictive analytics of Pittsburgh, PA, in solving urban poverty, homelessness, and violence problems, particularly by misdiagnosing child maltreatment and prescribing the wrong solutions [57]. This issue of automating inequality is discussed in length in the seminal work of Eubanks [58]. (b) Bias algorithmic decision-making has become one of the major unintended negative externalities, and the examples range from excluding women [59] to excluding people of colour [60], and from excluding religious minorities [61] to excluding indigenous people [62]. (c) In most cases, the failure and bias of algorithmic decisions led to the abolition of AI adoption endeavours in local governments. A good example is scrapping of the use of algorithms in benefit and welfare decisions in 20 local councils in the UK [63]. Furthermore, as stated by McKnight [64, p. 1], "as AI has no a moral compass, OpenAI's managers originally refused to release GPT-3 (Generative Pre-trained Transformer 3)—an autoregressive language model that uses deep learning to produce human-like text—ostensibly because they were concerned about the generator being used to create fake material, such as reviews of products or election-related commentary. Similarly, AI writing bots may need to be eliminated by humans, as in the case of Microsoft's racist Twitter prototype AI chatbot—i.e., Tay". (d) The other negative externalities include "creating opaque decision-making processes, challenges in accountability and trust in AI-enabled decisions, and risks to privacy due to sensitive, granular and in-depth data collection practices" [65, p. 3].

Moreover, even the most celebrated smart city initiatives—that represent urban innovation in management and policy as well as technology—have failed to deliver their promises or have even been abandoned before project initiation—e.g., Songdo, Masdar, PlanIT Valley and Sidewalk Toronto [66,67]. The main reasons behind this failure include technology myopia, a top-down approach, solutionism, and the lack of clear objectives and socio-spatial responsibility [68]. Consequently, technology giants—e.g., Google's Sidewalk Labs and Cisco—have recently pulled back from the smart city push that did not practiced clear responsible urban innovation principles—including accountability, anticipation, reflexivity, transparency, responsiveness, inclusiveness, and sustainability [69–71]. As argued by Green [72, p. 1], we need to "recognize the complexity of urban life rather than merely see the city as something to optimize, truly smart cities are the ones that successfully incorporate technology into a holistic vision of justice and equity".

Responsible urban innovation is central to addressing the current and emerging challenges of cities characterised by complexity, uncertainty, risk, and myopia [73]. It encompasses a public and environmental value-sensitive approach to technology design and adoption, which makes environmental (e.g., eco-responsibility) and societal (e.g., social-responsibility) factors as relevant as the economic (e.g., frugality) ones in the urban innovation and development processes [74–76]. In other words, responsible urban innovation carefully considers the effects of innovation on the environment and society [77]. Responsible urban innovation, thus, is characterised by its sustainability, which is vital for generating long-lasting solutions [78].

Furthermore, as stated by Ziegler [79, p. 195], there are two roles for responsible urban innovation to play for socio-spatial justice. These are:

(a) to contribute to the long-term stability of the society, and thus to find creative responses to socio-spatial challenges such as climate change, and; (b) to find ideas that specifically improve the benefits for the least advantaged members of the community in the present.

3 LOCAL GOVERNMENT ARTIFICIAL INTELLIGENCE SYSTEMS

As stated by Das and Rad [80, p. 1], AI-based algorithms "are transforming the way we approach real-world tasks done by humans; where recent years have seen a surge in the use of these algorithms in automating various facets of science, business, and social workflow". In particular, government agencies are increasingly interested in using AI capabilities to deliver policy and generate efficiencies in high-uncertainty environments [81,82]. A study by De Sousa et al. [83] disclosed a growing trend of interest in AI in the public sector, with the US as the most active country. This is also the case with many local government agencies [84]. According to Wirtz et al. [85], the most common AI applications in government agencies are as follows: (a) AI-based knowledge management software; (b) AI process automation systems; (c) chatbots/virtual agents; (d) predictive analytics and data visualisation; (e) identity analytics; (f) cognitive robotics and autonomous systems; (g) recommendation systems; (h) intelligent digital assistants; (i) speech analytics; and (j) cognitive security analytics and threat intelligence.

AI offers urban innovation opportunities to generate novel solutions to the problems of our cities [86,87]. It has the potential to create a great impact on the way citizens experience and receive services and interact with their government, as recent advances in AI have resulted in an increasing number of decisions being handed over to algorithms [88,89]. This also applies to local government operations and services [90,91]. Today, AI is not only becoming an integral part of local government operations and services but is also impacting and shaping the future of our cities and societies [92]—e.g., the forthcoming autonomous vehicle disruption [93–95].

In the context of cities, AI systems were first introduced as part of smart city initiatives [96] (Ullah et al., 2020). Nonetheless, today AI is no longer exclusively associated with smart city projects. For instance, there is an increasing number of local governments, with no smart city agenda, which have utilised AI-driven chatbots in their customer and service delivery services [97,98]. This is because local government agencies are becoming more aware of the benefits of AI. According to the International City/County Management Association (ICMA) [99], these benefits include: (a) local governments can run more efficiently; (b) local governments can focus on their residents; (c) local governments can remove a great deal of bias; and (d) local governments can make data-smart decisions and gain that extra edge for under-resourced departments.

Besides AI-powered chatbots for engaging with the local community, local governments are using AI for automating routine tasks via self-service and enhancing public services with data and analytics [100,101]. Additionally, hyper-personalised services, predictive maintenance of assets, workforce, schedule and resource optimisation, reducing our carbon footprints, optimising energy usage, and combatting child abuse and financial fraud are among the applications of AI used in local governments [102].

Today, many cities around the world are trying to position themselves as leaders of urban innovation through the development and utilisation of AI [103]. Some of these cities that are experimenting with and adopting AI systems include, but are not exclusively limited to, New York, Washington, Los Angeles, San Antonio, Pittsburgh, Phoenix, London, Singapore, Barcelona, Oslo, Helsinki, Hong Kong, Beijing, Brisbane, Sydney, Melbourne, Bangalore, Dubai and Jeddah [104]. Moreover, the world's first 'AI City'—or artificially intelligent city—is being planned in Chongqing (China) with wired AI-IoT, robotics, networking, and big data [105].

While the popularity of AI is skyrocketing in the urban context, Allam [106, p. 31] warns us that,

> whilst AI stands as a potential savior and as its role is being accentuated in urban planning, governance and management, there are increasing concerns that its practical implications and planning principles are disconnected with sensibilities linked to the dimensions of sustainability.

This very issue has also been raised by other urban scholars [107].

Importantly, as today AI-based decision-making is in a trend to become commonplace, local government AI systems should be used "responsibly and ethically that extends beyond compliance with the narrow letter of the law. It also requires the system to be aligned with broadly-accepted social norms, and considerate of impact on individuals, communities and the environment" [108, p. 1].

Despite the use of AI in local governments being relatively new, it is already possible to find promising examples of responsible practices, which include, but are not limited to (a) AI-driven transportation analytics and decision-making systems to address the urban traffic problems of Austin, TX [109]; (b) autonomous shuttle buses are currently being used as first- and last-mile solutions to increase public transport patronage and/or to provide transport service to disadvantaged populations in cities, including Lyon (France), Geneva (Switzerland), Wien (Austria), Oslo (Norway), Las Vegas (USA), Masdar City (UAE), Thuwal (Saudi Arabia), West Kowloon (Hong Kong) and Renmark (Australia) [110,111]; and (c) computer vision and machine learning for robots to identify material characteristics while sorting waste and increasing the capacity of recycling of the San Francisco Bay Area [112].

Building on the aforementioned practices which are limited but promising practices, Schmelzer [113, p. 1] raised the following key issues that must be fully tackled so that many more local governments can successfully implement responsible AI practices:

> (a) identifying the unique challenges around data at the local government level; (b) determining the areas that AI has the biggest impact at the local level; (c) understanding the challenges local governments face around data privacy, transparency and security; (d) developing an AI-ready local government workforce, including upskilling the current workforce around data and AI skills;

and (e) having a responsive lens when deploying and managing AI technologies.

A study by Chen et al. [114] reveals the success factors for AI adoption as: (a) innovation attributes of AI; (b) organisational capability; and (c) external environment.

4 RESPONSIBLE LOCAL GOVERNMENT ARTIFICIAL INTELLIGENCE

Amid the global push for AI use at the local government level [115], there are growing concerns over AI uptake in the absence of an in-depth understanding of the implementation challenges, contextual local differences, and local government readiness [116]—as well as the lack of responsible urban innovation practices [117].

The upcoming digital urban infrastructure that will be supporting future societies will intrinsically be based on AI [118]. This is evident from the work on the sixth-generation (6G) networks with extreme-scale ubiquitous AI services through the next-generation softwarisation, platformisation, heterogeneity, and configurability of networks [119–121]. Such infrastructure will provide unimaginable opportunities for urban innovation. Nevertheless, a responsible approach is critical when it comes to eliminating the negative externalities of innovation which could otherwise have catastrophic consequences—e.g., worsening socioeconomic inequity, widening digital divides, devastating environmental externalities and increased bias.

Some promising developments being made at present. A good outcome of the academic discourse is the 'Montreal Declaration for Responsible AI', developed under the auspices of the University of Montreal, following the Forum on the Socially Responsible Development of AI of November 2017 [122]. Another notable example is AI regulation and ethics frameworks being rolled out in various countries. The most celebrated of these is the European Parliament's initiative on the guidelines for the European Union (EU) on ethics in AI [123]. While thus far, more than 50 countries have developed their national AI strategies—where the intention is mainly economic development and national security [124]—only a few have attempted to form their AI ethics frameworks that advocate responsible AI [125].

Despite the increasing interest in the scholarly debate, responsible urban innovation and responsible local government AI remain highly understudied, and thus far, there is a limited empirical evidence base and few conceptual expansions [126]. Thus, local governments' adoption and use of AI systems [127], in the context of urban innovation, is an important topic of scholarly research, and key to expanding our understanding of the pathways leading to the achievement of responsible urban innovation through these systems.

Moreover, in the absence of an abundance of responsible urban innovation with local government AI, the generation of new insights and evidence is of utmost importance. Prospective research in this interdisciplinary field will aid in conceptualising and providing a sound understanding of the most appropriate approaches for local governments engaging with AI to achieve responsible urban innovation. Such conceptualisation will also aid in the efforts to develop clear pathways for healthy AI design and deployment. Furthermore, the outcomes of these future studies will help urban policymakers, managers and planners better understand the crucial role played by local government AI systems in ensuring responsible outcomes. This consolidated understanding will guide the efforts in developing new coordination and delivery practices for local government AI systems to achieve the desired responsible urban innovation outcomes in their cities.

5 CONCEPTUAL FRAMEWORK

We have developed a conceptual framework of responsible urban innovation with local government AI. The main rationale behind the development of this high-level information, i.e., the conceptual framework, is to highlight the relationship between the key drivers, components, and the fundamental principles of responsible urban innovation in the context of local government AI. We believe this approach will create an interest and curiosity in the academic community to further investigate the topic and subsequently generate new research directions and practical solutions for the government and industry to adopt or benefit from. Hence, the framework, which is illustrated in Figure 5.1 and elaborated below, shall not be seen as an operational framework to directly guide the AI system development, deployment, and management practices of local government agencies.

In this framework, the responsible urban innovation phenomenon is conceptualised through a process that involves technology, policy, and community to produce strategies, action plans, and initiatives, with balanced costs, benefits, risks, and impacts of developing, adopting, deploying, and managing local government AI systems.

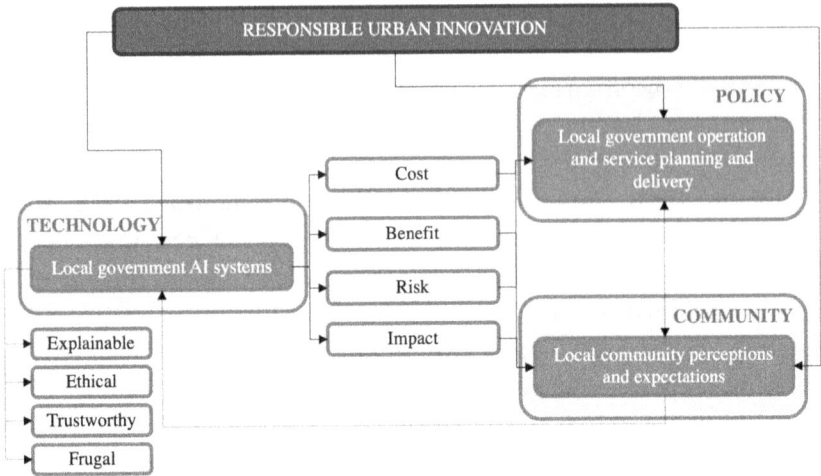

FIGURE 5.1 Conceptual framework of responsible urban innovation with local government artificial intelligence (AI).

First, technology, policy, and community are envisaged as the key drivers of responsible urban innovation; where local government AI systems represent 'technology', local government operation and service planning and delivery decisions represent 'policy', and local community perceptions and expectations represent the 'community' views and input. This is attributed to the fact that the trio of 'technology, policy and community' are being widely seen as the key drivers of smart cities—urban localities that use digital data and technology to create efficiencies for boosting economic development, enhancing the quality of life, and improving the sustainability of the city [128]—or technology-based urban growth practices [129–131] or technology-based public policy [132].

Second, 'cost', 'benefit', 'risk', and 'impact' are identified as the central foci of the framework for realising responsible outcomes concerning local government AI systems. This is attributed to the need for undertaking a cost-benefit analysis before developing, adopting, and deploying AI systems in order to ensure the worthiness of the investment [133]. Likewise, being aware of the potential hazards by undertaking a risk analysis, is critical when it comes to assuring the success of the AI system [134]. Similarly, forecasting the impact of the AI system on the society and the environment is crucial for identifying both positive and negative externalities [135]. Additionally, rather than only looking at costs and benefits overall, responsible use of AI should also consider the distribution of costs and benefits, and whether some groups or places (e.g., disadvantaged neighbourhoods) bear more risk. This is an ethical concern—equity as a public value.

Moreover, this conceptualisation advocates that local government AI systems should include the following characteristics: to be (a) explicable; (b) ethical; (c) trustworthy; and (d) frugal.

First, the effectiveness of AI systems is limited by the machine's inability to explain its thoughts and actions to human users. Hence, explainable AI (XAI)—which refers to methods and techniques that generate high-quality interpretable, intuitive, human-understandable explanations of AI decisions—is essential for operators and users to understand, trust, and effectively manage local government AI systems [136,137].

Second, the ethical considerations made by the designers and adopters of AI systems, are critical when it comes to avoiding the unethical consequences of AI systems [138,139]. As stated by Floridi et al. [122, p. 694],

> ensuring socially preferable outcomes of AI relies on resolving the tension between incorporating the benefits and mitigating the potential harms of AI, in short, simultaneously avoiding the misuse and underuse of these technologies. In this context, the value of an ethical approach to AI technologies comes into starker relief.

Third, today, we are seeing an increasing trend of autonomous decision-making, but we are not there yet. This is especially true for most government services [140,141]. In particular, local governments around the world are looking at technological solutions, including decisions being made autonomously by AI systems with limited or null human involvement, to address a range of social and policy challenges. While this handover of decisions to AI provides benefits—e.g., reduction in human error, faster decision-making, 24/7 availability, and completion of repetitive tasks—it also creates risks, e.g., algorithmic bias caused by bad/limited data and training, privacy violations, removing human responsibility, and a lack of transparency [142]. The public is already becoming increasingly distrustful of many AI decisions; robodebt and systematic racism traumas caused by AI systems in Australia and the US are among the recent examples [143–146]. In order to prevent these risks from occurring and gaining public confidence, AI must be trustworthy [147–149].

Fourth, at present comprehensive AI systems are significant investments for local governments, with many of these organisations having limited budgets and responsibilities for justifying the investment and management cost to their citizens and taxpayers. For a wide-scale AI adoption in local governments, AI systems should become accessible and affordable, or in other words frugal [150]. In this context, frugality is the minimum use of scarce resources—e.g., capital, time, workforce, and energy. Alternatively, the resources can be leveraged in new ways or other solutions can be found that do not jeopardise the delivery of high-value outputs [151].

The framework, illustrated in Figure 5.1 and elaborated above, under-lines the key drivers, components, and fundamental principles of respon-sible urban innovation with local government AI. It sheds light on the overall principles of the development and deployment of AI systems that assure the delivery of not only the desired urban outcomes but also the desired urban outcomes for all citizens, stakeholders, users, and the envi-ronment. The approach outlined in this framework is invaluable for local government agencies when it comes to being aware of the issues around responsible innovation; as local governments continue in the phase of experimentation with AI technologies, they need guidance on how best to design, develop, and deploy these solutions in a responsible manner that advances public value. Nevertheless, it should be noted that the framework presented here is a conceptual one developed with the goal of contributing to the academic discourse on the topic and generating directions for future research agendas. Thus, prospective research is needed to develop more operational frameworks involving specific guidelines for local govern-ments so that they can make informed decisions regarding their invest-ments in AI systems.

6 CONCLUSION

With the advent of powerful technologies and the strong short-term profit-at-any-cost mindset of many disruptive technology companies, ensuring responsible urban innovation will continue to be a major challenge in the third decade of the 21st century as well [152]. In particu-lar, there are some important issues, in the context of responsible urban innovation with local government AI, that require urgent attention.

With the aforementioned urgency in mind, in this perspective chap-ter, we underlined the importance of local government agencies making informed decisions while developing, adopting, deploying, and managing AI systems. This is becoming a highly critical issue, as stated by Arrieta et al. [137, p. 82], in recent years, the

> sophistication of AI systems has increased to such an extent that almost no human intervention is required for their design and

deployment. When decisions derived from such systems ultimately affect humans' lives, there is an emerging need for understanding how such decisions are furnished

and how responsible they are. Moreover, this also increases the danger that unintended consequences can grow and multiply through policy inattention or delayed recognition.

Nevertheless, at present, there are no operational frameworks and clear guidelines to assist local governments in achieving responsible innovation through AI practices. In the absence of such guidelines, the conceptual framework illustrated in Figure 5.1 is a step towards increasing awareness on the matter and triggering potential research agendas for the development of operational frameworks and guidelines. In this instance, we underline some of the fundamental issues, as listed below, where focusing on these could pave the way for a new research agenda concerning responsible urban innovation with local government AI:

- How can local governments utilise AI systems effectively, what are the requirements for making them responsible, and how can local government AI support responsible urban innovation efforts?

- Why do some local governments experiment with and adopt AI systems when the risks are not clearly known, while the others prefer to take a wait-and-see approach before making this decision?

- How can local governments conduct trade-offs between costs, risks, benefits, and impacts, and utilise AI systems in their municipal operations and services, and what externalities do these trade-offs generate?

- How can the AI systems' costs, risks, benefits, and impacts be distributed across the local government service users and communities, and how can the equity concerns be addressed?

- Further dwelling on the above issue, if some groups or geographic areas bear greater risk or costs, should AI be used anyway, should there be a just compensation in some manner, and if yes, how?

- How can local governments align the public perceptions and expectations of AI, and mitigate the impact of consequential negative externalities of AI systems on the environment and society?

- How can local governments support and ensure high levels of trust, transparency, and openness in the local community culture and extend these concepts of digital trust to AI?

- How can local governments successfully adopt, deploy, and manage AI systems to generate responsible urban innovation in their cities, and how can the guidelines for successful adoption be developed?

- How will local government AI systems shape the future of our cities and impact the lives of citizens, and how can the negative externalities of the disruption be alleviated?

- How can operational frameworks and guidelines be developed for local governments in regard to adopting, deploying, and managing AI systems to achieve responsible urban innovation outcomes?

There is very limited to no coverage on either theoretical or applied aspects concerning the aforementioned questions in the literature and practice—as the responsible local government AI notion is still cutting-edge but incipient in nature. Most studies reported in the literature, thus far, are drawn from the intuitions and predictions of scholars rather than hard evidence obtained through exhaustive empirical research. We strongly believe that investigation of these issues in prospective research projects, by the scholars of this highly interdisciplinary (more correctly transdisciplinary) field, will shed light on the better conceptualisation and practice of responsible urban innovation with local government AI.

Thus, the future research agenda should focus on tackling the aforementioned issues via carefully designed empirical studies in international contexts. In this regard, the conceptual framework, presented in Figure 5.1, is a useful compass to guide in the design of new empirical investigations. For instance, an example of a prospective research agenda, concerning responsible urban innovation with local government AI, could be as follows.

At present, AI is not only becoming an integral part of urban services but also is impacting and shaping the future of living and cities [153,154]. Nevertheless, the current AI practice has shown that urban innovation without responsibility generates more problems than it solves. In particular, the absence of a deep understanding of the costs, benefits, risks, and impacts of deploying local government AI systems creates concerns. Focusing on the local government case studies, prospective projects can generate new knowledge on the local government use of AI systems, and expand our understanding of the pathways for responsible urban innovation. These studies could also produce invaluable outcomes that include the local government AI adoption and implementation of guiding principles for informed decisions.

In conclusion, the AI revolution is already under way and its disruption will likely be comparable to that of the agricultural and industrial revolutions, which changed the course of human civilisation radically [155]. Nevertheless, it should not be forgotten that the AI revolution contains an equal measure of opportunities and challenges [156,157]. According to Walsh [158], the ethical challenges posed by AI—e.g., fairness, transparency, trustworthiness, protection of privacy and respect, and many other fundamental rights—are the biggest issues, and they must be addressed with utmost care and urgency. Importantly, the actions to address these challenges should be adopted before the local government AI systems are actually in use [159–161].

ACKNOWLEDGEMENTS

This chapter, with permission from the copyright holder, is a reproduced version of the following journal article: Yigitcanlar, T., Corchado, J., Mehmood, R., Li, R., Mossberger, K., & Desouza, K. (2021). Responsible urban innovation with local government artificial intelligence (AI): a conceptual framework and research agenda. *Journal of Open Innovation: Technology, Market, and Complexity, 7*(1), 71.

REFERENCES

1. Yun, J.J.; Zhao, X.; Yigitcanlar, T.; Lee, D.; Ahn, H. Architectural design and open innovation symbiosis: Insights from research campuses, manufacturing systems, and innovation districts. *Sustainability* 2018, 10, 4495.
2. Yun, J.J.; Jung, K.; Yigitcanlar, T. Open innovation of James Watt and Steve Jobs: Insights for sustainability of economic growth. *Sustainability* 2018, 10, 1553.
3. Beltagui, A.; Rosli, A.; Candi, M. Exaptation in a digital innovation ecosystem. *Res. Policy* 2020, 49, 103833.
4. Barnatt, C. The second digital revolution. *J. Gen. Manag.* 2001, 27, 1–16.
5. Jarach, D. The digitalisation of market relationships in the airline business. *J. Air Transp. Manag.* 2002, 8, 115–120.
6. Yun, J.J.; Zhao, X.; Jung, K.; Yigitcanlar, T. The culture for open innovation dynamics. *Sustainability* 2020, 12, 5076.
7. Flores, P.; Carrillo, F.J.; Robles, J.G.; Leal, M.A. Applying open innovation to promote the development of a knowledge city: The Culiacan experience. *Int. J. Knowl. Based Dev.* 2018, 9, 312–335.

8. Rindfleisch, A. The second digital revolution. *Mark. Lett.* 2020, 31, 13–17.
9. Makridakis, S. The forthcoming artificial intelligence (AI) revolution: Its impact on society and firms. *Futures* 2017, 90, 46–60.
10. Alam, F.; Mehmood, R.; Katib, I.; Albogami, N.; Albeshri, A. Data fusion and IoT for smart ubiquitous environments: A survey. *IEEE Access* 2017, 5, 9533–9554.
11. Alotaibi, S.; Mehmood, R.; Katib, I.; Rana, O.; Albeshri, A. Sehaa: A big data analytics tool for healthcare symptoms and diseases detection using. *TwitterApache SparkMach. Learn. Appl. Sci.* 2020, 10, 1398.
12. Yigitcanlar, T.; Butler, L.; Windle, E.; Desouza, K.; Mehmood, R.; Corchado, J. Can building 'artificially intelligent cities' protect humanity from natural disasters, pandemics and other catastrophes? *Sensors* 2020, 20, 2988.
13. Alaa, M.; Zaidan, A.; Zaidan, B.; Talal, M.; Kiah, M. A review of smart home applications based on Internet of Things. *J. Netw. Comput. Appl.* 2017, 97, 48–65.
14. Sodhro, A.; Gurtov, A.; Zahid, N.; Pirbhulal, S.; Wang, L.; Rahman, M.; Abbasi, Q. Towards convergence of AI and IoT for energy efficient communication in smart homes. *IEEE Internet Things J.* 2020, doi:10.1109/JIOT.2020.3023667.
15. Kumar, S.; Raut, R.; Narkhede, B. A proposed collaborative framework by using artificial intelligence-internet of things (AI-IoT) in COVID-19 pandemic situation for healthcare workers. *Int. J. Healthc. Manag.* 2020, 13, 337–345.
16. Dalenogare, L.; Benitez, G.; Ayala, N.; Frank, A. The expected contribution of Industry 4.0 technologies for industrial performance. *Int. J. Prod. Econ.* 2018, 204, 383–394.
17. Millar, C.; Lockett, M.; Ladd, T. Disruption: Technology, innovation and society. *Technol. Forecast. Soc. Chang.* 2018, 129, 254–260.
18. Yigitcanlar, T.; Cugurullo, F. The sustainability of artificial intelligence. *Sustainability* 2020, 12, 8548.
19. Yun, Y.; Lee, M. Smart city 4.0 from the perspective of open innovation. *J. Open Innov. Technol. Mark. Complex.* 2019, 5, 92.
20. Yigitcanlar, T.; Foth, M.; Kamruzzaman, M. Towards post-anthropocentric cities. *J. Urban Technol.* 2019, 26, 147–152.
21. Faisal, A.; Yigitcanlar, T.; Kamruzzaman, M.; Currie, G. Understanding autonomous vehicles. *J. Transp. Land Use* 2019, 12, 45–72.
22. Yigitcanlar, T.; Wilson, M.; Kamruzzaman, M. Disruptive impacts of automated driving systems on the built environment and land use: An urban planner's perspective. *J. Open Innov. Technol. Mark. Complex.* 2019, 5, 24.
23. Butler, L.; Yigitcanlar, T.; Paz, A. Barriers and risks of mobility-as-a-service (MaaS) adoption in cities. *Cities* 2021, 109, 103036.
24. Butler, L.; Yigitcanlar, T.; Paz, A. Smart urban mobility innovations. *IEEE Access* 2020, 8, 196034–196049.
25. Goddard, M.; Davies, Z.; Guenat, S.; Ferguson, M.; Fisher, J.; Akanni, A.; Bates, A. A global horizon scan of the future impacts of robotics and autonomous systems on urban ecosystems. *Nat. Ecol. Evol.* 2020, doi:10.1038/s41559-020-01358-z.

26. Liu, Z.; Ma, L.; Huang, T.; Tang, H. Collaborative governance for responsible innovation in the context of sharing economy. *J. Open Innov. Technol. Mark. Complex.* 2020, 6, 35.
27. Gonzales-Gemio, C.; Cruz-Cázares, C.; Parmentier, M. Responsible innovation in SMEs. *Sustainability* 2020, 12, 10232.
28. Von Schomberg, R. A vision of responsible research and innovation. In Richard Owen, John Bessant, Maggy Heintz (Eds.) *Responsible Innovation*; Wiley: London, 2013; pp. 51–74.
29. Owen, R.; Stilgoe, J.; Macnaghten, P.; Gorman, M.; Fisher, E.; Guston, D. A framework for responsible innovation. In Richard Owen, John Bessant, Maggy Heintz (Eds.) *Responsible Innovation*; Wiley: London, 2013; pp. 27–50.
30. Zhao, Y.; Liao, M. Chinese perspectives on responsible innovation. In Von Schomberg, R., Hankins, J. (Eds.) *International Handbook on Responsible Innovation*; Edward Elgar: Cheltenham, 2019; pp. 426–440.
31. Yigitcanlar, T. *Technology and the City: Systems, Applications and Implications*; Routledge: New York, NY, 2016.
32. Cugurullo, F. Urban artificial intelligence. *Front. Sustain. Cities* 2020, 2, 1–14.
33. Batty, M. Artificial intelligence and smart cities. *Environ. Plan. B* 2018, doi:10.1177/2399808317751169.
34. Luckey, D.; Fritz, H.; Legatiuk, D.; Dragos, K.; Smarsly, K. Artificial intelligence techniques for smart city applications. In *International Conference on Computing in Civil and Building Engineering*; Springer: Cham, Switzerland, 2020; pp. 3–15.
35. Desouza, K.; Swindell, D.; Smith, K.; Sutherland, A.; Fedorschak, K.; Coronel, C. Local government 2035: Strategic trends and implications of new technologies. *Issues Technol. Innov.* 2015, 27, 27.
36. Vogl, T.; Seidelin, C.; Ganesh, B.; Bright, J. Smart technology and the emergence of algorithmic bureaucracy. *Public Adm. Rev.* 2020, 80, 946–961.
37. Wang, Y.; Zhang, N.; Zhao, X. Understanding the determinants in the different government AI adoption stages. *Soc. Sci. Comput. Rev.* 2020, doi:10.1177/0894439320980132.
38. Cooke, P. Regionally asymmetric knowledge capabilities and open innovation: Exploring 'Globalisation 2'—A new model of industry organisation. *Res. Policy* 2005, 34, 1128–1149.
39. Belussi, F.; Sammarra, A.; Sedita, S.R. Learning at the boundaries in an "open regional innovation system": A focus on firms' innovation strategies in the Emilia Romagna life science industry. *Res. Policy* 2010, 39, 710–721.
40. Voegtlin, C.; Scherer, A. Responsible innovation and the innovation of responsibility. *J. Bus. Ethics* 2017, 143, 227–243.
41. Joss, S. *Sustainable Cities: Governing for Urban Innovation*; Palgrave: London, 2015.
42. Yun, J.J.; Lee, D.; Ahn, H.; Park, K.; Yigitcanlar, T. Not deep learning but autonomous learning of open innovation for sustainable artificial intelligence. *Sustainability* 2016, 8, 797.

43. Burget, M.; Bardone, E.; Pedaste, M. Definitions and conceptual dimensions of responsible research and innovation. *Sci. Eng. Ethics* 2017, 23, 1–19.

44. Dijk, M.; De Kraker, J.; Hommels, A. Anticipating constraints on upscaling from urban innovation experiments. *Sustainability* 2018, 10, 2796.

45. Yigitcanlar, T.; Kamruzzaman, M. Smart cities and mobility: Does the smartness of Australian cities lead to sustainable commuting patterns? *J. Urban Technol.* 2019, 26, 21–46.

46. Gibson, E. Sidewalk Labs Creates Machine-Learning Tool for Designing Cities. 2020. Available online: https://www.dezeen.com/2020/10/20/delve-sidewalk-labs-machine-learning-tool-cities (accessed on 10 February 2021).

47. Heaven, W.D. 2019. AI Planners in Minecraft could Help Machines Design Better Cities. 2020. Available online: https://www.technologyreview.com/2020/09/22/1008675/ai-planners-minecraft-urban-design-healthier-happier-cities (accessed on 10 February 2021).

48. University of Chicago What is Array of Things? 2020. Available online: https://arrayofthings.github.io/faq.html (accessed on 14 February 2021).

49. Hawthorne, M. Citizen Devices Tracking Chicago's Pollution Hot Spots. 2017. Available online: https://www.chicagotribune.com/news/breaking/ct-chicago-air-quality-testing-met-20171111-story.html (accessed on 14 February 2021).

50. Vallance, P.; Tewdwr-Jones, M.; Kempton, L. Building collaborative platforms for urban innovation: Newcastle City Futures as a quadruple helix intermediary. *Eur. Urban Reg. Stud.* 2020, 27, 325–341.

51. Wilson, A.; Tewdwr-Jones, M. Let's draw and talk about urban change: Deploying digital technology to encourage citizen participation in urban planning. *Environ. Plan. B* 2020, 47, 1588–1604.

52. Angelidou, M.; Psaltoglou, A. An empirical investigation of social innovation initiatives for sustainable urban development. *Sustain. Cities Soc.* 2017, 33, 113–125.

53. Fleming, J. *Profit at Any Cost?* Baker Books: Grand Rapids, MI, 2003.

54. Spiegel, J. The ethics of virtual reality technology. *Sci. Eng. Ethics* 2018, 24, 1537–1550.

55. De Saille, S. Innovating innovation policy: The emergence of 'responsible research and innovation'. *J. Responsible Innov.* 2015, 2, 152–168.

56. Selby, J.; Desouza, K. Fragile cities in the developed world. *Cities* 2019, 91, 180–192.

57. Eubanks, V. A child Abuse Prediction Model Fails Poor Families. 2018. Available online: https://www.wired.com/story/excerpt-from-automating-inequality (accessed on 10 February 2021).

58. Eubanks, V. *Automating Inequality: How High-Tech Tools Profile, Police, and Punish the Poor*; St. Martin's Press: New York, NY, 2018.

59. Eveleth, B. How Self-Tracking Apps Exclude Women. 2014. Available online: https://www.theatlantic.com/technology/archive/2014/12/how-self-tracking-apps-exclude-women/383673 (accessed on 10 February 2021).

60. Ledford, X. Millions of Black People Affected by Racial Bias in Health-Care Algorithms. 2019. Available online: https://www.nature.com/articles/d41586-019-03228-6 (accessed on 10 February 2021).

61. Bousquet, C. Algorithmic Fairness: Tackling Bias in City Algorithms. 2018. Available online: https://datasmart.hks.harvard.edu/news/article/algorithmic-fairness-tackling-bias-city-algorithms (accessed on 10 February 2021).

62. Smith, C. Dealing with Bias in Artificial Intelligence. 2020. Available online: https://www.nytimes.com/2019/11/19/technology/artificial-intelligence-bias. html (accessed on 10 February 2021).

63. Marsh, S. Councils Scrapping Use of Algorithms in Benefit and Welfare Decisions. 2020. Available online: https://www.theguardian.com/ society/2020/aug/24/councils-scrapping-algorithms-benefit-welfare-decisions-concerns-bias (accessed on 10 February 2021).

64. McKnight, L. To Succeed in an AI World, Students must Learn the Human Traits of Writing. 2021. Available online: https://theconversation.com/ to-succeed-in-an-ai-world-students-must-learn-the-human-traits-of-writing-152321 (accessed on 10 February 2021).

65. Van Noordt, C.; Misuraca, G. Exploratory insights on artificial intelligence for government in Europe. *Soc. Sci. Comput. Rev.* 2020, doi:10. 1177/0894439320980449.

66. Goodman, E.; Powles, J. Urbanism under Google. *Law Rev.* 2019, 88, 457–498.

67. Desouza, K.; Hunter, M.; Jacob, B.; Yigitcanlar, T. Pathways to the making of prosperous smart cities. *J. Urban Technol.* 2020, 27, 3–32.

68. Yigitcanlar, T.; Han, H.; Kamruzzaman, M.; Ioppolo, G.; Sabatini-Marques, J. The making of smart cities. *Land Use Policy* 2019, 88, 104187.

69. Stilgoe, J.; Owen, R.; Macnaghten, P. Developing a framework for responsible innovation. *Res. Policy* 2013, 42, 1568–1580.

70. Deschamps, T. Google's Sidewalk Labs Pulls out of Toronto Smart City Project. 2020. Available online: https://www.theguardian.com/technology/ 2020/may/07/google-sidewalk-labs-toronto-smart-city-abandoned (accessed on 8 January 2021).

71. Tilley, A. Cisco Systems Pulls Back from Smart City Push. 2020. Available online: https://www.wsj.com/articles/cisco-turns-off-lights-on-smart-city-push-11609178895 (accessed on 8 January 2021).

72. Green, B. *The Smart Enough City: Putting Technology in its Place to Reclaim our Urban Future*; MIT Press: Boston, MA, 2019.

73. Blasko, B.; Lukovics, M.; Buzás, N. Good practices in responsible innovation. In Richard Owen, John Bessant, Maggy Heintz (Eds.) *Responsible Innovation*; University of Szeged: Szeged, Hungary, 2014; pp. 179–192.

74. Albert, M. Sustainable frugal innovation: The connection between frugal innovation and sustainability. *J. Clean. Prod.* 2019, 237, 117747.

75. Hadj, T. Effects of corporate social responsibility towards stakeholders and environmental management on responsible innovation and competitiveness. *J. Clean. Prod.* 2020, 250, 119490.

76. Umbrello, S. Imaginative value sensitive design. *Sci. Eng. Ethics* 2020, 26, 575–595.

77. Martinuzzi, A.; Blok, V.; Brem, A.; Stahl, B.; Schönherr, N. Responsible research and innovation in industry: Challenges, insights and perspectives. *Sustainability* 2018, 10, 702.

78. Van den Buuse, D.; Van Winden, W.; Schrama, W. Balancing exploration and exploitation in sustainable urban innovation. *J. Urban Technol.* 2020, doi:10.1080/10630732.2020.1835048.

79. Ziegler, R. Justice and innovation–towards principles for creating a fair space for innovation. *J. Responsible Innov.* 2015, 2, 184–200.

80. Das, A.; Rad, P. Opportunities and challenges in explainable artificial intelligence (XAI): A survey. arXiv 2020, arXiv:2006.11371.

81. Mikhaylov, S.; Esteve, M.; Campion, A. Artificial intelligence for the public sector. *Philos. Trans. R. Soc. A* 2018, 376, 20170357.

82. Dwivedi, Y.K.; Hughes, L.; Ismagilova, E.; Aarts, G.; Coombs, C.; Crick, T.; Williams, M.D. Artificial intelligence (AI): Multidisciplinary perspectives on emerging challenges, opportunities, and agenda for research, practice and policy. *Int. J. Inf. Manag.* 2019, doi:10.1016/j.ijinfomgt.2019.08.002.

83. De Sousa, W.; De Melo, E.; Bermejo, P.; Farias, R.; Gomes, A. How and where is artificial intelligence in the public sector going? *Gov. Inf. Q.* 2019, 36, 101392.

84. Swindell, D.; Desouza, K.; Hudgens, R. Dubai Offers Lessons for Using Artificial Intelligence in Local Government. 2018. Available online: https://www.brookings.edu/blog/techtank/2018/09/28/dubai-offers-lessons-for-using-artificial-intelligence-in-local-government/ (accessed on 10 January 2021).

85. Wirtz, B.W.; Weyerer, J C.; Geyer, C. Artificial intelligence and the public sector: Applications and challenges. *Int. J. Public Adm.* 2019, 42, 596–615.

86. D'Amico, G.; L'Abbate, P.; Liao, W.; Yigitcanlar, T.; Ioppolo, G. Understanding sensor cities. *Sensors* 2020, 20, 4391.

87. Macrorie, R.; Marvin, S.; While, A. Robotics and automation in the city. *Urban Geogr.* 2020, doi:10.1080/02723638.2019.1698868.

88. Mehr, H.; Ash, H.; Fellow, D. Artificial Intelligence for Citizen Services and Government. 2017. Available online: https://ash.harvard.edu/files/ash/files/artificial_intelligence_for_citizen_services.pdf (accessed on 11 January 2021).

89. Walsh, T. Australia's AI future. *J. Proc. R. Soc. New South Wales* 2019, 152, 101–105.

90. Andreasson, U.; Stende, T. *Nordic Municipalities' Work with Artificial Intelligence; Nordic Council of Ministers*: Copenhagen, Denmark, 2019.

91. Androutsopoulou, A.; Karacapilidis, N.; Loukis, E.; Charalabidis, Y. Transforming the communication between citizens and government through AI-guided chatbots. *Gov. Inf. Q.* 2019, 36, 358–367.

92. Yigitcanlar, T.; Desouza, K.; Butler, L.; Roozkhosh, F. Contributions and risks of artificial intelligence (AI) in building smarter cities. *Energies* 2020, 13, 1473.

93. Faisal, A.; Yigitcanlar, T.; Kamruzzaman, M.; Paz, A. Mapping two decades of autonomous vehicle research: A systematic scientometric analysis. *J. Urban Technol.* 2020, doi:10.1080/10630732.2020.1780868.

94. Golbabaei, F.; Yigitcanlar, T.; Paz, A.; Bunker, J. Individual predictors of autonomous vehicle public acceptance and intention to use: A systematic review of the literature. *J. Open Innov. Technol. Mark. Complex.* 2020, 6, 106.

95. Golbabaei, F.; Yigitcanlar, T.; Bunker, J. The role of shared autonomous vehicle systems in delivering smart urban mobility: A systematic review of the literature. *Int. J. Sustain. Transp.* 2020, doi:10.1080/15568318.2020.1798571.

96. Ullah, Z.; Al-Turjman, F.; Mostarda, L.; Gagliardi, R. Applications of artificial intelligence and machine learning in smart cities. *Comput. Commun.* 2020, 154, 313–323.

97. Adam, M.; Wessel, M.; Benlian, A. AI-based chatbots in customer service and their effects on user compliance. *Electron. Mark.* 2020, doi:10.1007/s12525-020-00414-7.

98. Luo, X.; Tong, S.; Fang, Z.; Qu, Z. Frontiers: Machines vs. humans: The impact of artificial intelligence chatbot disclosure on customer purchases. *Mark. Sci.* 2019, 38, 937–947.

99. ICMA Using Artificial Intelligence as a Tool for Your Local Government. 2019. Available online: https://icma.org/blog-posts/using-artificial-intelligence-tool-your-local-government (accessed on 9 January 2021).

100. Bright, J.; Ganesh, B.; Seidelin, C.; Vogl, T. Data Science for Local Government. 2019. Available online: https://papers.ssrn.com/sol3/papers.cfm?abstract_id=3370217 (accessed on 11 January 2021).

101. Alomari, E.; Katib, I.; Albeshri, A.; Mehmood, R. COVID-19: Detecting government pandemic measures and public concerns from Twitter Arabic data using distributed machine learning. *Int. J. Environ. Res. Public Health* 2021, 18, 282.

102. Quddus, J. Innovation in Local Government. 2020. Available online: https://methods.co.uk/blog/innovation-in-local-government (accessed on 3 January 2021).

103. Robinson, S. Trust, transparency, and openness: How inclusion of cultural values shapes Nordic national public policy strategies for artificial intelligence (AI). *Technol. Soc.* 2020, 63, 101421.

104. Kirwan, C.; Zhiyong, F. *Smart Cities and Artificial Intelligence: Convergent Systems for Planning, Design, and Operations*; Elsevier: London, 2020.

105. Designboom Bjarke Ingels Group Plans AI CITY in China to Advance Future of Artificial Intelligence. 2020. Available online: https://www.designboom.com/architecture/bjarke-ingels-group-ai-city-china-artificial-intelligence-terminus-group-09-29-2020 (accessed on 11 January 2021).

106. Allam, Z. Urban chaos and the AI Messiah. In Allam, Z. (Ed.) *Cities and the Digital Revolution*; Palgrave Pivot: Cham, Switzerland, 2020; pp. 31–60.

107. Khakurel, J.; Penzenstadler, B.; Porras, J.; Knutas, A.; Zhang, W. The rise of artificial intelligence under the lens of sustainability. *Technologies* 2018, 6, 100.

108. Caetano, T.; Simpson-Young, B. Artificial Intelligence can Deepen Social Inequality: Here are 5 Ways to Help Prevent This. 2021. Available online: https://theconversation.com/artificial-intelligence-can-deepen-social-inequality-here-are-5-ways-to-help-prevent-this-152226 (accessed on 13 January 2021).

109. Diamante, R. UT Creates Artificial Intelligence to Solve Austin's Traffic Problems. 2017 Available online: https://spectrumlocalnews.com/tx/south-texas-el-paso/news/2017/12/11/ut-creates-artificial-intelligence-to-solve-austin-s-traffic-problems (accessed on 10 February 2021).

110. Iclodean, C.; Cordos, N.; Varga, B.O. Autonomous shuttle bus for public transportation: A review. *Energies* 2020, 13, 2917.

111. Dennis, S.; Paz, A.; Yigitcanlar, T. Perceptions and attitudes towards the deployment of autonomous and connected vehicles: Insights from Las Vegas, Nevada. *J. Urban Technol.* 2021, doi:10.1080/10630732.2021.1879606.

112. Ioannou, L.; Petrova, M. America is Drowning in Garbage: Now Robots are Being Put on Duty to Help Solve the Recycling Crisis. 2019. Available online: https://www.cnbc.com/2019/07/26/meet-the-robots-being-used-to-help-solve-americas-recycling-crisis.html (accessed on 10 February 2021).

113. Schmelzer, R. AI is Here to Stay in Your City and Local Government. 2020. Available online: https://www.forbes.com/sites/cognitiveworld/2020/11/08/ai-is-here-to-stay-in-your-city-and-local-government/?sh=28f22e0a77a3 (accessed on 10 February 2021).

114. Chen, H.; Li, L.; Chen, Y. Explore success factors that impact artificial intelligence adoption on telecom industry in China. *J. Manag. Anal.* 2020, doi: 10.1080/23270012.2020.1852895.

115. Fatima, S.; Desouza, K.; Dawson, G. National strategic artificial intelligence plans. *Econ. Anal. Policy* 2020, 67, 178–194.

116. Butcher, J.; Beridze, I. What is the state of artificial intelligence governance globally? *RUSI J.* 2019, 164, 88–96.

117. Nagenborg, M. Urban robotics and responsible urban innovation. *Ethics Inf. Technol.* 2020, 22, 345–355.

118. Yigitcanlar, T.; Kankanamge, N.; Regona, M.; Maldonado, A.; Rowan, B.; Ryu, A.; Desouza, K.; Corchado, J.; Mehmood, R.; Li, R. Artificial intelligence technologies and related urban planning and development concepts. *J. Open Innov. Technol. Mark. Complex.* 2020, 6, 187.

119. Janbi, N.; Katib, I.; Albeshri, A.; Mehmood, R. Distributed artificial intelligence-as-a-service (DAIaaS) for smarter IoE and 6G environments. *Sensors* 2020, 20, 5796.

120. Corchado, J.; Chamoso, P.; Hernandez, G.; Gutierrez, A.; Camacho, A.; Briones, A.; Omatu, S. Deepint.net: A rapid deployment platform for smart territories. *Sensors* 2021, 21, 236.

121. Repette, P.; Sabatini-Marques, J.; Yigitcanlar, T.; Sell, D.; Costa, E. The evolution of City-as-a-Platform. *Land* 2021, 10, 33.

122. Floridi, L.; Cowls, J.; Beltrametti, M.; Chatila, R.; Chazerand, P.; Dignum, V.; Vayena, E. AI4People-an ethical framework for a good AI society: Opportunities, risks, principles, and recommendations. *Minds Mach.* 2018, 28, 689–707.

123. European Parliament EU Guidelines on Ethics in Artificial Intelligence: Context and Implementation. 2019. Available online: https://www.europarl. europa.eu/RegData/etudes/BRIE/2019/640163/EPRS_BRI (accessed on 11 January 2021).

124. Allen, G. Understanding China's AI Strategy: Clues to Chinese Strategic Thinking on Artificial Intelligence and National security. 2019. Available online: https://nsiteam.com/social/wp-content/uploads/2019/05/CNAS-Understanding-Chinas-AI-Strategy-Gregory-C.-Allen-FINAL-2.15.19.pdf (accessed on 10 January 2021).

125. Australian Government Artificial Intelligence: Australia's Ethics Framework. 2018. Available online: https://consult.industry.gov.au/strategic-policy/artificial-intelligence-ethics-framework (accessed on 11 January 2021).

126. Macnaghten, P.; Owen, R.; Stilgoe, J.; Wynne, B.; Azevedo, A.; de Campos, A.; Garvey, B. Responsible innovation across borders. *J. Responsible Innov.* 2014, 1, 191–199.

127. Desouza, K.; Dawson, G.; Chenok, D. Designing, developing, and deploying artificial intelligence systems. *Bus. Horiz.* 2020, 63, 205–213.

128. Mora, L.; Deakin, M.; Reid, A. Strategic principles for smart city development. *Technol. Forecast. Soc. Chang.* 2019, 142, 70–97.

129. Nam, T.; Pardo, T. Conceptualizing smart city with dimensions of technology, people, and institutions. In *12th Annual International Digital Government Research Conference; Association for Computing Machinery: New York, NY, 2011; pp. 282–291.*

130. Tassey, G. A technology-based growth policy. *Issues Sci. Technol.* 2017, 33, 80–89.

131. Yigitcanlar, T. Smart city beyond efficiency: Technology-policy-community at play for sustainable urban futures. *Hous. Policy Debate* 2021, 31, 88–92.

132. Valle-Cruz, D.; Criado, J.; Sandoval-Almazán, R.; Ruvalcaba-Gomez, E. Assessing the public policy-cycle framework in the age of artificial intelligence: From agenda-setting to policy evaluation. *Gov. Inf. Q.* 2020, 37, 101509.

133. Ivanov, S.; Webster, C. Adoption of robots, artificial intelligence and service automation by travel, tourism and hospitality companies–a cost-benefit analysis. In *International Scientific Conference on Contemporary Tourism; Sofia University: Sofia, Bulgaria, 2017; pp. 1–9.*

134. Müller, V. *Risks of Artificial Intelligence*; CRC Press: Boca Raton, FL, 2016.

135. Rouhiainen, L. *Artificial Intelligence: 101 Things you Must Know Today About our Future; Createspace Independent Publishing Platform: New York, NY, 2018.*

136. Gunning, D.; Aha, D. DARPA's explainable artificial intelligence program. *Ai Mag.* 2019, 40, 44–58.

137. Arrieta, A.; Díaz-Rodríguez, N.; Del Ser, J.; Bennetot, A.; Tabik, S.; Barbado, A.; Herrera, F. Explainable artificial intelligence (XAI): Concepts, taxonomies, opportunities and challenges toward responsible AI. *Inf. Fusion* 2020, 58, 82–115.

138. Torresen, J. A review of future and ethical perspectives of robotics and AI. *Front. Robot. AI* 2018, 4, 75.

139. Mittelstadt, B. Principles alone cannot guarantee ethical AI. *Nat. Mach. Intell.* 2019, 1, 501–507.

140. Sun, T.; Medaglia, R. Mapping the challenges of artificial intelligence in the public sector: Evidence from public healthcare. *Gov. Inf. Q.* 2019, 36, 368–383.

141. Aoki, N. An experimental study of public trust in AI chatbots in the public sector. *Gov. Inf. Q.* 2020, 37, 101490.

142. Osoba, O.; Welser, W. *An Intelligence in Our Image: The Risks of Bias and Errors in Artificial Intelligence; Rand Corporation:* Santa Monica, CA, 2017.

143. Park, S.; Humphry, J. Exclusion by Design: Intersections of Social, Digital and Data Exclusion. *Inf. Commun. Soc.* 2019, 22, 934–953.

144. Curtis, C.; Gillespie, N.; Lockey, S. Australians have Low Trust in Artificial Intelligence and Want it to be Better Regulated. 2020. Available online: https://theconversation.com/australians-have-low-trust-in-artificial-intelligence-and-want-it-to-be-better-regulated-148262 (accessed on 10 January 2021).

145. Harrison, T.M.; Luna-Reyes, L.F. Cultivating trustworthy artificial intelligence in digital government. *Soc. Sci. Comput. Rev.* 2020, doi:10.1177/0894439320980122.

146. Sander, M. From Robodebt to Racism: What can go Wrong When Governments Let Algorithms Make the Decisions. 2020. Available online: https://theconversation.com/from-robodebt-to-racism-what-can-go-wrong-when-governments-let-algorithms-make-the-decisions-132594 (accessed on 11 January 2021).

147. Feng, J.; Lansford, J.; Katsoulakis, M.; Vlachos, D. Explainable and trustworthy artificial intelligence for correctable modeling in chemical sciences. *Sci. Adv.* 2020, 6, eabc3204.

148. Janssen, M.; Brous, P.; Estevez, E.; Barbosa, L.S.; Janowski, T. Data governance: Organizing data for trustworthy Artificial Intelligence. *Gov. Inf. Q.* 2020, 37, 101493.

149. Kokuryo, J.; Walsh, T.; Maracke, C. *AI for Everyone: Benefitting from and Building Trust in the Technology; AI Access:* Sydney, Australia, 2020.

150. Alliance, A. Artificial intelligence and frugal innovation. In Stachowicz-Stanusch, A., Amann, W. (Eds.) *Management and Business Education in the Time of Artificial Intelligence;* Information Age Publishing: Charlotte, NC, 2019; pp. 55–76.

151. Agarwal, N.; Chung, K.; Brem, A. 8 new technologies for frugal innovation. In Adela J. McMurray & Gerrit A. de Waal (Eds.) *Frugal Innovation: A Global Research Companion;* Routledge: New York, NY, 2019; 137–149.

152. Fisher, E. Reinventing responsible innovation. *J. Responsible Innov.* 2020, 7, 1–5.

153. Kassens-Noor, E.; Hintze, A. Cities of the future? The potential impact of artificial intelligence. *AI* 2020, 1, 192–197.

154. Kassens-Noor, E.; Wilson, M.; Kotval-Karamchandani, Z.; Cai, M.; Decaminada, T. Living with autonomy: Public perceptions of an AI-mediated future. *J. Plan. Educ. Res.* 2021, doi:10.1177/0739456X20984529.

155. Miailhe, N.; Hodes, C. Making the AI Revolution Work for Everyone. 2017. Available online: https://thefuturesociety.org/wp-content/uploads/2019/08/Making-the-AI-Revolution-work-for-everyone.-Report-to-OECD.-MARCH-2017.pdf (accessed on 11 January 2021).

156. Davenport, T. *The AI Advantage: How to Put the Artificial Intelligence Revolution to Work*; MIT Press: Cambridge, MA, 2018.

157. Alonso, C.; Berg, A.; Kothari, S.; Papageorgiou, C.; Rehman, S. Will the AI Revolution Cause a Great Divergence? 2020. Available online: https://www.imf.org/~/media/Files/Publications/WP/2020/English/wpiea2020184-print-pdf.ashx (accessed on 11 January 2021).

158. Walsh, T. The AI Revolution. 2017. Available online: https://education.nsw.gov.au/content/dam/main-education/teaching-and-learning/education-for-a-changing-world/media/documents/The_AI_Revolution_TobyWalsh.pdf (accessed on 11 January 2021).

159. Gandhi, S. Social Concerns about Artificial Intelligence. 2018. Available online: https://medium.com/@sharad.gandhi/social-concerns-about-artificial-intelligence-93e939b88a8c (accessed on 15 January 2021).

160. Walsh, T. Artificial Intelligence Should Benefit Society, not Create Threats. 2015. Available online: https://theconversation.com/artificial-intelligence-should-benefit-society-not-create-threats-36240 (accessed on 15 January 2021).

161. Glance, D. What should Governments be doing about the Rise of Artificial Intelligence? 2017. Available online: https://theconversation.com/what-should-governments-be-doing-about-the-rise-of-artificial-intelligence-86561 (accessed on 15 January 2021).

CHAPTER 6

Conceptual Framework of Responsible Urban Artificial Intelligence

1 INTRODUCTION

Driven by advancements in science and technology, emerging innovations have offered significant societal benefits and new commercial opportunities for our societies and cities, especially offering invaluable disruptive technology prospects in agriculture, biological, medical, and urban domains [1–7]. Nevertheless, these disruptive technologies may also raise significant ethical, social, and regulatory challenges, such as the technological and digital divide, inequality and disruption, the misuse of data and information, and others [8–10].

In the context of 'responsible research and innovation (RRI)', which has become a popular concept during the last decade, the terms responsible innovation and responsible technology (collectively referred to as 'responsible innovation and technology (RIT)' in this chapter) have been increasingly mentioned and practiced in academia, industrial circles, and public sectors. They were recognised as having strong potential to address the grand societal challenges associated with innovations and contribute to shaping our (smart) cities—creating pleasant places to live [11–13]. RIT is conceptually regarded as a socially desired/expected technological outcome in the agenda of RRI, which represents innovation and technology's ability to fulfil moral and social responsibilities while achieving socially

 DOI: 10.1201/9781003521440-8

desirable goals in a responsible manner [14]. To a certain extent, RIT can be called the carrier of the RRI concept, reflecting the practical results of the RRI theory in our cities and societies—particularly in the context of smart cities and societies [15–18].

Initial discussions of the concepts of 'responsible' or 'responsibility' in science and technology can be traced back to the developments in research integrity and ethics beginning in the early 20th century [17,19,20]. Via the broader philosophical and sociological analysis of this concept, it has become gradually recognised that scientific research could be governed in socially responsible ways via multiple and overlapping methods to overcome the concomitant challenges [11,21]. With the increasing attention on the notions of the social responsibility of science and technology, the term RRI has emerged over the past decade, and since then, it has been an integral part of European research and innovation (R&I) policies [11,22,23].

RRI has often been described as a forward-looking and comprehensive approach to innovation and research activities, which aims to prudently manage innovations to allow them to be properly embedded in our society [22–24]. With the growing interest in RRI, the number of relevant academic articles has been rapidly increasing over recent years [24]. For instance, Burget et al. [22] reviewed over 200 relevant articles to provide a discussion on the definitions and conceptual dimensions of RRI; Thapa et al. [13] investigated applications of RRI to regional studies; Wiarda et al. [24] identified the commonalities of RI and RRI and expounded on the accumulation of their knowledge; Liu et al. [25] discovered the landscape and evolution of RRI and provided an understanding of existing research.

Although the number of publications with an RRI focus is steadily growing, the research on the topic is still relatively limited. Existing research tends to expound the concepts of RRI in a top-down manner to formulate standardised principles or frameworks guiding innovation towards producing the 'right impacts' during the creation and implementation process [14,26–28]. Nonetheless, the intended outcomes of RRI remain unclear and lack attention—i.e., what kind of innovation or technology can be considered responsible in the context of RRI? [22,23].

Hence, additional investigations and reviews are needed to capture the growing knowledge on this topic and to bridge the research gap. The difference from previous studies is that the chapter at hand focuses on investigating the expected outcomes in the existing RRI practices, i.e., RIT, that attempt to broaden the understanding of responsible research from a bottom-up perspective. Accordingly, the following research question was posed in the chapter: what are the key characteristics of RIT—that is an umbrella term inclusive of 'Responsible Artificial Intelligence (AI)'?

To tackle this question, the rest of this chapter is organised as follows: Following this introduction, Section 2 presents the practices of RIT. Then, Section 3 outlines the research methodology. Next, Section 4 presents the results of the analysis. Afterwards, Section 5 discusses the study findings. Lastly, Section 6 concludes the chapter.

2 LITERATURE BACKGROUND

Since the term RRI emerged in the European R&I policy discourse, increasing industries and scenarios have advocated and attempted to incorporate or embed the concept into the innovation creation and implementation process in emerging technological fields, especially those that are potentially controversial. Today, this concept is moving from the early theoretical stages to one in which it is embedded in specific practices [25,29]. Guided by the RRI framework, a growing number of actors are exploring the characteristics of RIT, that is, establishing what kind of innovations and technologies can meet both social expectations and ethical standards and are able to embed into our cities and societies responsibly [23,30].

For instance, in the field of urban transport, Singh et al. [31] analysed the implementation case of electrical rickshaws (e-rickshaw)—also known as e-tuk-tuks—in India using the RRI framework. The authors pointed out that some key dimensions of the RRI concept have been evidently deployed in this case, i.e., deliberation and participation dimensions, which are the critical factors facilitating the successful implementation of responsible mobility and transport innovation in India. In addition, the authors mentioned that imparting universal and culture-specific values in the technical product to increase its acceptability has acquired significance for shaping RIT. The e-rickshaw case imparted these values during its deliberation and participation process, which made this innovation acceptable in India, providing a brilliant example of RIT implementation in a developing country.

To provide another example, in the agriculture field, Eastwood et al. [4] denoted that accessibility, both technical and financial, is one of the core challenges associated with robotics and automation adoption in agricultural systems. To improve existing agricultural technology design and innovation practices, the authors proposed a design guide for responsible robotic applications in pasture-grazed dairy farming based on the concepts of RRI, systems thinking, and co-design. The guide identified the critical design factors for responsible robotics and automation in smart farming, which involved broader considerations of the impacts on work design, worker well-being and safety, changes to farming systems, and the influences of market and regulatory constraints. Based on the guide, the authors stated that the focus of the further development of robotics and automation in smart farming should be on improving their technical adaptability and financial feasibility, aiming to provide wider accessibility for innovation to meet the market needs adequately.

Similarly, Hussain et al. [32] employed a 'responsible thinking' case study design to investigate software practitioners' perceptions of human values in software engineering. The survey results demonstrated that almost all participants agreed that human values, such as privacy, transparency, integrity, social justice, diversity, and so on, need to be explicitly addressed during software development. However, software companies tend to consider values mainly in the early phases of a project. The authors emphasised that the value issues need to be considered throughout the whole software development lifecycle because stakeholder values may conflict at different phases. These conflicts may be due to the different prioritisation of values chosen by stakeholders, such as some prioritising climate change while others prioritise economic equality. The authors indicated that resolving tensions between different values and embedding values in software design contributes to ensuring that technological outcomes meet social expectations, i.e., that they are aligned with universal human values.

Moreover, in the medical science field, Sujan et al. [6] investigated stakeholders' perceptions of AI-based applications in healthcare. The authors hold that, although most existing healthcare AI applications have been evaluated retrospectively, they are still not sufficient to ensure that the use of AI in healthcare settings is safe and free from any subsequent socio-technical concerns, such as trust, skill erosion, and ethical issues around fairness. The authors suggested embedding the notion of RRI in the healthcare innovation process, especially embedding the debate about societal concerns to ensure the diversity of views from actors and stakeholders. Such inclusive dialogue can ensure the meaningful and safe integration of AI into healthcare systems, which contributes to providing trustworthy healthcare innovation and technology for users and increasing their willingness to accept care.

Concerning market regulation in food and agricultural commodities, Merck et al. [33] stated that traditional regulatory regimes may be insufficient to deal with the more complex ethical, legal, and social implications of novel products produced using nanotechnology. For example, although nanotechnology has the potential to improve the sustainability, safety, and availability of agri-food products, in many cases, there remain uncertainties in assessing the potential risks they may pose, which present not readily addressed challenges to existing regulatory frameworks. The authors suggested implementing the principle of RRI to improve existing regulatory regimes, allowing innovators and policymakers to prospectively evaluate the associated influences and be more responsive to the public's needs and concerns. Appropriate and adequate regulations contribute to shaping responsible innovation in nano-agri-foods and promote the future development of agricultural innovation.

In addition to the above cases, an increasing number of studies have been exploring the specific practices of RIT in various technospheres, such as gene drive technology in biology [5], deep synthesis application in digital media [30], information and communication technologies (ICT) in tourism [34], community energy storage (CES) in the energy field [35], and others. The participants from various industries are attempting to shape emerging innovations and technologies to be more 'responsible' by embedding the RRI concept, aiming to address the grand challenges accompanied by technological and scientific progress.

Against this backdrop, the key characteristics of RIT, based on previous research efforts, could in a nutshell be categorised as follows: (a) acceptable; (b) accessible; (c) aligned; (d) trustworthy; and (e) well governed. The summary descriptions of these key characteristics are provided in Table 6.1.

3 RESEARCH DESIGN

This chapter adopts a systematic literature review method with the Preferred Reporting Items for Systematic Reviews and Meta-Analyses (PRISMA) protocol to address the following research question: 'What are the key characteristics of responsible innovation and technology (RIT)?' This chapter applied a three-stage procedure as the methodology, i.e., Stage 1 (planning), Stage 2 (review), and Stage 3 (reporting), which has been proven to be feasible and reliable by previously conducted systematic literature reviews—e.g., Li et al. [36] and Li et al. [37].

The task of the planning stage (Stage 1) is to form a feasible research plan, including setting up a research objective to address the abovementioned research question, selecting search keywords for relevant article searching, and developing the criteria of exclusion and inclusion for article

TABLE 6.1   Key characteristics of responsible innovation and technology

Characteristic	Description	Exemplar Reference
Acceptable	Publicly acceptable, ethically unproblematic, and harmless, including being free of bias and deception. Devoted to delivering equitable products and encouraging fair technology use for achieving an overall state of well-being and the common good.	[31]
Accessible	Broaden the notions of accessibility to deliver culturally inclusive, technically adaptable, and financially affordable products. Devoted to spreading the benefits of digitisation across societies and cities without barriers.	[4]
Aligned	Deliberate in decision-making practices and aligned with societal desirability and human values. Devoted to achieving meaningful, positive, and sustainable outcomes to solve the accompanied challenges and improve the well-being of life on Earth.	[32]
Trustworthy	Handle greater informational transparency and technical security within designing, producing, implementing, and operating processes. Devoted to delivering human-understandable explanations of decisions to increase public understanding, trust, and confidence robustly.	[6]
Well governed	Adhere to statutory regulations and governance requirements and can be well governed by the broader stakeholder groups. Devoted to ensuring its dependability and accountability to maintain public support and trust, which leads to higher acceptance and further implementation.	[33]

screening. The research objective was framed to conceptualise the key characteristics of RIT. Therefore, the keywords were confirmed as 'responsible innovation' and 'responsible technology', which were used to search across the titles, abstracts, and keywords of available articles. The search task was conducted via an academic search engine, which covered approximately 400 different bibliographic repositories, including Directory of Open Access Journals, Web of Science, Wiley Online Library, Scopus, and ScienceDirect. The inclusion and exclusion criteria were developed to improve the efficiency of screening tasks (Table 6.2), which can assist in selecting suitable articles and reduce unnecessary efforts.

TABLE 6.2   Inclusion and exclusion criteria

Primary Criteria		Secondary Criteria	
Inclusionary	**Exclusionary**	**Inclusionary**	**Exclusionary**
Academic journal articles	Duplicate records	Responsible innovation and technology-related	Not responsible innovation or technology-related
Peer-reviewed	Books and chapters		
Full-text available online	Industry reports	Relevance to the research objective	Irrelevant to the research objective
Published in English	Government reports		

In the review stage (Stage 2), the reviewing task followed the PRISMA 2020 statement to ensure transparency, integrity, and accuracy of the article selecting and reviewing process. The search task was conducted in August 2022. The initial search did not include any restrictions for publication year so that we could inspect the suitability of all time periods covered by the academic search engine. However, in consideration of 'RIT' as an emerging concept that has grown rapidly during the last decade, most of the highly relevant articles were published in this period [38,39]. Therefore, the final search task developed a literature database with a limited publishing period, covering the articles published between January 2010 and August 2022. Additionally, a fuzzy format—'*'—was included in the query string to ensure the comprehensiveness of the obtained data.

The final query string of the search task was determined as follows: TITLE-ABS-KEY ('Responsible innovation' OR 'Responsible technolog*') AND (LIMIT-TO (DOCTYPE, 'ar')) AND (LIMIT-TO (LANGUAGE, 'English')). The result of the initial search returned a total of 1201 articles based on the primary criteria. After removing duplicates, the records went down to 1008 articles. Based on the secondary criteria (reviewing article titles and abstracts), the result of the second-round review recorded a total of 178 articles. In the third screening task, a full-text review of 178 articles was undertaken to evaluate the relevance, consistency, and reliability of these articles. The result of the third-round review returned a total of 51 articles.

In addition, the repetitive screening test and the snowballing strategy were used in this stage to ensure the comprehensiveness and validity of the final article selection, which additionally recorded 14 articles. Snowballing is a literature retrieval strategy that identifies additional relevant papers by tracking the reference list of articles, which is adopted in the complementary search task of this chapter, aiming to expand candidate articles at the

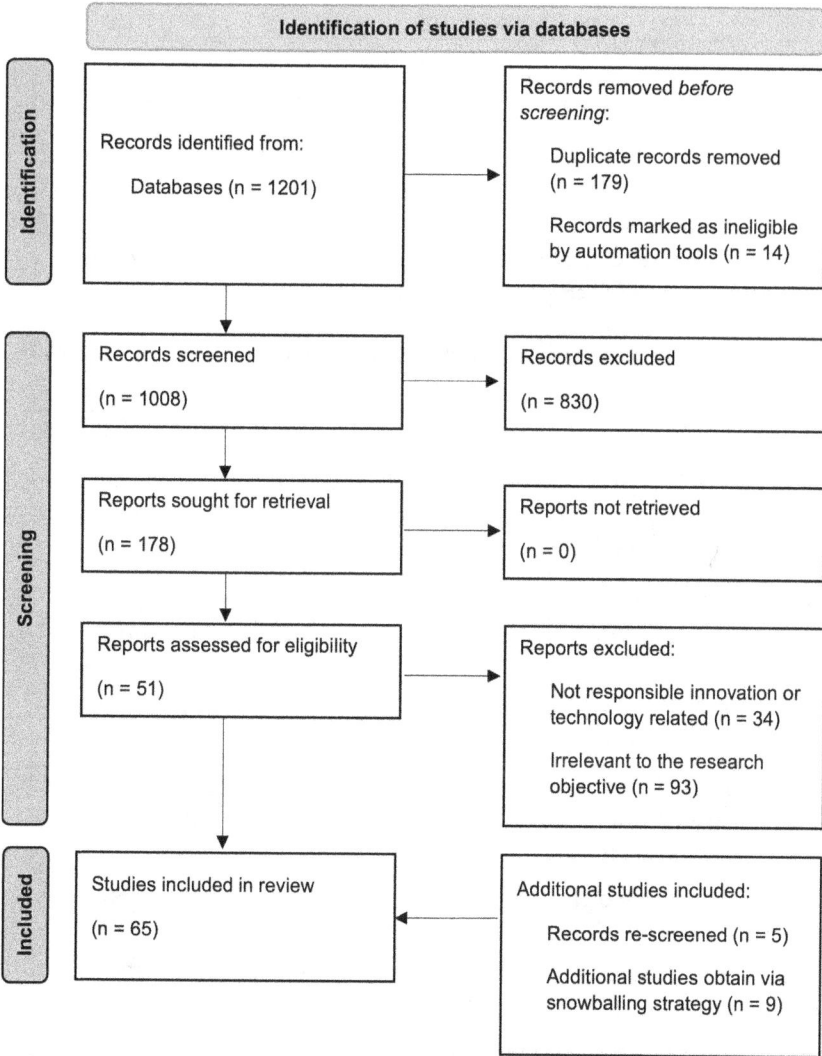

FIGURE 6.1   Literature selection procedure.

specific themes to discover additional insights [40,41]. Finally, a total of 65 articles included in the qualitative analysis were recorded (Figure 6.1).

In the reporting and dissemination stage (Stage 3), the insights were captured from recorded articles and sorted into specific themes via a qualitative analytical approach, focusing on understanding the characteristics of responsible innovation or technology. In this stage, the eye-balling technique was adopted to identify the commonalities and disparities of

TABLE 6.3   Categorisation criteria

Criteria
Identify the themes and contents associated with responsible innovation or technology in the articles;
Determine the domain of existing practices relevant to responsible innovation or technology in the selected articles;
Capture the insights about responsible innovation or technology in the selected articles;
Conceptualise the key characteristics of responsible innovation or technology;
Narrow down themes and crosscheck the consistency and reliability of themes against other published literature;
Conduct a final review of the selected and reviewed literature and reconsider the refined themes.

recorded articles, which helped in the categorisation of themes [36,42]. Lastly, the insights of articles were finally classified under five themes: 'aligned', 'accessible', 'acceptable', 'trustworthy', and 'well governed'. The detailed criteria for this categorisation work were developed and are shown in Table 6.3. The completed reporting table is presented in Appendix 6.A.

4 ANALYSIS AND RESULTS

4.1 General Observations

Based on the statistical data extracted from the reviewed articles ($n = 65$), the number of RIT studies has increased over time, reflecting the growing interest in this topic over the past decade. Since the European Union (EU) mainstreamed the notion of 'responsible' in the EU's R&I policy, it has evoked extensive discussion and reflection regarding the 'responsibility' of innovation and technology among various participators, including researchers, practitioners, and policymakers. This discussion and reflection recently gained momentum, especially in the emerging technological fields that are potentially controversial, such as artificial intelligence (AI), gene technology, and nanotechnology [43–45]. The reason might be the growing concerns over uncertainty about the potential consequences and opportunities presented by these promising but potentially disruptive technological advances. Figure 6.2 shows the publication trend of RIT studies during the last decade.

In addition, the statistical data indicated that RIT studies mainly focus on AI ($n = 10$), healthcare technology ($n = 6$), robotics ($n = 4$), nanotechnology ($n = 4$), information and communications technology (ICT) ($n = 4$), and gene technology ($n = 4$). The technology categories of articles

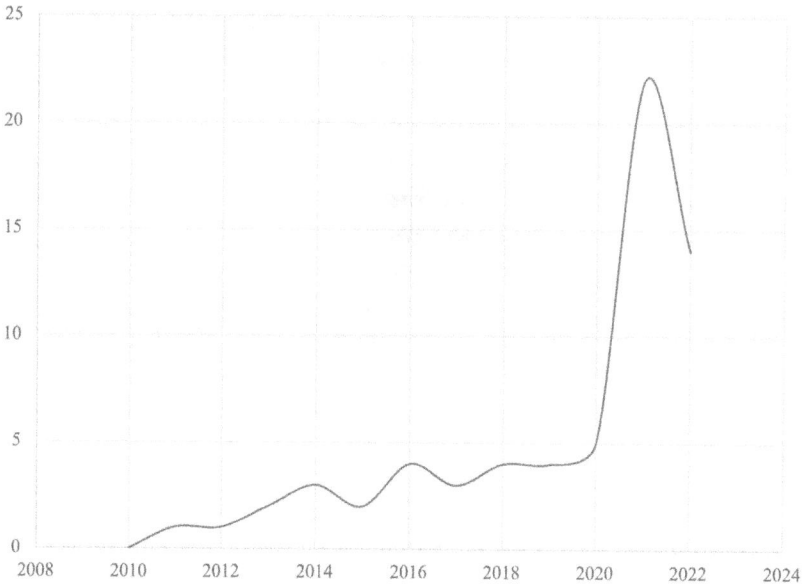

FIGURE 6.2   Publication trend of RIT studies.

were classified based on the keywords or explicit statements in the paper. The pieces without specific statements about technology were classified into the category of 'non-specific'. The articles that had less than two pieces but had specific statements about technology were classified into the category of 'others'. The result shows that RIT-related articles had a great interest in AI technology (over 15% of the total reviewed articles). The reason might be that the global proliferation and societal penetration of AI have raised widespread concerns regarding human autonomy, agency, fairness, and justice, and relevant sectors are attempting to introduce the concept of RIT in AI practices, aiming to offset these concerns and promote the development of responsible AI innovations [43,46]. Figure 6.3 shows the technology categories of RIT studies during the last decade.

Furthermore, the most mentioned characteristic of RIT is 'well governed', occupying 34% of the recorded articles ($n = 46$, 65 articles in total). The proportion of remaining characteristics are relatively average, namely, 'trustworthy' (20%, $n = 28$), 'acceptable' (17%, $n = 23$), 'accessible' (15%, $n = 21$), and 'aligned' (14%, $n = 19$). The reason might be that existing research tends to formulate standardised principles or frameworks in a top-down fashion to guide the development of innovation and technology towards a responsible direction [14,26–28]. Figure 6.4 shows the proportion

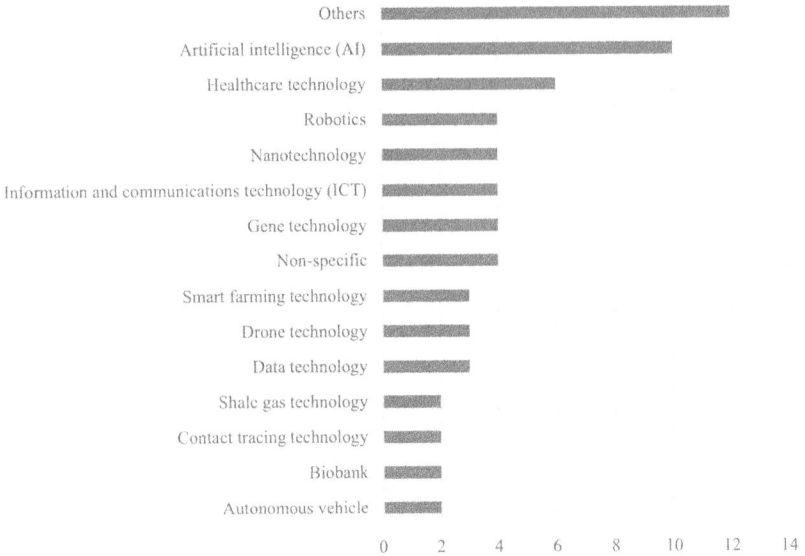

FIGURE 6.3   Technology categories of RIT studies.

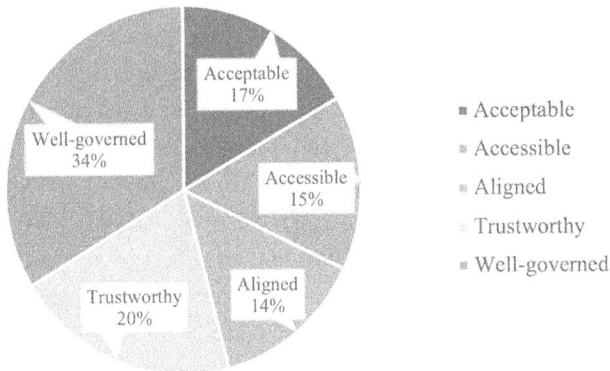

FIGURE 6.4   Proportion of characteristics mentioned in RIT studies.

of RIT characteristics mentioned in the recorded articles. The following five sub-sections will provide a detailed analysis of these characteristics.

4.2 Acceptable Innovation and Technology

The *acceptability of innovation and technology* has been often mentioned in studies on the topic, such as Owen et al. [47], Stilgoe and Guston [48], and Bacq and Aguilera [49]. Based on the reviewed articles, the 'acceptable' characteristics of RIT can be interpreted as follows.

Innovation and technology should be publicly acceptable, ethically unproblematic, and harmless, including being free of bias and deception, while ensuring they will not disrupt the existing social orders. Marketable products and services should be bound by the inherently safe design and meaningful control approach to ensure they do not harm human beings and the environment. The design principles should eliminate systematic stereotyping, encouraging more equitable innovation implementation for achieving overall well-being and the common good.

Based on the above description, the RIT characteristic of 'acceptable' comprises three keywords, namely, ethical, harmless, and equitable. Table 6.4 lists the keywords of this RIT characteristic and provides their summary descriptions.

From an ethical point of view, innovation and technology should follow the principle of people-centred development and use to be compatible with respect for freedoms and human rights, including autonomy, dignity, privacy, and so on, ensuring they are ethically acceptable [30,50–52]. Additionally, Foley et al. [53] stated that technological innovation should afford people freedom of expression and freedom from oppression while not reinforcing social orders that subjugate human beings but aiming to achieve the overarching aspirations of human flourishing.

TABLE 6.4   Keywords of RIT's acceptable characteristic

Keyword	Description	Exemplar Reference
Ethical	Afford and respect human rights and freedoms, including dignity and privacy, while ensuring innovations and technologies do not reinforce social orders that subjugate human beings, promoting autonomy and ethical acceptability to avoid unethical consequences.	[50,51,53]
Equitable	Eliminate systematic stereotyping to reduce the potential risks and impacts of perpetuated and/or increased inequalities between individuals and groups in society while encouraging more broad, democratic, and equitable innovation implementation.	[53,55,57]
Harmless	Ensure products do not harm human health (including physical and psychological) and/or the environment while reducing or eliminating the harmful effects of technology use by appropriate safeguards, e.g., inherently safe design, meaningful human control approach, and so on.	[3,4,62]

Therefore, broader and more open discussions regarding moral and societal values and potential ethical issues are needed to incorporate into the innovation process to make technological outcomes more ethical and democratic [54,55]. In addition, innovators and adopters should be especially prudent in considering potential ethical problems in highly sensitive settings to either avoid or fuel controversy. For example, in Europe, there is a highly valued animal health and welfare context, so it has a long historical arc of concerns about animal food safety [56].

The planning objective and design principle of innovation and technology should encourage a broad, democratic, and equitable implementation approach to avoid disadvantaging specific groups or individuals [53,55,57]. Li et al. [30] stated that innovation and technology should ensure fairness and justice during their entire lifecycle to avoid prejudice and discrimination. Li et al. [30] and Bunnik and Bolt [51] suggested applying a non-discriminatory and more inclusive design approach to eliminate systematic stereotyping during the innovation process, which may contribute to reducing the risks of perpetuated and/or increased inequalities between individuals and groups in society, such as marginalisation towards minority groups. Additionally, Brandao et al. [58] indicated that innovation and technology would face many unforeseen challenges regarding equity issues during real-world implementation, such as indirect discrimination, social inequalities produced by fairness-unrelated decision-making, and others. The authors suggested including realistic fairness models in the early stage of the innovation process, which are important to anticipate potential fairness conflicts or issues and optimise equity in realistic contexts [58].

Furthermore, Li et al. [30] stated that being harmless is one of the core conditions for the high acceptance of innovation and technology. Innovation and technology must do no harm to human beings' physical and psychological health, which is also considered to be the bottom line for enabling technological innovation attempts [3,4,30]. The specific practices and marketable products must not damage human abilities and must not subvert human statuses, such as disrupting interpersonal relationships or replacing human roles [4,6]. Additionally, technological practices and outcomes should avoid causing irreversible social or environmental damage to ensure the sustainability of society and the environment [3,30,59].

However, Boden et al. [60] highlighted that technological products are "just tools designed to achieve goals and desires that humans specify"; all the participants, including users, are responsible for ensuring their actions obey the rules humans have made. This is partly because users may make these products do things their designer did not foresee [60, p. 126]. Therefore, the appropriate safeguards should be embedded in the design process, such as inherently safe design, meaningful human control (MHC) approaches, eco-friendly design, and so on. These measurements are crucial to ensuring the right and proper human control over life and the surrounding environment and to reducing or eliminating the harmful effects during practice and use as far as possible [3,61,62].

4.3 Accessible Innovation and Technology

The second key RIT characteristic relates to the *accessibility of innovation and technology*, which numerous studies on the topic have mentioned [13,22,26]. Based on the reviewed articles, the 'accessible' characteristics of RIT can be interpreted as discussed below.

Innovation and technology should actively incorporate diversified considerations into the design and practice strategies to broaden the notions of accessibility. To create better conditions for widespread availability, marketable products and services should be technically adaptable, financially affordable, and culturally inclusive. The design principles should overcome technological and ideological lock-ins to minimise the digital divide's potential impacts, aiming to spread digitisation benefits across societies and cities without barriers.

Based on the above description, the RIT characteristic of 'accessible' comprises three keywords, namely, inclusive, adaptable, and affordable. Table 6.5 lists the keywords of this characteristic and provides their summary descriptions.

Technically, innovation and technology should be able to integrate with existing technologies and leverage new opportunities to capture all the potential benefits, such as increased work flexibility, productivity gains, and so on [4]. Marketable products must be reliable and robust to deal with complex operating environments under real-world conditions [63,64]. Additionally, it is essential to ensure that innovations and technology, as well as their marketable products and services, are easy to train, use, and maintain, which reduces the technical difficulty for adoption to provide more extensive adaptability for a broader range of people [4].

Economically, innovations and technology and their marketable products and services should ensure avoiding negative financial implications for individuals or users [65]. The design principles should ensure innovation and technology can deliver effective and efficient outcomes while balancing their technical performance and economic viability [38]. Alternatively, other

TABLE 6.5   Keywords of RIT's accessible characteristic

Keyword	Description	Exemplar Reference
Adaptable	Produce valid and reliable products adaptable to existing technologies and complex operating environments, ensuring they are easy to train, use, and maintain to increase the flexibility for application scenarios and the usability for a broader range of people.	[4,63,64]
Affordable	Ensure the delivery of high-value outputs while maintaining economic viability; alternatively, leverage resources in economic ways to avoid any negative financial implications for users, which creates better conditions for wider implementation scenarios.	[38,65,66]
Inclusive	Incorporate diversified cultures, knowledge, and values to align innovation more responsibly with practical societal contexts, aiming to overcome technological and ideological lock-ins and make technological trajectories more responsive to the needs of society.	[68,69,75]

solutions should be provided so as not to jeopardise the delivery of high-value outputs, such as improving the ways resources are leveraged and so on [66]. Some studies stated that improving the financial accessibility of innovation and technology is expected to create better conditions for wider implementation scenarios, which is one of the essential aspects of achieving sustainability outcomes [35,38,67].

Culturally, innovation and technology should adopt more inclusive strategies to incorporate diversified cultural considerations and social impacts into the technological design [68,69]. In some cross-cultural settings, the design and operational criteria should actively respect and include local values, needs, and preferences, aiming to ensure that the marketable products are able to recognise local knowledge and governance [35,69]. Additionally, the interdisciplinary dialogue between fields should be supported during the innovation process because it is essential for adding the richness of understanding to possible cultural impacts and ensuring an accommodation between public values and technological outcomes [57,70].

4.4 Aligned Innovation and Technology

The third characteristic of RIT responds to the initiative stated by former EU Research and Innovation Commissioner Máire Geoghegan-Quinn in supporting the Horizon 2020 Strategy for European R&I. Her opinion is that "innovation must respond to the needs and ambitions of society, reflect its values and be responsible" [71, p. 1]. In other words, innovation and technology must be *aligned with the social desirability* that responds to public needs and/or preferences. Based on the reviewed articles, the 'aligned' characteristics of RIT can be interpreted as follows.

Innovation and technology should always be thoughtful and careful in decision-making practices throughout their entire lifecycle to minimise irreversible social, health, and environmental consequences. Marketable products and services need to achieve a better alignment with societal desirability and/or preferences and human values of freedom, justice, privacy, and so on. The design principles should be devoted to delivering meaningful, positive, and sustainable outcomes to solve the challenges that accompany technological and scientific progress and improve the well-being of life on Earth.

Based on the above description, the RIT characteristic of 'aligned' comprises three keywords, namely, deliberate, meaningful, and sustainable. Table 6.6 lists the keywords of this characteristic and provides their summary descriptions.

As mentioned in multitudinous articles, the practices of innovation and technology could have various unforeseen consequences for society, the environment, and the economy due to the inevitable entry of provided products and services into the complex scenarios of human use [56,63,72].

TABLE 6.6   Keywords of RIT's aligned characteristic

Keyword	Description	Exemplar Reference
Deliberate	Carefully anticipate and assess the associated consequences and opportunities and exercise deliberation in decision-making practices to mitigate actual and potential negative impacts for life on earth to the extent feasible.	[44,63,72]
Meaningful	Achieve a better alignment between people's needs and/or preferences and innovative technologies and social practices to create expected and meaningful outcomes; e.g., address significant problems or societal needs and improve human well-being.	[59,73,77]
Sustainable	Taking the environment into consideration is part of the innovation to treat resources with respect and in the most responsible way throughout the entire lifecycle of innovation, which ensures broad sustainability outcomes while avoiding large and irreversible consequences for the earth.	[53,59,78]

Given this, the context and uses of innovation should be clearly investigated as early as possible, aiming to provide sufficient knowledge and information for participators to comprehensively anticipate and consider the future scenarios of innovation and technology practices, including the actual and potential near-term and longer-term risks and benefits [44,63,73]. Innovation and technology need to keep care and moderation in decision-making practices throughout their entire lifecycles to reduce unforeseen and undesirable consequences to the extent feasible, such as significant or irreversible consequences for life on Earth [72,73]. Therefore, they require stakeholders, including innovators, actors, and researchers, to keep humility, avoid easy judgment, and learn to hesitate during the innovation and practice processes [56,74].

In addition to minimising the undesirable consequences, Van den Hove et al. [75] stated that innovation and technology should be re-targeted, focusing not just on their technical characters or the potential for economic growth but also more directly on their roles in improving human or social well-being. Innovation and technology should be "designed to truly meet people's needs and to put the user at the center of service provision, with all the associated benefits" [76, p. 854]. Therefore, innovation and technology should actively seek to align their processes and expected outcomes with societal needs and/or preferences, aiming to create effective and efficient products to address significant problems or societal needs and positively impact social well-being [59,67,77]. Some studies suggested that embedding human values, including universal and culture-specific values, throughout the innovation process would contribute to achieving a better alignment between technological advancements and societal desirability and acceptability [31,32,70].

Moreover, innovation and technology should commit to achieving broader sustainability outcomes, which not only concentrate on social and economic sustainability but should incorporate environmental considerations as a critical part of the innovation process [4,78]. During the designing and practice process, the resources should be treated and used in a respectful and non-wasteful manner to foster the development of environmentally friendly innovation and technology [53,59,78]. Additionally, Eastwood et al. [4] and Middelveld and Macnaghten [56] posed an interesting point: innovation and technology should ensure not merely human well-being but should also take care of the welfare of other creatures on this planet. The implication for other creatures on this planet during the technical practices is one of the key considerations to ensure broad sustainability outcomes of innovation and technology [4].

4.5 Trustworthy Innovation and Technology

The fourth RIT characteristic relates to the *public trustworthiness in innovation and technology*. Asveld et al. [79] stated that RRI is a way to stimulate public trustworthiness in technological outcomes. Trustworthiness in innovation and technology is one of the desirable outcomes of RRI practices and is also a prerequisite for the successful adoption of technological achievements in our societies and cities [65,79,80]. Based on the reviewed articles, the 'trustworthy' characteristics of RIT can be interpreted as outlined below.

Innovation and technology should foster greater informational transparency throughout their entire lifecycle, especially if information and data are related to matters that affect human beings. Any decisions or acts made by participators or technology itself should be understandable and explainable. Marketable products and services should be physically and digitally secure to minimise the risks of harm or adverse consequences. The design principles should be devoted to enhancing public understanding, trust, and confidence in innovation and technology to increase public acceptance.

Based on the above description, the RIT characteristic of 'trustworthy' comprises three keywords, namely, transparent, secure, and explainable. Table 6.7 lists the keywords of this characteristic and provides their summary descriptions.

TABLE 6.7   Keywords of RIT's trustworthy characteristic

Keyword	Description	Exemplar Reference
Explainable	Make explicit the reason or standard for any decisions or acts made and be able to justify these choices to provide not just the experts but also the public with adequate understanding and trust, which is essential for effective implementation and management.	[3,50,76]
Secure	Assure the safety and security of innovation in society, both physically and digitally, to minimise the risks of harm or the adverse consequences these technologies may cause as possible, aiming to build the public's trust and confidence in them.	[60,66,84]
Transparent	Information and data regarding the design, production, implementation, operating processes, and future planning of innovation should be transparently disclosed to increase public understanding of innovation, including its opportunities, benefits, risks, and consequences.	[33,58,83]

According to Samuel et al. [52], the practice of innovation and technology should take place in contexts where public trust in relevant sectors is established and robust. Given this, the decisions or acts made by participators or technology itself during innovation practices should be understandable and explainable, which is essential for establishing solid public trust in technological outcomes [34,74]. Whether in the innovation process or in specific practice, the reason or standard for choosing any options in the decision process should be clearly made explicit and be able to be justified [50,81]. Professionals and participators should be able to explain the rationale and the strengths and weaknesses of innovation and technology to relevant audiences in an interpretable, intuitive, and human-understandable way [34,51,62]. Additionally, in addition to providing sufficient explanations, there need to be clear responses to audiences' suggestions and concerns [43]. Santoni de Sio [3] noted the need to create adequate social and legal spaces in which professionals and participators can provide the required explanations to audiences, and the audiences can be able to require further explanations and share their opinions with professionals and participators.

In addition, innovation and technology should ensure that the promise of safety and security is delivered to users in practical scenarios, thereby building and enhancing their trust and confidence [60,66,67]. Digitally, innovation and technology should minimise the inherent risks of algorithmic processing, including bias, privacy violation, and cybersecurity vulnerabilities, to reassure users that technological outcomes will not be misused, used to discriminate, or used to unjustifiably target any individual in any way [82,83]. The privacy-assuring methods should be applied in the personal information collection process for heightened data protection and processing while promoting the measures of informed consent and user control over data [3,58].

Physically, innovation and technology should provide greater stability and accuracy to reduce error and risk in practical applications, creating products that are safer than their conventional counterparts [6,44]. Given this, sufficient safeguards, rigorous safety studies, and assiduous protection mechanisms should be adopted in the designing process to increase safety and security and reduce the created risks or adverse consequences as possible [6,74,84]. Additionally, Sujan et al. [6] and Merck et al. [33] underlined that adopting independent oversight and third-party testing to provide sound safety evidence is vital to assure the safety and security of innovation and technology. Bunnik and Bolt [51] pointed out that security is a crucial condition that responsible innovation must provide in its practical application scenarios.

Furthermore, greater transparency during the entire lifecycle of innovation and technology should be implemented to increase people's understanding, which was deemed important to build people's trust in innovation and technology [85,86]. Greater transparency gives stakeholders as much information as possible about the matter involved to increase their understanding of innovation practices. Stakeholders can evaluate the issues from all possible viewpoints based on adequate information and communicate all their concerns [43,59,87]. This informed discourse may assist decision makers in fully considering all relevant matters, especially the uncertainties and limitations that may be relevant for various stakeholders, to direct technological developments towards more responsible goals [63,65,88]. In addition, with the growing public concerns about digital privacy, there is a growing call for transparency in information and data processing. Merck et al. [33], Akintoye et al. [82], and Ienca et al. [83] suggested that innovation and technology should contain a range of measures or processes to disclose how and for what purpose the information and data will be collected, managed, and used. Chamuah and Singh [67] stated that ensuring greater data transparency not only helps to build trust among the users but also would further make innovation and technology responsible.

4.6 Well-Governed Innovation and Technology

The last RIT characteristic relates to the governance of innovation and technology. According to Stilgoe et al. [14], RIT should "take care of the future through collective stewardship of science and innovation in the present" [14, p. 1570]. Innovation and technology, thus, should be well governed to ensure the desired outcomes can be delivered for our cities and society. Based on the reviewed articles, the 'well governed' characteristics of RIT can be interpreted as follows.

Innovation and technology must adhere to statutory regulations and governance requirements during the entire lifecycle while providing explicit accountability mechanisms to make participators consider and act upon the values of responsibility. Participatory governance approaches should be adopted to ensure innovation and technology can be well governed by a broader range of stakeholders, including the public and private sectors, communities, and other relevant entities. The design principles should be devoted to maintaining and strengthening public support and

TABLE 6.8   Keywords of the RIT characteristic well governed

Keyword	Description	Exemplar Reference
Accountable	The entities with legal responsibility for any decisions and acts made or other failures are identifiable, traceable, and held accountable, aiming to embed the values of responsibility in the overall innovation processes.	[50,54,83]
Participatory	Widen stakeholder groups, enhance participation level, and support mutually responsive relations to bring diversified public views and values into innovation, which maintains public support and trust, while embedding innovation successfully in the complex and dynamic societal context via participatory and responsive governance.	[65,91,92]
Regulated	Operated as far as is practicable to comply with existing regulations, fundamental rights, and freedoms, which helps address the complex challenges and reach a consensus that fosters and facilitates innovation and improves the wider implications to society.	[60,96,98]

trust in innovation practices and ensuring that the desirable technological outcomes can be delivered to our cities and societies.

Based on the above description, the RIT characteristic of 'well governed' comprises three keywords, namely, regulated, accountable, and participatory. Table 6.8 lists the keywords of this characteristic and provides their summary descriptions.

According to Hemphill [89], three basic methods can be taken to support the governance of innovation and technology—i.e., government regulation, self-regulation, and public regulation. The authors indicated that self-regulation and public regulation are promising supplementary approaches to improve the traditional regulatory method (government regulation), which may solve the shortcomings of traditional regulatory regimes in dealing with the more complex implications carried by rapid scientific and technological progress.

From the self-regulation perspective, innovation and technology should be auditable and accountable while ensuring that the locus of responsibility remains with the human participators, e.g., designer, operator, or other legal entity [30,50,90]. The intelligent system should provide traceable historical records of every action to identify specific responsibility ascriptions, which contribute to facilitating the clear incident investigation process [6,54]. Additionally, the participators responsible for different stages throughout the lifecycle of innovation practice should be identifiable and accountable for the results of decisions and acts made [60,62,83]. Mecacci and Santoni de Sio [90, p. 105] indicated that "only humans can be held responsible for unwanted actions or mistakes of a technical system". The authors suggested deploying the concept of MHC in the decisional chain of intelligent systems to promote a strong and clear connection between human agents and intelligent devices, thereby resulting in more transparent accountability [90].

From the public regulation perspective, an increasing number of studies have stated that the regulation of innovation and technology must include efforts to engage with the public, which makes it "a more inclusive, participatory, reflexive and responsive heralding responsible governance" [89,91, p. 637, 92]. According to Russell et al. [5] and MacDonald et al. [87], the decision maker and key participator of innovation should focus more on finding creative ways for public engagement, such as allowing for the bottom-up approach that considers the voices of wider stakeholder groups [57,93,94]; interaction with stakeholders in the early stage of the innovation process [33,59,95]; applying interdisciplinary approaches to embed public values and cultures into innovation [64,96,97]; and others. Similarly, Stemerding et al. [98] advocated that public engagement enhanced reflexivity about the different needs and that the interests of stakeholders should be considered in shaping the responsible innovation agenda.

Additionally, considering that some implications of innovation practices may be controversial, innovators, decision makers, and regulators should carefully consider the relevant audience's concerns, insights, and feedback to shape, modify, and restrain innovation, which supports mutually responsive relations in the innovation practices [65,99]. This responsive relationship would assist different stakeholders in reaching a consensus on potential conflicts in the context of the complex and dynamic embedding of technology in society [54,100,101]. The enhanced participation level allows wider stakeholder groups to establish a common ground of innovation practices, which can, along with other positives, increase user acceptance of technological outcomes and maintain public support and trust [31,35,52].

Consequently, a formal and inclusive cooperative mechanism needs to be established that allows wider stakeholders institutionalised access to deliberative settings and provides them with sustained engagement in innovation practices, which ensures they can identify their respective obligations and can voice opinions throughout the innovation process [3,43,73]. Hemphill

[89, p. 242] stated that "participatory public regulation might be a far more thoughtful, efficient, and effective approach to ensuring responsible innovation and technology, which could act as an alternative, complementary, or hybrid form to improve traditional regulation mechanisms".

Lastly, from the traditional regulatory perspective (government regulation), the entire lifecycle of innovation and technology must adhere to existing laws and statutory regulations, while ensuring technological outcomes align with academic discipline, research integrity, and ethics [44,60,102]. On that point, Samanta and Samanta [88] stated that the combination of ethicolegal principles and statutory regulations would enable innovation and technology to maximise its benefits in a responsible way in practical applications. Hence, professionals and legislature should cooperatively establish a more sound and clearer regulatory framework or guideline to address the emerging challenges of defining ethics and reaching a consensus [54,69,96]. Meanwhile, the comprehensive and explicit legislation would contribute to addressing increasing concerns of privacy and safety, while helping to facilitate public acceptance and foster future innovative development [82,89,98].

Yet, Merck et al. [33] indicated that relevant legislation should not appear as a regulatory barrier to innovation. The innovation trajectories should be flexibly steered within a highly regulated environment without generating potential safety or efficacy issues [73]. Given this, regulators should appropriately balance the technological viability (what can be done), statutory permissibility (what may be done), and ethical acceptance (what should be done) during the development of relevant provisions to ensure the appropriateness of regulations [84]. Moreover, Koirala et al. [35] and Leenes et al. [103] suggested that the legislature can adopt socio-technical models specific to local, social, and physical conditions to develop flexible and adaptive legislation programmes better suited to specific circumstances.

5 FINDINGS AND DISCUSSION

5.1 Key Findings

This chapter reviewed studies ($n = 65$) with a focus on RIT, which were published between January 2010 and August 2022 and aimed to conceptualise the key characteristics of RIT and to broaden the understanding of responsible research in the technosphere, particularly from a bottom-up perspective. The findings of this review disclosed the following: (a) the number of RIT studies has increased over time, reflecting the growing interest in this topic over the past decade; (b) RIT studies mainly focus on AI, healthcare technology, robotics, nanotechnology, ICT, and gene technology; (c) RIT is characterised as acceptable, accessible, aligned, trustworthy, and well governed; and (d) these characteristics may be shaped and influenced by cultures, values, social norms, and virtues. The key findings of this chapter are summarised and presented in Tables 6.1 and 6.4–6.8. In accordance with the above efforts, we conceptualise and define RIT in this chapter as follows:

> Responsible innovation and technology is an approach to deliver acceptable, accessible, trustworthy, and well governed technological outcomes, while ensuring these outcomes are aligned with societal desirability and human values and can be responsibly integrated into our cities and societies.

In other words, technological advancements should be developed and integrated into cities and societies in a way that is aligned with societal needs, values, cultures, and ethics. The technological outcomes and practices should ensure accessibility, acceptability, and trustworthiness for relevant audiences and be appropriately managed to ensure its 'right' impact on society. The goal of RIT should pursue a balance between the promising opportunities of technology and its potentially negative consequences, ensuring that it can spread the benefits of technological progress across our societies and cities in a responsible manner.

In terms of practical design and practices, a first suggestion is that decision makers and key participators of innovation should apply a series of formal and evidence-based procedures to evaluate the impacts in different phases of the innovation process, e.g., impact prediction, assessment, and monitoring. Impact prediction helps to identify the potential risks, unintended consequences, and negative impacts of the technology before it is developed and implemented. It allows for early consideration and provides participators with first-mover advantages to take proactive steps to deal with, mitigate, or solve problems; minimise harm; and maximise benefit.

Impact assessment can assist innovators in balancing the opportunities and consequences of innovation and technology, ensuring that technological outcomes deliver the desired outcomes and positively impact society. Additionally, regular impact monitoring should be incorporated into the entire process and make improvements continuously to the technological product as needed to ensure innovation and technology progresses towards more responsible outcomes. The decision makers and key participators of innovation should maintain 'long-term thinking' in the whole evaluation process to consider the long-term implications of innovation and technology and ensure that improved results are resilient and sustainable. A thorough impact evaluation procedure contributes to promoting RIT development.

Our second suggestion is that a broader and clearer ethical framework should be adopted to guide decision making, ensuring that innovation and technology are developed and deployed in line with ethical considerations. The framework should involve two key components. The first involves core and common ethical principles, such as respect for human rights and dignity, ensuring fairness and non-discrimination, and being harmless to human beings and the environment. The second component is being flexible depending on specific cultures, values, and industries, such as geo-cultural characteristics and values, different industry standards, and requirements for sensitive industries. Additionally, industry-specific ethical guidelines can be incorporated into the framework to improve its completeness, such as the EU's Ethical Framework for Trustworthy AI and the IEEE's Ethically Aligned Design for autonomous and intelligent systems. Incorporating a comprehensive and adaptable ethical framework helps build public trust in innovation and technology, which can also promote their responsible implementation in our cities and societies.

Third, the regulatory sector plays a crucial role in promoting RIT development by setting laws, policies, and regulations that ensure legal innovation practices. The regulatory sector should formulate a clear and effective regulatory framework to ensure that innovation practices are carried out under the constraints of the law. The framework can incentivise participators to prioritise responsible innovation by setting clear expectations and consequences for non-compliance. Additionally, effective regulation and governance also facilitates transparency and accountability in innovation practices, encouraging stakeholders to take responsibility for their decisions and actions. A well governed innovation practice can ensure that the technological outcomes are trustworthy and help to ensure that the outcomes are developed and used in a way that benefits society while avoiding severe consequences.

Lastly, an effective participation mechanism is critical in developing RIT. The desirable characteristics of RIT should be realised via active engagement with a broader range of stakeholders, such as policymakers, industry, civil society, academic communities, and underrepresented groups. The decision makers and key participators of innovation should actively consider the diverse perspectives of stakeholders to understand their concerns and opinions, ensuring that innovation and technology will deliver the best possible outcomes. The mechanism can be built by establishing interdisciplinary research programmes, partnerships between industry and academia, and public engagement initiatives.

Additionally, a supportive environment for ongoing dialogue and collaboration should be created, including but not limited to providing practical funding and regulatory support. Incorporating participation mechanisms into the innovation process may contribute to broadening the perspectives and knowledge base of the innovation sector to help them make the right decision, such as acquiring more comprehensive background information regarding local cultures, values, and potential influences. Moreover, an effective participation approach can increase the transparency of the innovation process, which helps to enhance public trust and confidence in technology and increase their acceptance.

5.2 Conceptual Framework

Based on the systematic literature review, this section conceptualises a broad framework outlining the fundamental results and key findings, aiming to explore the possible design procedure of RIT and assist its future research, development, and practices towards more responsible outcomes. The framework (Figure 6.5) is invaluable for governments, companies, practitioners, researchers, and other stakeholders as a tool to address the grand challenges that accompany technological and scientific progress— especially given the context of smart and sustainable cities. The framework also informs science, technology, and innovation policy.

The conceptual framework (Figure 6.5) indicates that ensuring a responsible innovation environment is essential to the delivery of a socially desirable technological outcome, i.e., RIT. Innovation practices should be conducted within the statutory regulatory framework and ensure it adheres to the core ethical principle, which should be considered as the bottom line enabling technological innovation attempts. During the innovation process, the innovation sector should consider adopting the concept of RRI and applying specific RRI tools to ensure the technological outcomes can satisfy the 'responsible' characteristics. These characteristics are as follows: (a) acceptable; (b) accessible; (c) aligned; (d) trustworthy; and (e) well governed.

According to Stilgoe et al. [14] and Burget et al. [22], the RRI concept is an attempt to govern the process of emerging science and innovation by anticipating and discerning opportunities and consequences of innovation within a broader social context. The existing concept pays more attention to describing the processual elements required for a 'responsible' innovation process rather than certain outcomes—e.g., anticipation, reflexivity, inclusion, and responsiveness [22]. Thus, specific tools are needed in this process to achieve concrete results or outcomes. Thapa et al. [13] presented some of the most frequently applied RRI tools in practices, such

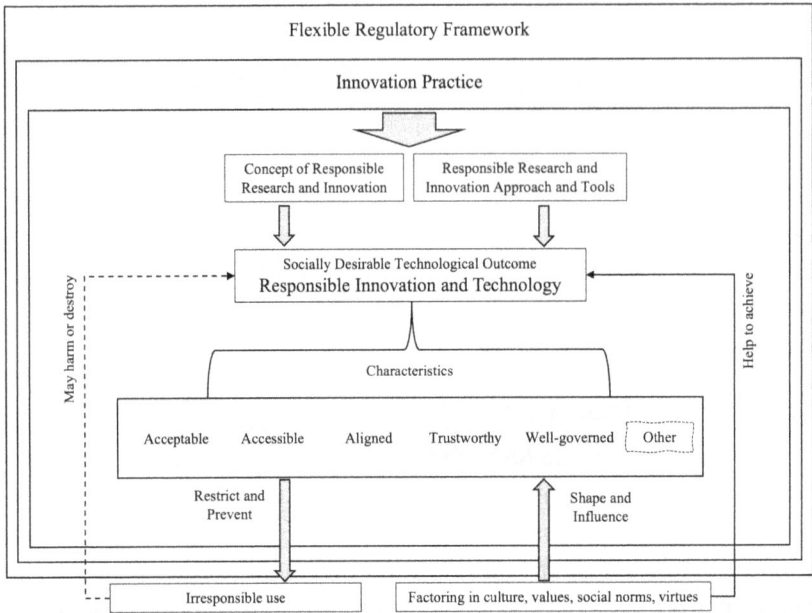

FIGURE 6.5 Conceptual framework of RIT design and implementation.

as comprehensive and acceptability analysis, participatory appraisal, technology assessment, etc., which may assist the innovation sector in designing and developing responsible products to meet social expectations.

In addition, cultures, values, social norms, and virtues may exert significant influence in shaping the characteristics of RIT. According to Von Schomberg [15], the vision of RRI is to embed fundamental social values into the R&I process to ensure that technological outcomes are socially desirable. Nonetheless, culture is considered one of the important determinants in the innovation process [104,105,106]. Although certain core human values are universal but may be interpreted in different ways by different cultures, even within the same society, "the values appreciated by Western society or the developed part of the world may be disliked by developing parts of the world, and vice versa" [107, p. 2]. Therefore, the social perceptions and expectations of 'responsible' outcomes might exhibit differences in different cultural contexts, assuming that 'one size fits all' is unwise. RIT should respect and recognise 'other' (as shown in Figure 6.5) possible characteristics shaped by specific cultural settings, e.g., local social norms and virtues, and approach sensitively when values from different societies and cultures conflict with each other.

Finally, irresponsible use may remove some of the characteristics of RIT, making it unable to meet social expectations. As Boden et al. [60, p. 126] underlined, innovation and technology are "just tools designed to achieve goals and desires that humans specify". Although RIT is expected to provide certain safeguards to restrict or prevent irresponsible use, the awareness of 'responsible use' still needs to be promoted, especially in practical application scenarios. The 'responsible' notion in innovation and technology requires a collective effort to sustain, not only by innovators but also by users.

6 CONCLUSION

This chapter investigated the reported practices of the RRI concept in our societies and cities and conceptualised the key characteristics of its expected outcomes—i.e., RIT—that is an umbrella term inclusive of 'Responsible Artificial Intelligence'. Additionally, a focused discussion has been provided on possible solutions for realising these characteristics. The conceptual framework has been developed, and the possible design procedure of RIT has been outlined, which will broaden the understanding of responsible research from a bottom-up perspective, especially regarding its application to specific practices. It sheds light on the overall design principles of RIT that assure emerging innovations and technologies to be more 'responsible' by embedding the RRI concept. The framework could be used by the government, practitioners, researchers, and other stakeholders as a tool to address the grand challenges accompanied by technological and scientific progress and ensure innovation and technology can be responsibly embedded into our cities and societies.

Although the interest in the theme of 'being responsible' in technospheres is growing, specific ways to achieve desired outcomes still need further research. Given this, this section underlines some issues to be further explored/studied, which may pave the way for a new research agenda concerning achieving the development and practices of responsible technological outcomes in our cities and societies. The following issues/questions are important for prospective research to focus on and address:

- How can we decide if innovation and technology contain the responsible characteristics, and what are the specific evaluation criteria?

- What are the specific scales in defining responsible characteristics in innovation and technology? For example, when a technological innovation contains many characteristics of RIT but is missing some, should it still be considered 'responsible'?

- How should RIT be compatible with specific values in complex cultural contexts, and how should innovation and technology sectors balance the possible conflict of values from different levels of cultural contexts, such as individual, collective, community, and national levels?

- How can we define the social desirability of innovation and technology that may vary with different values or specific innovation purposes, and how can we make choices when different social expectations or needs conflict?

Lastly, this chapter offers invaluable insights into responsible research in the technosphere—particularly from a bottom-up perspective. However, it is essential to acknowledge several limitations that may influence the interpretation of the findings. First, the chapter did not apply any automated analysis tools or techniques to conduct the qualitative analysis. Second, the selected search keywords may not cover all studies relevant to the research objective. Third, the literature selection only records available online and peer-reviewed academic journal articles, which may omit some additional insights from other forms of studies. Fourth, the findings may be influenced by the unconscious biases of the authors.

Despite these limitations, our research lays the groundwork for future investigations in this area. The topic of responsible research in the technosphere is relatively new, and there are still significant research gaps that need to be bridged, especially the uncertain pathway between theoretical study and practical applications. This chapter advocates that future research should open new discussions on responsible research in the context of specific practices and scenarios, with the aim of making the concept more responsive to specific settings—e.g., innovation and technology. On that very point, our prospective studies will concentrate on providing more clarity and measurability to RIT—that is inclusive of responsible AI.

ACKNOWLEDGEMENTS

This chapter, with permission from the copyright holder, is a reproduced version of the following journal article: Li, W., Yigitcanlar, T., Browne, W., & Nili, A. (2023). The making of responsible innovation and technology: an overview and framework. *Smart Cities*, 6(4), 1996–2034.

APPENDIX 6.A   SUMMARY OF THE REVIEWED LITERATURE

No	Author	Journal	Title	Year	Aspect	Innovation	Characteristic	Keyword	Finding
1	Sujan et al.	*Safety Science*	Stakeholder perceptions of the safety and assurance of artificial intelligence in healthcare.	2022	Healthcare	Artificial intelligence	Acceptable	Harmless	1. The use of innovations should not disrupt the relationship between patients and clinicians.
							Trustworthy	Secure	1. Innovation should provide greater efficiency and accuracy to reduce error and to make care safer. 2. Need for rigorous approaches, sound safety evidence, and independent oversight.
							Well governed	Participatory	1. Diversity of views can support responsible innovation.
								Accountable	1. Provide auditable and traceable history of every action that the AI did to facilitate the incident investigation process.

(Continued)

APPENDIX 6.A Continued

No	Author	Journal	Title	Year	Aspect	Innovation	Characteristic	Keyword	Finding
2	Li et al.	Computers in Human Behavior	What drives the ethical acceptance of deep synthesis applications? A fuzzy set qualitative comparative analysis.	2022	Digital media	Artificial intelligence	Aligned	Meaningful	1. Should promote the progress of society and human civilisation, create a more intelligent way of work and life, improve people's well-being, and benefit all humanity, including future generations.
							Acceptable	Ethical	1. Should follow the principle of people-centred development and use, based on the principle of respecting human autonomy.
								Harmless	1. Well-being and the common good, justice, and a lack of harm are the core conditions of high ethical acceptance.
								Equitable	1. Should ensure fairness and justice, avoid prejudice and discrimination against specific groups or individuals, and avoid disadvantaging vulnerable groups.

3	Eastwood et al.	*Frontiers in Robotics and AI*	2022	Agriculture	Robotics	Responsible robotics design–A systems approach to developing design guides for robotics in pasture-grazed dairy farming.	Trustworthy	Transparent	1. Accountability and transparency are the most important guiding principles in AI development.
							Well governed	Accountable	1. AI should be auditable and accountable. The people responsible for different stages of the life cycle of an AI system should be identifiable and responsible for the results of the AI system.
							Aligned	Meaningful	1. Robotics should provide positive impacts on social well-being.
								Sustainable	1. To ensure broad sustainability outcomes with milking robotics. A key consideration is the implications for animal well-being using robots.
							Accessible	Adaptable	1. Should be able to integrate with existing technologies and leverage new opportunities for productivity gains.

(Continued)

APPENDIX 6.A   Continued

No	Author	Journal	Title	Year	Aspect	Innovation	Characteristic	Keyword	Finding
									2. Must be robust to deal with complex operating environments.
									3. Must be easy to train, use, and maintain to increase job flexibility and ability for a wider range of people.
								Affordable	1. High-throughput robotics may be using high-cost technology to perform low-cost jobs.
							Acceptable	Harmless	1. Robotics should reduce injuries and physical demands on people and avoid negative psychological impacts, such as changes to the self-identity of staff if robotics replaces their roles.
							Well governed	Regulated	1. Potential regulatory barriers also need to be assessed in robotics development.

Conceptual Framework of Responsible Urban Artificial Intelligence ■ 209

No.	Author	Title	Year	Field	Technology	Category	Principle	Description
4	Ienca et al.	Towards a Governance Framework for Brain Data.	2022	Medical science	Data technology	Trustworthy	Secure	1. Brain data should be considered a special category of personal data that warrants heightened protection during collection and processing. 2. Usage of brain data should consider and prevent inherent risks of algorithmic processing including bias, privacy violation, and cybersecurity vulnerabilities.
							Transparent	1. The process of collecting, managing, and/or processing identifiable brain data should be transparently disclosed.
						Well governed	Accountable	1. Legal entities responsible for data breaches and other regulatory failures should be identifiable and held accountable.
5	Bao et al.	Whose AI? How different publics think about AI and its social impacts.	2022	Technology	Artificial intelligence	Well governed	Participatory	1. AI development and regulation must include efforts to engage with the public in order to account for the varied perspectives that different social groups hold concerning the risks and benefits of AI.

(Continued)

APPENDIX 6.A   Continued

No	Author	Journal	Title	Year	Aspect	Innovation	Characteristic	Keyword	Finding
6	Townsend and Noble	*Sociologia Ruralis*	Variable rate precision farming and advisory services in Scotland: Supporting responsible digital innovation?	2022	Agriculture	Smart farming technology	Well governed	Participatory	1. Design and implementation of innovation should allow for a more bottom-up approach that considers the voices of a broader stakeholder group.
7	Stahl	*International Journal of Information Management*	Responsible innovation ecosystems: Ethical implications of the application of the ecosystem concept to artificial intelligence.	2022	Technology	Artificial intelligence	Aligned	Meaningful	1. Innovations should actively seek to align their processes and expected outcomes with societal needs and/or preferences.
8	Middelveld et al.	*Public Understanding of Science*	Imagined futures for livestock gene editing: Public engagement in the Netherlands.	2022	Agriculture	Gene technology	Aligned	Deliberate Sustainable	1. We need to exercise care (and moderation) in decision-making practices of innovation to avoid large and irreversible consequences for life on Earth.
9	Merck et al.	*Bulletin of Science, Technology & Society*	What Role Does Regulation Play in Responsible Innovation of Nanotechnology in Food and Agriculture? Insights and Framings from US Stakeholders.	2022	Agriculture	Nanotechnology	Trustworthy	Transparent	1. Should provide a range of processes to ensure transparency (via processes, safety studies, and assessments), requiring disclosure, and changing the use of confidential business information.

							Well governed	Secure	1. Should conduct more safety studies to strengthen safety and ensure independent third-party testing.
								Regulated	1. Require basic and appropriate regulations to ensure safety and efficacy but ensure that they do not appear as a barrier to innovation or a guardrail against the risks of novel products.
								Participatory	1. Should interact with stakeholders early in innovation processes, involving the public and engaging stakeholders, to strengthen community and stakeholder engagement.
10	Russell et al.	Journal of Responsible Innovation	Opening up, closing down, or leaving ajar? How applications are used in engaging with publics about gene drive.	2022	Biology	Gene technology	Well governed	Participatory	1. Should devote more attention to finding creative ways for public engagement, which can create fresh perspectives, engagements, and collective actions.

(Continued)

APPENDIX 6.A Continued

No	Author	Journal	Title	Year	Aspect	Innovation	Characteristic	Keyword	Finding
11	Foley et al.	*Journal of Responsible Innovation*	Innovation and equality: an approach to constructing a community governed network commons.	2022	Technology	Information and communication technologies	Well governed	Participatory	1. Apply more interdisciplinary approaches to bring the public's values into innovation, aiming to align technological and societal research for equitable outcomes.
12	Donnelly et al.	*AI & Society*	Born digital or fossilised digitally? How born digital data systems continue the legacy of social violence towards LGBTQI+ communities: a case study of experiences in the Republic of Ireland.	2022	Technology	Data technology	Acceptable	Equitable	1. Developers should eliminate systematic stereotyping during the innovation process and take a more inclusive approach within the original software design to reduce marginalisation towards minority groups in society.
13	Samuel et al.	*Critical Public Health*	COVID-19 contact tracing apps: UK public perceptions.	2022	Healthcare	Contact tracing technology	Acceptable	Ethical	1. New health innovation practices should be compatible with respect to personal privacy and autonomy.
							Trustworthy	Explainable	1. Innovation practices should be within contexts where public trust in government and institutions is established and robust.

14	MacDonald et al.	*Journal of the Royal Society of New Zealand*	Conservation pest control with new technologies: public perceptions.	2022	Biology	Gene technology	Well governed — Trustworthy	Participatory — Transparent		1. Effective communication and engagement are helpful in maintaining public support and trust. 1. Innovation design and processes need to involve and communicate/be transparent with the public.
15	Bunnik and Bolt	*Epigenetics Insights*	Exploring the Ethics of Implementation of Epigenomics Technologies in Cancer Screening: A Focus Group Study.	2021	Medical science	Epigenomics technology	Well governed — Acceptable	Participatory — Ethical; Equitable		1. From an ethical point of view, innovation and its implementation should be respected for autonomy and ethically acceptable. 1. Should grant all categories of road users the same level of protection, which aims to redress inequalities in vulnerability among road users. 2. Should adopt non-discriminatory and more inclusive designs to reduce the risks of perpetuated and increased inequalities between individuals and groups in society. 3. Should avoid blaming the victim, stigmatisation, and discrimination.

(Continued)

APPENDIX 6.A Continued

No	Author	Journal	Title	Year	Aspect	Innovation	Characteristic	Keyword	Finding
							Trustworthy	Transparent	1. Informed consent was deemed important because people need to understand 'the whole chain' of events or decisions they may be confronted with based on the possible outcomes of screening.
								Explainable	1. Professionals should be able to explain what epigenomic screening entails and what the results might mean.
								Secure	1. The security of data and samples and the protection of the privacy of screening participants were crucial conditions for the responsible implementation of epigenomics technologies in public health settings.
16	Santoni de Sio	*Ethics and Information Technology*	The European Commission report on ethics of connected and automated vehicles and the future of ethics of transportation.	2021	Urbanology	Autonomous vehicle	Acceptable	Harmless	1. CAVs should prevent unsafe use by inherently safe design and meaningful human control approaches. 2. Products do not harm human health and/or the environment.

Trustworthy	Secure	1. The first and most obvious is the safeguard of informational privacy, in line with some basic principles of GDPR such as data minimisation, storage limitation, and strict necessity requirements, as well as the promotion of informed consent of informed consent practices and user control over data.
	Explainable	1. Create adequate social and legal spaces where questions about the design and use choices of CAVs can be posed and answered, making the relevant people aware, willing, and able to provide the required explanations to the relevant audience and the relevant audiences able and willing to require and understand the explanations.
Well governed	Participatory	1. Create the institutional, social, and educational environment to ensure that all key stakeholders can discuss, identify, decide, and accept their respective obligations.

(Continued)

APPENDIX 6.A Continued

No	Author	Journal	Title	Year	Aspect	Innovation	Characteristic	Keyword	Finding
								Accountable	1. Should address the following often-posed question: who is to blame (and held legally culpable) for accidents involving CAVs?
17	Kokotovich et al.	*NanoEthics*	Responsible innovation definitions, practices, and motivations from nanotechnology researchers in food and agriculture.	2021	Agriculture	Nanotechnology	Aligned	Meaningful	1. Create effective and efficient products to improve human well-being and solve societal problems.
								Deliberate	1. Should carefully and comprehensively consider the potential social, health, and environmental impacts associated with the innovation and ensure that actual and potential near-term and longer-term negative impacts are mitigated to the extent feasible.
							Acceptable	Harmless	1. Develop products that are publicly acceptable because of their potential to impact the uptake of technology.

18	Grieger et al.	*NanoImpact*	Responsible innovation of nano-agrifoods: Insights and views from US stakeholders.	2021	Agriculture	Nanotechnology			
							Trustworthy	Secure	1. Use nanotechnologies and/or engineered nanomaterials to create agrifood products that were more safe than conventional counterparts.
							Well governed	Regulated	1. Adhere to regulations to ensure agrifood products align with the mission of academic discipline and research integrity and ethics.
								Participatory	1. Engage stakeholders and collaborate interdisciplinarily to determine what specific products to pursue.
							Aligned	Meaningful	1. Create effective and efficient products to address a significant problem or societal need and improve the world.
								Sustainable	1. Should treat resources with respect and in the most responsible way and use the resource in a non-wasteful manner.
							Acceptable	Harmless	1. Do nothing that could cause irreversible harm to public health or the environment.
							Trustworthy	Transparent	1. Improve transparency and communication.

(Continued)

APPENDIX 6.A   Continued

No	Author	Journal	Title	Year	Aspect	Innovation	Characteristic	Keyword	Finding
							Well governed	Participatory	1. Consider the impacts of an innovative development on the different stakeholders and engage them in the early stage of innovation process.
								Regulated	1. Adhere to regulations.
19	Laursen and Meijboom	*Journal of Agricultural and Environmental Ethics*	Between Food and Respect for Nature: On the Moral Ambiguity of Norwegian Stakeholder Opinions on Fish and Their Welfare in Technological Innovations in Fisheries.	2021	Agriculture	Fishery technology	Well governed	Regulated	1. Governance regulation becomes a key outcome of the innovation process to make innovation acceptable in India.
20	Singh et al.	*Technological Forecasting and Social Change*	Analysing acceptability of E-rickshaw as a public transport innovation in Delhi: A responsible innovation perspective.	2021	Urbanology	E-rickshaw	Aligned	Sustainable	1. Universal and culture-specific values should be embedded in innovation, which can make the product socially, economically, and environmentally sustainable.
							Accessible	Inclusive	

21	*Stitzlein et al.*	*Reputational risk associated with big data research and development: an interdisciplinary perspective.*	2021	Agriculture	Data technology	Well governed	Participatory	1. Engage stakeholders via a representative and inclusive process to establish common ground or consensus, which can, along with other positives, make innovation acceptable and workable.
								2. Some implications of technological innovation may be controversial, and thus, public research and development must reconcile the possible repercussions of participating in their development.
						Sustainability	Equitable	1. Encourage fairer and more equitable technology use.
						Acceptable	Ethical	1. Incorporate moral and societal values into the design processes for making emerging technologies more ethical and more democratic.
						Trustworthy	Transparent	1. Handle greater transparency about data collection, reuse, consent, and custodianship.

(*Continued*)

APPENDIX 6.A Continued

No	Author	Journal	Title	Year	Aspect	Innovation	Characteristic	Keyword	Finding
22	Christodoulou and Iordanou	*Frontiers in Political Science*	Democracy under attack: challenges of addressing ethical issues of AI and big data for more democratic digital media and societies.	2021	Digital media	Artificial intelligence	Well governed	Participatory	1. Should involve even more stakeholders (public) to identify common ground across countries and regions, as well as cultural-specific challenges that need to be addressed.
						Data technology		Regulated	1. Legislation should help to address the challenge of defining ethics and reaching a consensus that improves the wider implications to society.
23	Middelveld and Macnaghten	*Elem Sci Anth*	Gene editing of livestock: Sociotechnical imaginaries of scientists and breeding companies in the Netherlands.	2021	Agriculture	Gene technology	Aligned	Deliberate	1. Require stakeholders—including agricultural scientists—to exercise humility, avoid easy judgment, and learn to hesitate.
								Meaningful	1. The purpose of livestock gene editing applications is represented as that of producing 'better' animals, with 'improved' animal welfare and 'increased' disease resistance, that contribute a vital role to play in 'solving' the global food challenge.

								Acceptable	Ethical	1. Highly valued animal health and welfare and the long historical arc of concerns about animal food safety in unison create a high ethical sensitivity to animals in Europe. Innovation should be particularly careful to avoid or fuel future controversy.
24	Mladenović and Haavisto	*Case Studies on Transport Policy*	Interpretative flexibility and conflicts in the emergence of Mobility as a Service: Finnish public sector actor perspectives.	2021	Urbanology	Mobility-as-a-Service		Aligned	Meaningful	1. Mobility as a Service (MaaS) should be designed to truly meet people's needs and to put the user at the centre of transport service provision, with all the associated benefits.
									Sustainable	1. MaaS would reduce the percentage of car use, by freeing people from car dependency to increase the share of sustainable modes.
								Accessible	Inclusive	1. Embrace the inherent conflict in the value-laden mobility domain, which paves the way for a culture of technological innovation.

(Continued)

APPENDIX 6.A Continued

No	Author	Journal	Title	Year	Aspect	Innovation	Characteristic	Keyword	Finding
							Well governed	Participatory	1. The public and private sectors need to find a way to cooperate and begin dialogue, and information sharing are important in a fast-moving field.
25	Rochel and Evéquoz	*AI & Society*	Getting into the engine room: a blueprint to investigate the shadowy steps of AI ethics.	2021	Technology	Artificial intelligence	Trustworthy	Explainable	1. Innovator should be able to make explicit their reasons or standard for choosing option A over other existing options and be able to justify these choices.
26	Pickering	*Future Internet*	Trust, but Verify: Informed Consent, AI Technologies, and Public Health Emergencies.	2021	Healthcare	Artificial intelligence	Acceptable	Harmless	1. Reduce or eliminate the harmful effects of technology use and achieve an overall state of well-being.
							Trustworthy	Explainable	1. Should explain the rationale to users, characterise the strengths and weaknesses, and convey an understanding of how they will behave in the future.
							Well governed	Accountable	1. Understand the actors who will often be regulated with specific obligations. These actors would all influence the trust context.

#	Author	Journal	Title	Year	Domain	Technology			Description
27	Stankov and Gretzel	*Information Technology & Tourism*	Digital well-being in the tourism domain: mapping new roles and responsibilities.	2021	Tourism	Information and communication technologies	Trustworthy	Explainable	1. Generating high-quality interpretable, intuitive, human-understandable explanations of AI decisions is essential for operators and users to understand, trust, and effectively manage local government AI systems.
28	Yigitcanlar et al.	*Journal of Open Innovation: Technology, Market, and Complexity*	Responsible urban innovation with local government artificial intelligence (AI): A conceptual framework and research agenda.	2021	Urbanology	Artificial intelligence	Acceptable	Ethical	1. The ethical considerations made by the designers and adopters of AI systems are critical when it comes to avoiding the unethical consequences of AI systems.
							Accessible	Affordable	1. AI systems should be accessible and affordable. Alternatively, the resources can be leveraged in new ways, or other solutions can be found that do not jeopardise the delivery of high-value outputs.

(Continued)

APPENDIX 6.A Continued

No	Author	Journal	Title	Year	Aspect	Innovation	Characteristic	Keyword	Finding
							Trustworthy	Transparent	1. Reduce the created risks or adverse consequences as much as possible and build users' trust and confidence via increasing transparency and security of the system.
								Secure	
29	Iakovleva et al.	*Sustainability*	Changing Role of Users—Innovating Responsibly in Digital Health.	2021	Healthcare	Healthcare technology	Well governed	Participatory	1. Should carefully consider the ability and willingness of users to get involved and contribute their insights and absorb this type of feedback to shape and modify innovation in response to their insights.
30	Buhmann and Fieseler	*Technology in Society*	Towards a deliberative framework for responsible innovation in artificial intelligence.	2021	Technology	Artificial intelligence	Accessible	Inclusive	1. The inclusion of all arguments constitutes the main precondition of the rationality of the process of deliberation.

No.	Author	Year	Domain	Subdomain			Description
31	*Health Services Management Research*	2021	Healthcare	Healthcare technology	Trustworthy	Transparent	1. Stakeholders need to have as much information as possible about the issues at stake, the various suggestions for their solution, and the ramifications of these proposed solutions.
	Lehoux et al.					Explainable	1. Needs to be clearly responsive to stakeholders' suggestions and concerns.
	Responsible innovation in health and health system sustainability: Insights from health innovators' views and practices.				Well governed	Participatory	1. Stakeholders need institutionalised access to deliberative settings to ensure they have a chance to voice their concerns, opinions, and arguments.
					Accessible	Affordable	1. Address specific system-level benefits but often struggle with the positioning of their solution within the health system.
					Acceptable	Ethical	1. Increase general practitioners' capacity or patients and informal caregivers' autonomy.

(Continued)

APPENDIX 6.A   Continued

No	Author	Journal	Title	Year	Aspect	Innovation	Characteristic	Keyword	Finding
32	Akintoye et al.	*Journal of Information, Communication and Ethics in Society*	Understanding the perceptions of UK COVID-19 contact tracing app in the BAME community in Leicester.	2021	Healthcare	Contact tracing technology	Well governed	Participatory	1. Engage stakeholders at an early ideation stage using context-specific methods combining both formal and informal strategies.
							Acceptable	Ethical	1. Reassure users that this technology will not target them and will not be misused in any way. The collected data will securely be processed and will not be used in any way to discriminate against them or unjustifiably target them.
							Trustworthy	Secure Transparent	1. Commit to full transparency in the implementation of the technology to provide clear information on how and what the data will be used for in the future.
							Well governed	Regulated	1. Clear regulation or policy to prevent misuse or dual use of concern.

#	Author	Journal	Year	Field	Technology			Recommendations
33	Ten Holter et al.	*Technology Analysis & Strategic Management*	2021	Technology	Quantum computing	Accessible	Inclusive	1. Support interdisciplinary dialogue between fields to empower researchers, which is essential to add richness of understanding to possible impacts.
						Acceptable	Equitable	1. Ensure wide, democratic access to technologies.
						Well governed	Participatory	1. Generate more frequent, more detailed conversations with society to increase public understanding of quantum technologies. 2. Widen the pool of stakeholders consulted to incorporate the views of wider groups of stakeholders.
34	Macdonald et al.	*Journal of Responsible Innovation*	2021	Urbanology	Drone technology	Accessible	Inclusive	1. Incorporate diversified ethical considerations and social impacts into the technological design, especially in some cross-cultural settings, to ensure products match local users' needs and preferences and recognise local knowledge and governance.

Indigenous-led responsible innovation: lessons from co-developed protocols to guide the use of drones to monitor a biocultural landscape in Kakadu National Park, Australia.

Reading the road: challenges and opportunities on the path to responsible innovation in quantum computing.

(Continued)

APPENDIX 6.A Continued

No	Author	Journal	Title	Year	Aspect	Innovation	Characteristic	Keyword	Finding
35	Chamuah and Singh	*Aircraft Engineering and Aerospace Technology*	Responsibly regulating the civilian unmanned aerial vehicle deployment in India and Japan.	2021	Urbanology	Drone technology	Well governed	Participatory Regulated	1. Participation of both internal and external stakeholders in regulations would make it more inclusive, participatory, reflexive, and responsive, heralding responsible governance that is suitable for robust policymaking.
								Accountable	1. Have to keep a strategy and plan and identify essential values to ensure the accountability of new and emerging technology.
							Well governed	Regulated	1. Establish guidelines for technological systems to remain human-centric, serving humanity's values and ethical principles, including ensuring humans remain in the loop about drone development and use.

36	Rose et al.	*Land use policy*	Agriculture 4.0: Making it work for people, production, and the planet.	2021	Agriculture	Smart farming technology	Accessible	Inclusive Participatory	1. Having open conversations about the future of agriculture should include the crucial views of marginalised individuals who might possess differing opinions.
							Well governed	Regulated	1. Require updates to legislation, guidelines, and possible support for various technologies in the form of skills training, improved infrastructure, or perhaps funding.
37	Hussain et al.	*IEEE Transactions on Software Engineering*	Human values in software engineering: Contrasting case studies of practice.	2020	Technology	Software engineering	Aligned	Meaningful	1. Technology should be socially desirable and aligned with human values of freedom, justice, privacy, and so on.
38	Lockwood	*IET Smart Cities*	Bristol's smart city agenda: vision, strategy, challenges and implementation.	2020	Urbanology	Smart city technology	Accessible	Inclusive	1. Reduce the impacts of the digital divide on the most deprived areas or more vulnerable groups to spread the benefits of digitisation spreading across the city.

(Continued)

APPENDIX 6.A   Continued

No	Author	Journal	Title	Year	Aspect	Innovation	Characteristic	Keyword	Finding
							Trustworthy	Secure	1. It is vital that smart technologies are developed with sufficient safeguards to minimise the risk of harm these technologies may cause, be that data protection and privacy breaches or biased, discriminatory outcomes.
							Well governed	Regulated	1. Regulators in developing relevant policy frameworks, regulations, and standards should appropriately balance what can be done (what is technologically viable), what may be done (from a legal perspective), and what should be done (what is ethical and acceptable).
39	Mecacci and Santoni de Sio	*Ethics and Information Technology*	Meaningful human control as reason-responsiveness: the case of dual-mode vehicles.	2020	Urbanology	Autonomous vehicle	Trustworthy Well governed	Secure Accountable	1. Promote a strong and clear connection between human agents and intelligent systems, thereby resulting in better safety and clearer accountability.

#	Author	Journal	Title	Year	Domain	Technology			Description
40	Brandao et al.	*Artificial Intelligence*	Fair navigation planning: A resource for characterising and designing fairness in mobile robots.	2020	Urbanology	Robotics	Acceptable	Equitable	1. Include realistic fairness models within the planning objectives of innovation.
							Trustworthy	Transparent	1. Provide transparency to let users understand and control the impact of the technology in terms of the values of interest.
								Secure	1. Require data collection to go together with privacy-assuring methods.
41	Chamuah and Singh	*SN Applied Sciences*	Securing sustainability in Indian agriculture through civilian UAV: a responsible innovation perspective.	2020	Agriculture	Drone technology	Aligned	Meaningful	1. Have the ability or power to address existing problems or societal needs.
							Accessible	Affordable	1. The economic viability is one of the essential aspects of sustainability for civilian UAVs in India.
							Trustworthy	Secure	1. Should maintain the safety, security and privacy rights of the people while deploying the drone.

(Continued)

APPENDIX 6.A Continued

No	Author	Journal	Title	Year	Aspect	Innovation	Characteristic	Keyword	Finding
								Transparent	1. The transparency and traceability of data provided by civil UAVs further make them accountable and entwine the values of responsibility in the overall civilian UAV innovations.
							Well governed	Accountable	
42	Rivard et al.	*BMJ Innovations*	Double burden or single duty to care? Health innovators' perspectives on environmental considerations in health innovation design.	2019	Healthcare	Healthcare technology	Aligned	Sustainable	1. Taking the environment into consideration is part of responsible practice in health innovation to foster environmentally friendly health innovations, which realise supporting patient care while reducing environmental impacts.
43	Hemphill	*Journal of Responsible Innovation*	'Techlash', responsible innovation, and the self-regulatory organisation.	2019	Digital media	Information and communication technologies	Well governed	Regulated	1. Implement a regulatory regime to address policy concerns of privacy, public safety, and national security. 2. Except for the traditional approach (government regulation), self-regulation and public regulation can be well-reasoned alternatives.

No.	Author	Journal	Title	Year	Field	Technology			Notes
44	Stemerding et al.	*Futures*	Future making and responsible governance of innovation in synthetic biology.	2019	Biology	Synthetic technology	Well governed	Regulated	1. Need to foster and facilitate innovation via more generic institutional, regulatory, and pricing measures.
								Participatory	1. Stakeholder engagement enhanced reflexivity about the different needs and interests that should be considered in shaping the innovation agenda.
45	Samuel and Prainsack	*New Genetics and Society*	Forensic DNA phenotyping in Europe: Views 'on the ground' from those who have a professional stake in the technology.	2019	Medical science	DNA phenotyping	Accessible	Adaptable	1. Had to meet two criteria: be valid and reliable; be ethically unproblematic.
							Acceptable	Ethical	2. It is important only to use tests that are deemed ethically 'safe'.
							Well governed	Participatory	1. Must engage both professional and public stakeholder views regarding future policy decisions.
46	Rose and Chilvers	*Frontiers in Sustainable Food Systems*	Agriculture 4.0: Broadening responsible innovation in an era of smart farming.	2018	Agriculture	Smart farming technology	Accessible	Inclusive	1. Broadening notions of 'inclusion' that open up to wider 'ecologies of participation', which change public opinion to accept technologies rather than making technological trajectories more responsive to the needs of society.

(Continued)

APPENDIX 6.A   Continued

No	Author	Journal	Title	Year	Aspect	Innovation	Characteristic	Keyword	Finding
47	Koirala, et al.	*Applied Energy*	Community energy storage: A responsible innovation towards a sustainable energy system?	2018	Energy	Community energy storage	Aligned	Meaningful	1. Provide higher flexibility as well as accommodate the needs and expectations of citizens and local communities.
								Sustainable	1. Guarantee socially and technologically acceptable transformation towards an inclusive and sustainable energy system.
							Accessible	Inclusive	1. Allow stakeholders to express their values and design operational criteria to respect and include them.
								Affordable	1. Decentralised markets for flexibility, ease of market participation, and community empowerment are expected to create better conditions for its implementation.

						Well governed	Participatory
1. Enhance participation level to allow local communities to provide important feedback to the technology providers, which leads to higher acceptance and further technological innovation.
							Regulated
1. Flexible legislation programme for the experimentation and development of socio-technical models specific to the local, social, and physical conditions.
48	Winfield and Jirotka	*Physical and Engineering Sciences*	2018	Technology	Artificial intelligence	Ethical governance is essential to building trust in robotics and artificial intelligence systems.	Acceptable	Harmless
1. Do no harm, including being free of bias and deception.
							Ethical
1. Respect human rights and freedoms, including dignity and privacy, while promoting well-being.
						Trustworthy	Explainable
1. Should be explainable or even capable of explaining their own actions (to non-experts) and being transparent (to experts).

(Continued)

APPENDIX 6.A Continued

No	Author	Journal	Title	Year	Aspect	Innovation	Characteristic	Keyword	Finding
49	Pacifico Silva et al.	Health Research Policy and Systems	Introducing responsible innovation in health: a policy-oriented framework.	2018	Healthcare	Healthcare technology		Transparent	1. Be transparent and dependable while ensuring that the locus of responsibility and accountability remains with their human designers or operators.
							Well governed	Accountable	1
							Aligned	Sustainable	. Need to reduce, as much as possible, the negative environmental impacts of health innovations throughout their entire lifecycle.
							Accessible	Affordable	1. Deliver both high-performing products as well as affordable ones to support equity and sustainability.
							Acceptable	Equitable	1. Increase our ability to attend to collective needs whilst tackling health inequalities.
50	Sonck et al.	Life Sciences, Society and Policy	Creative tensions: mutual responsiveness adapted to private sector research and development.	2017	Business	Non-specific	Well governed	Participatory	1. Support mutually responsive relations in the innovation process, which assist innovators and stakeholders reach some form of joint understanding about how the innovation is shaped and eventually applied.

51	Boden et al.	*Connection Science*	Principles of robotics: regulating robots in the real world.	2017	Technology	Robotics	Acceptable	Harmless	1. Robots should not be designed solely or primarily to kill or harm humans, except for national security interests.
							Trustworthy	Secure	1. Robots should be designed using processes that assure their safety and security and make sure that the safety and security of robots in society are assured so that people can trust and have confidence in them.
								Transparent	1. Robots should not be designed in a deceptive way to exploit vulnerable users; instead, their machine nature should be transparent.
							Well governed	Regulated	1. Robots should be designed and operated as far as is practicable to comply with existing laws, fundamental rights, and freedoms, including privacy.

(Continued)

APPENDIX 6.A Continued

No	Author	Journal	Title	Year	Aspect	Innovation	Characteristic	Keyword	Finding
								Accountable	1. The person with legal responsibility for a robot should be attributed.
52	Leenes et al.	Law, Innovation and Technology	Regulatory challenges of robotics: some guidelines for addressing legal and ethical issues.	2017	Technology	Robotics	Well governed	Regulated	1. Develop a method, a framework, or guidelines that can be used to make innovation in a certain context more responsible. 2. Develop self-learning procedures that can be used to make innovation in a certain context more responsible.
53	Demers-Payette et al.	Journal of Responsible Innovation	Responsible research and innovation: a productive model for the future of medical innovation.	2016	Medical science	Healthcare technology	Aligned	Deliberate	1. Carefully anticipate the consequences and opportunities associated with medical innovations to generate a clear understanding of the uses of medical innovation and of its context.

54	Foley et al.	*Journal of Responsible Innovation*	2016	Technology	Non-specific	Towards an alignment of activities, aspirations and stakeholders for responsible innovation.			

Meaningful	1. Ensure potential innovations align with clinical and healthcare system challenges and needs to achieve a better alignment between health and innovation value systems and social practices.
Participatory	Well governed	1. Use formal deliberative mechanisms to provide a sustained engagement way for stakeholders in the innovation process.
Regulated		1. Flexible steering of innovation trajectories within a highly regulated environment without compromising the safety of new products.
Meaningful	Aligned	1. Support the creation of technologies that contribute to the stewardship of planetary systems identified.
Sustainable		1. Does not interfere with access to basic resources critical to a healthy human life.

(Continued)

APPENDIX 6.A Continued

No	Author	Journal	Title	Year	Aspect	Innovation	Characteristic	Keyword	Finding
							Acceptable	Ethical	1. Affords people freedom of expression and freedom from oppression and does not reinforce social orders that subjugate human beings.
								Equitable	1. Ensure that select groups of people are not inequitably burdened by negative impacts.
55	Dignum et al.	*Science and engineering Ethics*	Contested technologies and design for values: The case of shale gas.	2016	Energy	Shale gas technology	Well governed	Participatory	1. To create and implement a technological design, we must look beyond technology itself and iteratively include institutions and stakeholder interactions to acknowledge the complex and dynamic embedding of a (new) technology in a societal context.

56	Arentshorst et al.	*Technology in Society*	Exploring responsible innovation: Dutch public perceptions of the future of medical neuroimaging technology.	2016	Medical science	Neuroimaging technology	Accessible	Affordable	
							Trustworthy	Transparent	1. Freedom of choice, guaranteed privacy, the right to know or to be kept in ignorance, and informed consent should be self-evident prerequisites. 2. The acts, competencies, and knowledge of experts developing and working with neuroimaging can be trusted in terms of doing good and determining the correct treatment plan.
							Well governed	Participatory	1. Relevant actors need to become mutually responsive, and participants' concerns should be taken seriously in order to promote responsible embedding of neuroimaging.

1. Should never result in negative social or economic implications for individuals/patients.

(Continued)

APPENDIX 6.A Continued

No	Author	Journal	Title	Year	Aspect	Innovation	Characteristic	Keyword	Finding
57	Fisher et al.	*Journal of Responsible Innovation*	Mapping the integrative field: Taking stock of socio-technical collaborations.	2015	Technology	Non-specific	Accessible	Inclusive	1. Adopt more inclusive strategies to integrate wider stakeholders to align science, technology, and innovation more responsibly with their broader societal contexts.
58	Samanta and Samanta	*Journal of Medical Ethics*	Quackery or quality: the ethicolegal basis for a legislative framework for medical innovation.	2015	Healthcare	Healthcare technology	Trustworthy	Transparent Accountable	1. At the heart of the regulation of medical innovation is care delivered by a process that is accountable and transparent and that allows full consideration of all relevant matters.
							Well governed	Regulated	1. A combination of ethicolegal principles and statutory regulations would permit responsible medical innovation and maximise benefits in terms of therapy and patient-centred care.

59	Toft et al.	*Applied Energy*	Responsible technology acceptance: Model development and application to consumer acceptance of Smart Grid technology.	2014	Energy	Smart grid technology	Accessible	Adaptable	1. Acceptance of a new technology depends on believing that the technology is easy to use and useful for achieving a personal goal.
60	Wickson and Carew	*Journal of Responsible Innovation*	Quality criteria and indicators for responsible research and innovation: Learning from transdisciplinarity.	2014	Environment conservation	Nanotechnology	Aligned	Meaningful	1. Addressing a grand social challenge.
								Deliberate Sustainable	1. Generating a range of positive and negative future scenarios and identifying and assessing associated risks and benefits of these for social, environmental, and economic sustainability.
							Accessible	Inclusive	1. Openly and actively seeking ongoing critical input, feedback, and feed-forward from a range of stakeholders.

(Continued)

APPENDIX 6.A Continued

No	Author	Journal	Title	Year	Aspect	Innovation	Characteristic	Keyword	Finding
								Adaptable	1. Outcomes work reliably under real-world conditions. 2. Resources are carefully considered and allocated to efficiently achieve maximum utility and impact.
							Trustworthy	Transparent	1. Transparent identification of a range of uncertainties and limitations that may be relevant for various stakeholders.
								Explainable	1. Openly communicated lines of delegation and ownership able to respond to process dynamics and contextual change.
							Well governed	Regulated	1. Documented compliance with highest-level governance requirements, research ethics, and voluntary codes of conduct, which are all actively monitored throughout.

#	Author	Journal	Title	Year	Field					Description
61	Taebi et al.	*Journal of Responsible Innovation*	Responsible innovation as an endorsement of public values: The need for interdisciplinary research.	2014	Energy	Shale gas technology	Accessible	Inclusive	Accountable	1. Preparedness to accept accountability for both potentially positive and negative impacts. 1. Responsible innovation as an accommodation of public values, which requires undertaking interdisciplinary research and interaction between innovators and other stakeholders in conjunction with the early assessment of ethical and societal desirability.
62	Lauss et al.	*Biopreservation and Biobanking*	Towards biobank privacy regimes in responsible innovation societies: ESBB conference in Granada 2012.	2013	Biology	Biobank	Trustworthy Well governed	Secure Regulated		1. Biobank privacy regimes presuppose knowledge of and compliance with legal rules, professional standards of the biomedical community, and state-of-the-art data safety and security measures.

(Continued)

APPENDIX 6.A   Continued

No	Author	Journal	Title	Year	Aspect	Innovation	Characteristic	Keyword	Finding
63	Gaskell et al.	*European Journal of Human Genetics*	Publics and biobanks: Pan-European diversity and the challenge of responsible innovation.	2013	Biology	Biobank	Aligned	Deliberate	1. Lying behind European diversity is a number of common problems, issues, and concerns—many of which are not set in stone and can be addressed by informed and prudent actions on the part of biobank developers and researchers.
							Trustworthy	Secure	1. Assiduous mechanisms for the protection of privacy and personal data should be given careful consideration.
								Explainable	1. Need to consider how to explain to the public the rationale for cooperation with other actors that can help to increase people's trust.
							Well governed	Participatory	1. Stakeholder engagement relates to readiness to participate in biobank research and to agree to broad consent.

64	Van den Hove et al.	*Environmental Science & Policy*	The Innovation Union: a perfect means to confused ends?	2012	Technology	Non-specific	Aligned	Meaningful	1. Innovation should be re-targeted to deliver better health and well-being, improved quality of life, and sustainability.
							Accessible	Inclusive	1. A broader concept of innovation must be deployed, aiming to overcome technological and ideological lock-ins.
65	Stahl	*Journal of Information, Communication and Ethics in Society*	IT for a better future: how to integrate ethics, politics and innovation.	2011	Technology	Information and communication technologies	Acceptable	Ethical	1. Incorporate ethics into ICT research and development to engage in discussion of what constitutes ethical issues and be open to incorporation of gender, environmental, and other issues.
								Regulated	1. Provide a regulatory framework that will support ethical impact assessment for ICTs to proactively consider solutions to foreseeable problems that will likely arise from the application of future and emerging technologies.

(Continued)

APPENDIX 6.A   Continued

No	Author	Journal	Title	Year	Aspect	Innovation	Characteristic	Keyword	Finding
							Well governed	Participatory	1. To allow and encourage stakeholders to exchange ideas, to express their views, and to reach a consensus concerning good practices in the area of ethics and ICT.
								Accountable	1. Ensure that specific responsibility ascriptions are realised within technical work and further sensitise possible subjects of responsibility to some of the difficulties of discharging their responsibilities.

REFERENCES

1. Yigitcanlar, T.; Guaralda, M.; Taboada, M.; Pancholi, S. Place making for knowledge generation and innovation: Planning and branding Brisbane's knowledge community precincts. *J. Urban Technol.* 2016, 23, 115–146.
2. Yigitcanlar, T.; Sabatini-Marques, J.; da-Costa, E.; Kamruzzaman, M.; Ioppolo, G. Stimulating technological innovation through incentives: Perceptions of Australian and Brazilian firms. *Technol. Forecast. Soc. Change* 2019, 146, 403–412.
3. Santoni de Sio, F. The European Commission report on ethics of connected and automated vehicles and the future of ethics of transportation. *Ethics Inf. Technol.* 2021, 23, 713–726.
4. Eastwood, C.; Rue, B.; Edwards, J.; Jago, J. Responsible robotics design–A systems approach to developing design guides for robotics in pasture-grazed dairy farming. *Front. Robot. AI* 2022, 9, 914850.
5. Russell, A.; Stelmach, A.; Hartley, S.; Carter, L.; Raman, S. Opening up, closing down, or leaving ajar? How applications are used in engaging with publics about gene drive. *J. Responsible Innov.* 2022, 9, 151–172.
6. Sujan, M.; White, S.; Habli, I.; Reynolds, N. Stakeholder perceptions of the safety and assurance of artificial intelligence in healthcare. *Saf. Sci.* 2022, 155, 105870.
7. David, A.; Yigitcanlar, T.; Li, R.; Corchado, J.; Cheong, P.; Mossberger, K.; Mehmood, R. Understanding Local Government Digital Technology Adoption Strategies: A PRISMA Review. *Sustainability* 2023, 15, 9645.
8. Ocone, R. Ethics in engineering and the role of responsible technology. *Energy AI* 2020, 2, 100019.
9. Yigitcanlar, T., & Cugurullo, F. The sustainability of artificial intelligence: An urbanistic viewpoint from the lens of smart and sustainable cities. *Sustainability* 2020, 12, 8548.
10. Regona, M.; Yigitcanlar, T.; Xia, B.; Li, R. Opportunities and adoption challenges of AI in the construction industry: A PRISMA review. *J. Open Innov. Technol. Mark. Complex.* 2022, 8, 45.
11. Ribeiro, B.; Smith, R.; Millar, K. A mobilising concept? Unpacking academic representations of responsible research and innovation. *Sci. Eng. Ethics* 2017, 23, 81–103.
12. Bashynska, I.; Dyskina, A. The overview-analytical document of the international experience of building smart city. *Bus. Theory Pract.* 2018, 19, 228–241.
13. Thapa, R.; Iakovleva, T.; Foss, L. Responsible research and innovation: A systematic review of the literature and its applications to regional studies. *Eur. Plan. Stud.* 2019, 27, 2470–2490.
14. Stilgoe, J.; Owen, R.; Macnaghten, P. Developing a framework for responsible innovation. *Res. Policy* 2013, 42, 1568–1580.
15. Von Schomberg, R. A vision of responsible research and innovation. In *Responsible Innovation: Managing the Responsible Emergence of Science and Innovation in Society*; Owen, R., Bessant, J., Heintz, M., Eds.; Wiley: London, 2013; pp. 51–74.

16. Koops, B. The concepts, approaches, and applications of responsible innovation. In Bert-Jaap Koops; Ilse Oosterlaken; Henny Romijn; Tsjalling Swierstra; Jeroen van den Hoven (Eds.) *Responsible Innovation 2*; Springer: Cham, Switzerland, 2015; pp. 1–15.

17. Genus, A.; Iskandarova, M. Responsible innovation: Its institutionalisation and a critique. *Technol. Forecast. Soc. Change* 2018, 128, 1–9.

18. Dudek, M.; Bashynska, I.; Filyppova, S.; Yermak, S.; Cichoń, D. Methodology for assessment of inclusive social responsibility of the energy industry enterprises. *J. Clean. Prod.* 2023, 394, 136317.

19. Resnik, D. *The Ethics of Science: An Introduction*; Routledge: London, 2005.

20. Owen, R.; Macnaghten, P.; Stilgoe, J. Responsible research and innovation: From science in society to science for society, with society. *Sci. Public Policy* 2012, 39, 751–760.

21. Glerup, C.; Horst, M. Mapping 'social responsibility' in science. *J. Responsible Innov.* 2014, 1, 31–50.

22. Burget, M.; Bardone, E.; Pedaste, M. Definitions and conceptual dimensions of responsible research and innovation: A literature review. *Sci. Eng. Ethics* 2017, 23, 1–19.

23. Jakobsen, S.; Fløysand, A.; Overton, J. Expanding the field of Responsible Research and Innovation (RRI)–from responsible research to responsible innovation. *Eur. Plan. Stud.* 2019, 27, 2329–2343.

24. Wiarda, M.; van de Kaa, G.; Yaghmaei, E.; Doorn, N. A comprehensive appraisal of responsible research and innovation: From roots to leaves. *Technol. Forecast. Soc. Change* 2021, 172, 121053.

25. Liu, J.; Zhang, G.; Lv, X.; Li, J. Discovering the landscape and evolution of responsible research and innovation (RRI). *Sustainability* 2022, 14, 8944.

26. Von Schomberg, R. (2011). Research and Innovation in the Information and Communication Technologies and Security Technologies Fields: A Report from the European Commission Services; European Union, Publications Office of the European Union: Luxembourg.

27. De Saille, S. Innovating innovation policy: The emergence of 'Responsible Research and Innovation'. *J. Responsible Innov.* 2015, 2, 152–168.

28. Voegtlin, C.; Scherer, A. Responsible innovation and the innovation of responsibility: Governing sustainable development in a globalised world. *J. Bus. Ethics* 2017, 143, 227–243.

29. Loureiro, P.; Conceição, C.P. Emerging patterns in the academic literature on responsible research and innovation. *Technol. Soc.* 2019, 58, 101148.

30. Li, M.; Wan, Y.; Gao, J. What drives the ethical acceptance of deep synthesis applications? A fuzzy set qualitative comparative analysis. *Comput. Hum. Behav.* 2022, 133, 107286.

31. Singh, R., Mishra, S., & Tripathi, K. Analysing acceptability of E-rickshaw as a public transport innovation in Delhi: A responsible innovation perspective. *Technological Forecasting and Social Change* 2021, 170, 120908.

32. Hussain, W.; Perera, H.; Whittle, J.; Nurwidyantoro, A.; Hoda, R.; Shams, R.; Oliver, G. Human values in software engineering: Contrasting case studies of practice. *IEEE Trans. Softw. Eng.* 2020, 48, 1818–1833.

33. Merck, A.; Grieger, K.; Cuchiara, M.; Kuzma, J. What role does regulation play in responsible innovation of nanotechnology in food and agriculture? Insights and framings from U.S. *Stakeholders. Bull. Sci. Technol. Soc.* 2022, 42, 85–103.

34. Stankov, U.; Gretzel, U. Digital well-being in the tourism domain: Mapping new roles and responsibilities. *Inf. Technol. Tour.* 2021, 23, 5–17.

35. Koirala, B.; van Oost, E.; van der Windt, H. Community energy storage: A responsible innovation towards a sustainable energy system? *Appl. Energy* 2018, 231, 570–585.

36. Li, W.; Yigitcanlar, T.; Erol, I.; Liu, A. Motivations, barriers and risks of smart home adoption: From systematic literature review to conceptual framework. *Energy Res. Soc. Sci.* 2021, 80, 102211.

37. Li, F.; Yigitcanlar, T.; Nepal, M.; Thanh, K.; Dur, F. Understanding urban heat vulnerability assessment methods: A PRISMA review. *Energies* 2022, 15, 6998.

38. Pacifico Silva, H.; Lehoux, P.; Miller, F.; Denis, J. Introducing responsible innovation in health: A policy-oriented framework. *Health Res. Policy Syst.* 2018, 16, 90.

39. Ulnicane, I.; Eke, D.; Knight, W.; Ogoh, G.; Stahl, B. Good governance as a response to discontents? Déjà vu, or lessons for AI from other emerging technologies. *Interdiscip. Sci. Rev.* 2021, 46, 71–93.

40. Sayers, A. Tips and tricks in performing a systematic review. *Br. J. Gen. Pract.* 2008, 58, 136.

41. Wohlin, C.; Kalinowski, M.; Felizardo, K.; Mendes, E. Successful combination of database search and snowballing for identification of primary studies in systematic literature studies. *Inf. Softw. Technol.* 2022, 147, 106908.

42. Yigitcanlar, T.; Desouza, K.; Butler, L.; Roozkhosh, F. Contributions and risks of artificial intelligence (AI) in building smarter cities: Insights from a systematic review of the literature. *Energies* 2020, 13, 1473.

43. Buhmann, A.; Fieseler, C. Towards a deliberative framework for responsible innovation in artificial intelligence. *Technol. Soc.* 2021, 64, 101475.

44. Kokotovich, A.; Kuzma, J.; Cummings, C.; Grieger, K. Responsible innovation definitions, practices, and motivations from nanotechnology researchers in food and agriculture. *NanoEthics* 2021, 15, 229–243.

45. Owen, R.; von Schomberg, R.; Macnaghten, P. An unfinished journey? Reflections on a decade of responsible research and innovation. *J. Responsible Innov.* 2021, 8, 217–233.

46. Yigitcanlar, T.; Agdas, D.; Degirmenci, K. Artificial intelligence in local governments: Perceptions of city managers on prospects, constraints and choices. *AI Soc.* 2023, 38, 1135–1150.

47. Owen, R.; Stilgoe, J.; Macnaghten, P.; Gorman, M.; Fisher, E.; Guston, D. A framework for responsible innovation. In Richard Owen; John Bessant; Maggy Heintz (Eds.) *Responsible Innovation: Managing the Responsible Emergence of Science and Innovation in Society*; John Wiley & Sons: Hoboken, NJ, 2013; Volume 31, pp. 27–50.

48. Stilgoe, J.; Guston, D. *Responsible Research and Innovation*; MIT Press: Cambridge, MA, 2016.

49. Bacq, S.; Aguilera, R. Stakeholder governance for responsible innovation: A theory of value creation, appropriation, and distribution. *J. Manag. Stud.* 2022, 59, 29–60.

50. Winfield, A.; Jirotka, M. Ethical governance is essential to building trust in robotics and artificial intelligence systems. *Philos. Trans. R. Soc. A Math. Phys. Eng. Sci.* 2018, 376, 20180085.

51. Bunnik, E.; Bolt, I. Exploring the Ethics of Implementation of Epigenomics Technologies in Cancer Screening: A Focus Group Study. *Epigenetics Insights* 2021, 14, 25168657211063618.

52. Samuel, G.; Roberts, S.L.; Fiske, A.; Lucivero, F.; McLennan, S.; Phillips, A.; Hayes, S.; Johnson, S.B. COVID-19 contact tracing apps: UK public perceptions. *Crit. Public Health* 2022, 32, 31–43.

53. Foley, R.; Bernstein, M.; Wiek, A. Towards an alignment of activities, aspirations and stakeholders for responsible innovation. *J. Responsible Innov.* 2016, 3, 209–232.

54. Stahl, B. IT for a better future: How to integrate ethics, politics and innovation. *J. Inf., Commun. Ethics in Soc.* 2011, 9, 140–156.

55. Stitzlein, C.; Fielke, S.; Waldner, F.; Sanderson, T. Reputational risk associated with big data research and development: An interdisciplinary perspective. *Sustainability* 2021, 13, 9280.

56. Middelveld, S.; Macnaghten, P. Gene editing of livestock: Sociotechnical imaginaries of scientists and breeding companies in the Netherlands. *Elem. Sci. Anthr.* 2021, 9, 00073.

57. Ten Holter, C.; Inglesant, P.; Jirotka, M. Reading the road: Challenges and opportunities on the path to responsible innovation in quantum computing. *Technol. Anal. Strateg. Manag.* 2021, 35, 844–856.

58. Brandao, M.; Jirotka, M.; Webb, H.; Luff, P. Fair navigation planning: A resource for characterizing and designing fairness in mobile robots. *Artif. Intell.* 2020, 282, 103259.

59. Grieger, K.; Merck, A.; Cuchiara, M.; Binder, A.; Kokotovich, A.; Cummings, C.; Kuzma, J. Responsible innovation of nano-agrifoods: Insights and views from US stakeholders. *NanoImpact* 2021, 24, 100365.

60. Boden, M., Bryson, J., Caldwell, D., Dautenhahn, K., Edwards, L., Kember, S., ... & Winfield, A. Principles of robotics: regulating robots in the real world. *Connect. Sci.* 2017, 29(2), 124–129.

61. Arbolino, R.; Carlucci, F.; De Simone, L.; Ioppolo, G.; Yigitcanlar, T. The policy diffusion of environmental performance in the European countries. *Ecol. Indic.* 2018, 89, 130–138.

62. Pickering, B. Trust, but Verify: Informed Consent, AI Technologies, and Public Health Emergencies. *Future Internet* 2021, 13, 132.

63. Wickson, F.; Carew, A. Quality criteria and indicators for responsible research and innovation: Learning from transdisciplinarity. *J. Responsible Innov.* 2014, 1, 254–273.

64. Samuel, G.; Prainsack, B. Forensic DNA phenotyping in Europe: Views "on the ground" from those who have a professional stake in the technology. *New Genet. Soc.* 2019, 38, 119–141.

65. Arentshorst, M.; de Cock Buning, T.; Broerse, J. Exploring responsible innovation: Dutch public perceptions of the future of medical neuroimaging technology. *Technol. Soc.* 2016, 45, 8–18.

66. Yigitcanlar, T.; Corchado, J.; Mehmood, R.; Li, R.; Mossberger, K.; Desouza, K. Responsible urban innovation with local government artificial intelligence (AI): A conceptual framework and research agenda. *J. Open Innov. Technol. Mark. Complex.* 2021, 7, 71.

67. Chamuah, A.; Singh, R. Securing sustainability in Indian agriculture through civilian UAV: A responsible innovation perspective. *SN Appl. Sci.* 2020, 2, 106.

68. Fisher, E.; O'Rourke, M.; Evans, R.; Kennedy, E.; Gorman, M.; Seager, T. Mapping the integrative field: Taking stock of socio-technical collaborations. *J. Responsible Innov.* 2015, 2, 39–61.

69. Macdonald, J.M.; Robinson, C.J.; Perry, J.; Lee, M.; Barrowei, R.; Coleman, B.; Markham, J.; Barrowei, A.; Markham, B.; Ford, H.; et al. Indigenous-led responsible innovation: Lessons from co-developed protocols to guide the use of drones to monitor a biocultural landscape in Kakadu National Park, Australia. *J. Responsible Innov.* 2021, 8, 300–319.

70. Taebi, B.; Correlje, A.; Cuppen, E.; Dignum, M.; Pesch, U. Responsible innovation as an endorsement of public values: The need for interdisciplinary research. *J. Responsible Innov.* 2014, 1, 118–124.

71. Geoghegan-Quinn, M. Commissioner Geoghegan-Quinn Keynote Speech at the "Science in Dialogue" Conference. In *Proceedings of the Science in Dialogue Conference*, Odense, Denmark, 23–25 April 2012.

72. Middelveld, S.; Macnaghten, P.; Meijboom, F. Imagined futures for livestock gene editing: Public engagement in the Netherlands. *Public Underst. Sci.* 2022, 32, 143–158.

73. Demers-Payette, O.; Lehoux, P.; Daudelin, G. Responsible research and innovation: A productive model for the future of medical innovation. *J. Responsible Innov.* 2016, 3, 188–208.

74. Gaskell, G.; Gottweis, H.; Starkbaum, J.; Gerber, M.M.; Broerse, J.; Gottweis, U.; Hobbs, A.; Helén, I.; Paschou, M.; Snell, K.; Soulier, A. Publics and biobanks: Pan-European diversity and the challenge of responsible innovation. *Eur. J. Hum. Genet.* 2013, 21, 14–20.

75. Van den Hove, S.; McGlade, J.; Mottet, P.; Depledge, M. The Innovation Union: A perfect means to confused ends? *Environ. Sci. Policy* 2012, 16, 73–80.

76. Mladenović; M.; Haavisto, N. Interpretative flexibility and conflicts in the emergence of Mobility as a Service: Finnish public sector actor perspectives. *Case Stud. Transp. Policy* 2021, 9, 851–859.

77. Stahl, B. Responsible innovation ecosystems: Ethical implications of the application of the ecosystem concept to artificial intelligence. *Int. J. Inf. Manag.* 2022, 62, 102441.

78. Rivard, L.; Lehoux, P.; Miller, F. Double burden or single duty to care? Health innovators' perspectives on environmental considerations in health innovation design. *BMJ Innov.* 2019, 6, 4–9.

79. Asveld, L.; Ganzevles, J.; Osseweijer, P. Trustworthiness and responsible research and innovation: The case of the bio-economy. *J. Agric. Environ. Ethics* 2015, 28, 571–588.

80. Papenbrock, J. Explainable, Trustworthy, and Responsible AI for the Financial Service Industry. *Front. Artif. Intell.* 2022, 5, 902519.

81. Rochel, J.; Evéquoz, F. Getting into the engine room: A blueprint to investigate the shadowy steps of AI ethics. *AI Soc.* 2021, 36, 609–622.

82. Akintoye, S.; Ogoh, G.; Krokida, Z.; Nnadi, J.; Eke, D. Understanding the perceptions of UK COVID-19 contact tracing app in the BAME community in Leicester. *J. Inf. Commun. Ethics Soc.* 2021, 19, 521–536.

83. Ienca, M.; Fins, J.J.; Jox, R.J.; Jotterand, F.; Voeneky, S.; Andorno, R.; Ball, T.; Castelluccia, C.; Chavarriaga, R.; Chneiweiss, H.; Ferretti, A. Towards a Governance Framework for Brain Data. *Neuroethics* 2022, 15, 20.

84. Lockwood, F. Bristol's smart city agenda: Vision, strategy, challenges and implementation. *IET Smart Cities* 2020, 2, 208–214.

85. Yigitcanlar, T.; Mehmood, R.; Corchado, J. Green artificial intelligence: Towards an efficient, sustainable and equitable technology for smart cities and futures. *Sustainability* 2021, 13, 8952.

86. Oldeweme, A.; Märtins, J.; Westmattelmann, D.; Schewe, G. The role of transparency, trust, and social influence on uncertainty reduction in times of pandemics: Empirical study on the adoption of COVID-19 tracing apps. *J. Med. Internet Res.* 2021, 23, e25893.

87. MacDonald, E.; Neff, M.; Edwards, E.; Medvecky, F.; Balanovic, J. Conservation pest control with new technologies: Public perceptions. *J. R. Soc. New Zealand* 2022, 52, 95–107.

88. Samanta, J.; Samanta, A. Quackery or quality: The ethicolegal basis for a legislative framework for medical innovation. *J. Med. Ethics* 2015, 41, 474–477.

89. Hemphill, T. 'Techlash', responsible innovation, and the self-regulatory organization. *J. Responsible Innov.* 2019, 6, 240–247.

90. Mecacci, G.; Santoni de Sio, F. Meaningful human control as reason-responsiveness: The case of dual-mode vehicles. *Ethics Inf. Technol.* 2020, 22, 103–115.

91. Chamuah, A.; Singh, R. Responsibly regulating the civilian unmanned aerial vehicle deployment in India and Japan. *Aircr. Eng. Aerosp. Technol.* 2021, 93, 629–641.

92. Bao, L.; Krause, N.M.; Calice, M.N.; Scheufele, D.A.; Wirz, C.D.; Brossard, D.; Newman, T.P.; Xenos, M.A. Whose AI? How different publics think about AI and its social impacts. *Comput. Hum. Behav.* 2022, 130, 107182.

93. Rose, D.; Wheeler, R.; Winter, M.; Lobley, M.; Chivers, C. Agriculture 4.0: Making it work for people, production, and the planet. *Land Use Policy* 2021, 100, 104933.

94. Townsend, L.; Noble, C. Variable rate precision farming and advisory services in Scotland: Supporting responsible digital innovation? *Sociol. Rural.* 2022, 62, 212–230.

95. Lehoux, P.; Silva, H.; Rocha de Oliveira, R.; Sabio, R.; Malas, K. Responsible innovation in health and health system sustainability: Insights from health innovators' views and practices. *Health Serv. Manag. Res.* 2022, 35, 196–205.

96. Christodoulou, E.; Iordanou, K. Democracy under attack: Challenges of addressing ethical issues of AI and big data for more democratic digital media and societies. *Front. Political Sci.* 2021, 3, 682945.

97. Foley, R.; Sylvain, O.; Foster, S. Innovation and equality: An approach to constructing a community governed network commons. *J. Responsible Innov.* 2022, 9, 49–73.

98. Stemerding, D.; Betten, W.; Rerimassie, V.; Robaey, Z.; Kupper, F. Future making and responsible governance of innovation in synthetic biology. *Futures* 2019, 109, 213–226.

99. Iakovleva, T.; Oftedal, E.; Bessant, J. Changing Role of Users—Innovating Responsibly in Digital Health. *Sustainability* 2021, 13, 1616.

100. Dignum, M.; Correljé; A.; Cuppen, E.; Pesch, U.; Taebi, B. Contested technologies and design for values: The case of shale gas. *Sci. Eng. Ethics* 2016, 22, 1171–1191.

101. Sonck, M.; Asveld, L.; Landeweerd, L.; Osseweijer, P. Creative tensions: Mutual responsiveness adapted to private sector research and development. *Life Sci. Soc. Policy* 2017, 13, 14.

102. Lauss, G.; Schröder, C.; Dabrock, P.; Eder, J.; Hamacher, K.; Kuhn, K.; Gottweis, H. Towards biobank privacy regimes in responsible innovation societies: ESBB conference in Granada 2012. *Biopreserv. Biobanking* 2013, 11, 319–323.

103. Leenes, R.; Palmerini, E.; Koops, B.; Bertolini, A.; Salvini, P.; Lucivero, F. Regulatory challenges of robotics: Some guidelines for addressing legal and ethical issues. *Law Innov. Technol.* 2017, 9, 1–44.

104. Westwood, R.; Low, D. The multicultural muse: Culture, creativity and innovation. *Int. J. Cross Cult. Manag.* 2003, 3, 235–259.

105. Zhu, B.; Habisch, A.; Thøgersen, J. The importance of cultural values and trust for innovation: A European study. *Int. J. Innov. Manag.* 2018, 22, 1850017.

106. Kim, S.; Parboteeah, K.; Cullen, J.; Liu, W. Disruptive innovation and national cultures: Enhancing effects of regulations in emerging markets. *J. Eng. Technol. Manag.* 2020, 57, 101586.

107. Setiawan, A. The influence of national culture on responsible innovation: A case of CO2 utilisation in Indonesia. *Technol. Soc.* 2020, 62, 101306.

CHAPTER 7

Assessment Framework of Responsible Urban Artificial Intelligence

1 INTRODUCTION

Against the backdrop of ever-changing and exponentially advancing science and technology, along with digital transformation pressures on our institutions, many people are worried about the undesired effects of unparalleled technological developments [1,2]. These technological advances are not only shaping our cities to be digital and smart but also significantly affecting our society [3–5]. While the advancements are undeniably ground-breaking, particularly developments in artificial intelligence (AI) field, they introduce multiple potential futures and uncertainties [6–9].

The inherent complexity of innovation practices has sparked broader discussions about their alignment with societal values, ethical standards, and collective aspirations, especially concerning disruptive innovations and technologies [10–12]. As advanced digital technologies increasingly permeate into our daily lives, it is imperative to ensure that their development trajectories address not only economic or functional objectives but also uphold societal values, demonstrate ethical integrity, and promote long-term sustainability [13–15].

Amid growing concerns regarding aligning technological achievements with societal needs, the dialogue around 'responsible innovation and technology' (RIT)—that is an umbrella term inclusive of 'Responsible

DOI: 10.1201/9781003521440-9

Artificial Intelligence'—has gained increasing attention recently [16–18]. Such dialogue emphasises that modern innovation practice should deliver responsible technological outcomes and ensure these outcomes can be rationally integrated into our cities and societies, thereby mitigating potential repercussions of contemporary technological leaps [19–22]. As the major actors in innovation practices, technology companies, especially those with global influence, find themselves at the nexus of this transformative dialogue [23,24]. They realise that adopting a responsible approach in the innovation process is not only beneficial for society but also crucial for maintaining a positive brand image and securing long-term success in an ever-changing market [25–27]. As stakeholders demand greater accountability and consumers seek out ethically sound products and services, the RIT concept has become a key driver in guiding corporates' future technological development direction [10,28–31].

Although the interest of academia and the policy community in RIT is growing, and some effort has been put into RIT practices, research from the business community's perspective remains limited [10,32]. Existing research and discussion regarding 'responsibility' in the business setting are more concerned with widely recognised concepts such as 'corporate social responsibility' (CSR) and 'corporate sustainability' (CS) [23–25,33]. However, research focused on corporates' specific insights into the RIT concept remains relatively underexplored. Although RIT, CSR, and CS are all concepts that can be used to address the ethical, social, and environmental responsibilities of businesses, each concept has its unique emphasis.

RIT is distinct in its focus on innovation practices, ensuring that new products, services, or technologies are developed and introduced responsibly [32,34]. On the other hand, CSR and CS have broader scopes, addressing the wider social and environmental responsibilities and sustainability of businesses, respectively [35–38]. As RIT practices continue to be launched by leading high-tech companies, additional investigations and reviews are needed to capture the growing knowledge on RIT from corporations' perspective. This will help bridge the research gap and may contribute to making the concept more responsive and adaptable to real-world scenarios.

In addressing this gap, the chapter at hand conducts a comprehensive policy review to investigate insights into RIT from the top 100 global high-tech companies (ranked by market capitalisation)—so-called tech giants. This study aims to clarify companies' guidance on integrating the RIT concept into their innovation practices. Accordingly, the following research question was posed in this chapter: How do technology companies guide their products and services to respond to the concept of RIT? The findings of this chapter contribute to a richer understanding of how these highly innovative and successful technology companies interpret and enact the RIT concept in their innovative pursuits. By synthesising insights from technology companies, this chapter provides valuable insights for academic and industry stakeholders as well as social stakeholders, e.g., customers and local communities, to practically foster more responsible and socially beneficial technological outcomes.

This chapter's key contribution is that it develops an indicative framework based on corporate insights and established viewpoints from the policy community and academia. Unlike previous approaches, this covers a comprehensive range of considerations essential for RIT. This framework provides a systematic approach to holistically evaluate technological outcomes against a broad spectrum of responsible innovation principles. By transforming an abstract concept into specific objectives and statements, the framework operationalises the goals of RIT, making them more actionable and measurable within organisations.

2 LITERATURE BACKGROUND

The dialogue around RIT originated in the European research and innovation (R&I) policy—the Horizon 2020 framework programme [39,40]. Since this concept earned a prominent position in policy spheres, the policy community has begun to emphasise that the modern innovation process should be open, interactive, and transparent [13,41,42]. This approach enables social actors to participate in the innovation process and share the responsibilities with innovation actors to shape RIT, aiming to ensure the technological outcomes can both meet societal needs and address the potential or unanticipated social impacts and challenges that accompany them [16,19,43,44]. To this end, European Commission officer Rene von Schomberg preliminarily proposed specific criteria to elucidate the somewhat abstract notion of RIT, which encompasses 'ethically acceptable', 'sustainable', and 'societal desirable' [44].

On this basis, Li et al. [22] (p. 1) further identified the key characteristics of RIT from an academic perspective, which they defined as 'acceptable', 'accessible', 'aligned', 'trustworthy', and 'well governed'. The authors emphasised that "RIT should deliver acceptable, accessible, trustworthy, and well governed technological outcomes, while ensuring these outcomes are aligned with societal desirability and human values and should also be responsibly integrated into our cities and societies". Moreover, to clarify these characteristics in greater detail, the authors provided detailed descriptions for each characteristic. For example, they explained that the 'acceptable' characteristic covers the 'ethical', 'equitable', and 'harmless' aspects. Their study provides a conceptual framework of RIT design and implementation, assisting innovation actors, policymakers, and social stakeholders to ensure emerging innovations and technologies are more 'responsible'.

It is noteworthy that commercialisation is a pivotal phase in innovation, and most innovations are founded and produced by the private sector [23,45]. With growing interest in the theme of 'being responsible' in technospheres, some influential technology companies have embraced the RIT concept to respond to societal needs and challenges by creating responsible products and services [24,28,46]. These companies have established various principles based on their own perspectives, missions, or specific business backgrounds to help guide their product and services to respond to the concept of RIT, e.g., Microsoft's Responsible AI, Amazon's Responsible Use of Machine Learning, Samsung's Responsible Production and Sales, and Atlassian's Responsible Technology.

RIT in a business context refers to the emerging concept where businesses actively consider the broader societal, ethical, and environmental implications of their innovations and technological deployments [10,23,47]. Li et al. [24] and Jarmai [48] posited that the impetus for corporates to adopt a responsible approach in their innovation process primarily stems from both internal factors, such as company vision and the pursuits of key individuals, and external factors, including market demand, policy pressure, and the expectations of civil society. Some studies indicated that incorporating the RIT concept into business strategies contributes to leading long-term benefits, including enhanced brand reputation, increased trust among consumers and stakeholders, and a more sustainable and resilient business model [27,49–51].

Nonetheless, Boenink and Kudina [15] pointed out that the interpretations of RIT characteristics are not universal and eternal but vary by region, time, or other specific factors. The characteristics appreciated by academia and the policy community may be disliked or overlooked by industry or consumer groups, and vice versa [10,23].

Against this backdrop, this chapter undertakes a policy review supported by the previous research efforts. The aim of this study is to glean complementary insights about RIT from a corporate perspective and to integrate existing views from the high-tech policy community and academia, ultimately attempting to formulate a comprehensive and implementable RIT framework.

3 RESEARCH DESIGN

This chapter concentrates on addressing the following research question: How do technology companies guide their products and services to respond to the concept of RIT? To address this, this study conducts a policy analysis, defined by the Centers for Disease Control and Prevention (CDC) [52] as "the process of identifying potential policy options that could address the problem and then comparing those options to choose the most effective, efficient, and feasible ones".

To conduct a policy analysis, this study adopts a qualitative and quantitative thematic analysis of corporate documents using the NVivo software (v.14). Content analysis of corporate documents has been highly praised in previous academic papers as a promising approach to understanding corporate practices [53]. Despite NVivo being renowned for analysing the contents of interviews, increasing studies have used it as a tool for content analysis of documentation, e.g., CSR-related reporting documents [53], strategy documents of local governments [54], and healthcare planning documents [55].

Different from previous content analysis studies, this chapter applies a targeted search strategy to identify RIT-related documents that are publicly accessible and written in English by searching via the internal search engine of corporations' official websites. In addition, for those websites without internal search engines or displaying ambiguous RIT-related content, we conducted complementary search tasks through the Google search engine to ensure the comprehensiveness of our database. In some search results, content that does not provide clear information, leaves room for multiple interpretations, omits crucial details necessary for a comprehensive understanding, makes broad statements without specifics, or presents information without the necessary context is identified as ambiguous content.

The list of technology companies selected for this purpose was sourced from CompaniesMarketCap [56], which offers a daily updated global ranking of technology companies by market capitalisation (https://companies marketcap.com/tech/largest-tech-companies-by-market-cap (accessed on 1 December 2023)). The reliability of this website has been verified by numerous organisations in the public and private sectors, and the data provided have been incorporated into their official reports [57–59]. The

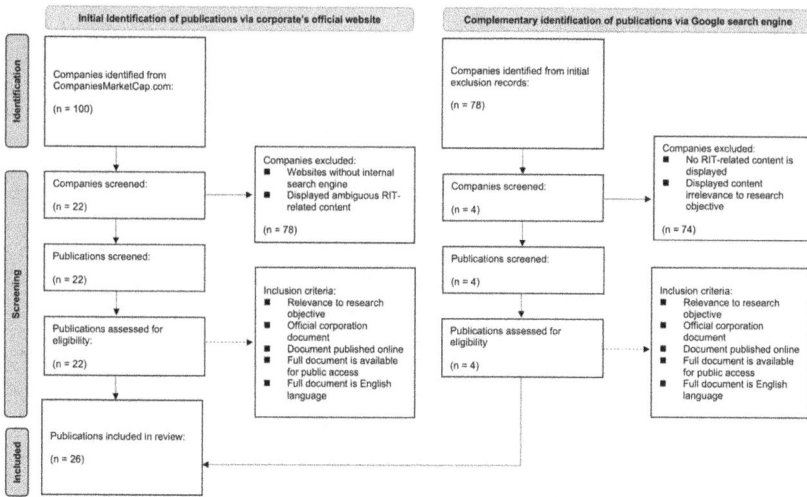

FIGURE 7.1   Policy document selection process.

query string of the search task included a fuzzy format '*' to assure further the comprehensiveness of the obtained data, which was determined as follows: ('innovation' OR 'technolog*') AND ('responsible' OR 'ethic*' OR 'explainable' OR 'trust*' OR 'transparen*'). Figure 7.1 shows the specifics of the corporate document selection process.

The search results show that, out of the top 100 global technology companies by market capitalisation, 26 companies provide RIT-related documents that are publicly accessible online. These documents include corporate reporting, website content, official blog posts, and internal policies and guidelines. Table 7.1 provides a brief profile of these companies.

After identifying companies that have directly included RIT-related documents on their websites, we downloaded and saved all the documents ($n = 26$) that were highly relevant to our research objectives from their official websites and uploaded them to NVivo. To accurately address the research question, we established five nodes (acceptability goals, accessibility goals, alignment goals, trustworthiness goals, and well-governance goals) and 15 sub-nodes drawing from Li et al. [22], who conducted a specific analysis of the corporate's RIT guidelines. In addition, to ensure the conceptual framework developed by Li et al. [22] is suitable for this study, we made necessary improvements to the node establishment work, e.g., modifying titles and keywords and clarifying specific content categories. Table 7.2 lists the nodes and the associated sub-nodes for each analysis task.

TABLE 7.1   Profiles of the case companies

Company	Region	Profile
Microsoft	USA	Microsoft is a technology company specialising in software, hardware, cloud services, and digital solutions, driving innovation in numerous sectors, from computing to business applications.
Alphabet (Google)	USA	Alphabet is the parent company of Google, focusing on search, advertising, cloud computing, AI, and digital services, with ventures in healthcare, autonomous vehicles, and other technological innovations.
Amazon	USA	Amazon is an e-commerce company providing cloud services via AWS, streaming with Prime Video, and branching into AI, devices, and retail, driving transformative consumer and business solutions.
NVIDIA	USA	NVIDIA is a technology company renowned for its graphics processing units (GPUs) for gaming, also venturing into AI, deep learning, automotive AI solutions, and data centre advancements.
Meta (Facebook)	USA	Meta Platforms, formerly Facebook, focuses on social media services, augmented and virtual reality, advertising, and digital communication tools, aspiring to build a comprehensive metaverse for global users.
Samsung	Republic of Korea	Samsung is an electronics company specialising in smartphones, TVs, semiconductors, and home appliances, while also venturing into software, digital services, and cutting-edge technology innovations.
Oracle	USA	Oracle is a technology company specialising in database software, cloud solutions, enterprise software products, and hardware systems, serving businesses with integrated technology stacks.
Adobe	USA	Adobe is a software company known for creative and multimedia solutions, digital marketing tools, and document management, driving digital content creation and optimisation across platforms.
Salesforce	USA	Salesforce is a cloud-based software company specialising in customer relationship management (CRM) solutions and offering a suite of enterprise applications for marketing, sales, service, and analytics.
Cisco	USA	Cisco is a technology company focusing on networking hardware, software, telecommunications equipment, and cybersecurity solutions, enabling seamless connectivity and digital transformation for businesses.

(Continued)

TABLE 7.1 Continued

Company	Region	Profile
Intel	USA	Intel is a semiconductor manufacturer specialising in microprocessors, chipsets, and integrated solutions, driving advancements in computing, data centres, AI, and broader technology ecosystems.
Qualcomm	USA	Qualcomm is a semiconductor manufacturer specialising in wireless technology innovations, designing chips for smartphones, and pioneering advances in 5G, IoT, and AI across various platforms.
IBM	USA	IBM is a technology company focusing on cloud computing, AI, enterprise software, and hardware, offering integrative business solutions and IT consultancy.
Sony	Japan	Sony is an electronics company specialising in electronics, entertainment, gaming (PlayStation), music, film production, and professional broadcasting solutions, driving innovation in media and consumer technologies.
Schneider Electric	France	Schneider Electric is a global specialist in energy management and automation, offering solutions for homes, buildings, data centres, infrastructure, and industries, driving sustainable and integrated efficiency.
Automatic Data Processing	USA	Automatic Data Processing (ADP) is a global provider specialising in human capital management solutions, offering payroll, tax, HR services, and analytics to businesses of varying sizes.
Airbnb	USA	Airbnb is a global online platform connecting travellers with hosts, specialising in unique accommodations, experiences, and evolving into travel services, redefining how people experience new destinations.
Equinix	USA	Equinix is a technology company specialising in data centre services, connecting businesses to their customers and partners inside interconnected data centres, driving digital business performance through platform solutions.
VMware	USA	VMware is a software company specialising in cloud infrastructure, virtualisation, networking, security, and digital workspace technology, empowering businesses with integrated IT solutions for modern computing.
Workday	USA	Workday is a cloud-based software provider focusing on human capital management, financial management, and enterprise planning, offering adaptive solutions for business insights and growth.

(Continued)

TABLE 7.1 Continued

Company	Region	Profile
Baidu	China	Baidu is a technology company specialising in internet services, AI research, autonomous driving, and digital advertising, often referred to as China's premier search engine platform.
NXP Semiconductors	The Netherlands	NXP Semiconductors is a technology company specialising in secure connectivity solutions for embedded applications, driving innovations in automotive, industrial, and IoT markets.
Atlassian	Australia	Atlassian is a software company providing collaboration and productivity tools for teams, including Jira, Confluence, and Bitbucket, serving developers and businesses to enhance workflow and project management.
Dell	USA	Dell is a technology company specialising in personal computers, servers, storage solutions, and network devices, also offering software and IT services to drive digital transformation for businesses.
Xiaomi	China	Xiaomi is an electronics company, known for smartphones, smart home devices, and IoT products, emphasising innovative technology, design, and cost-effective solutions for a connected lifestyle.
Palantir	USA	Palantir is a software company specialising in big data analytics, offering platforms for data integration, decision-making, and operational intelligence, serving government agencies and private sectors.

TABLE 7.2 Coding of the corporate document data

Node	Sub-Node
Acceptability Goals	Equitability considerations, Ethics considerations, Harmlessness considerations
Accessibility Goals	Adaptability considerations, Affordability considerations, Inclusiveness considerations
Alignment Goals	Deliberateness considerations, Meaningfulness considerations, Sustainability considerations
Trustworthiness Goals	Explainability considerations, Security considerations, Transparency considerations
Well-governance Goals	Accountability considerations, Participation considerations, Regulatory considerations

4 ANALYSIS AND RESULTS

4.1 Quantitative Content Analysis

Corporate document data underwent rigorous analysis using NVivo (v.14). Word clouds visualised word frequency, with larger font sizes indicating a higher recurrence in corporate documents, enabling a concise representation of key thematic areas. Figure 7.2 displays word frequency across 26 corporations' RIT guidance documents, whereas Figure 7.3 showcases word frequency within the specifically coded segments. The visualisation results revealed that corporations' RIT guidelines predominantly pertain to AI. This emphasis likely stems from concerns regarding AI's profound potential to transform various sectors [60], coupled with ethical dilemmas related to 'black box' decision-making [61], inherent biases during model training [62], and data privacy issues regarding model construction [63].

Given AI's vast potential and the range of associated challenges, these leading technology companies are endeavouring to embrace RIT concepts to navigate the transformative impact of AI technology in our cities and societies [64,65]. For instance, these companies emphasise enhancing the explainability and transparency of AI systems [66–68]. They also prioritise embedding human-in-the-loop mechanisms, striking a balance between efficiency and ethical considerations [69–71]. Furthermore, there's a concerted effort to integrate robust privacy and security measures into AI-driven practices and applications [72–74]. These views also coincide with those emphasised by the academic community [75–77]. Notably, keywords linked to these measures frequently appear in their RIT guidelines, as illustrated in Figure 7.3, e.g., explainable, privacy, security, transparency, control, oversight, and others.

Table 7.3 lists the analysed nodes, sub-nodes, number of sub-nodes mentioned within the guidance documents, sub-node frequency, and total frequencies. Figure 7.4 presents a hierarchy chart to visualise the information and data in Table 7.4. Within the same colour setting, larger rectangles (nodes) represent major themes. Smaller rectangles (sub-nodes) within larger rectangles (nodes) represent sub-categories of the larger theme. The size of each rectangle indicates the volume of instances coded to that node, with larger rectangles indicating a greater volume of instances.

In Figure 7.4, there are five nodes representing five primary areas of focus for technology companies: trustworthiness goals, well-governance goals, alignment goals, acceptability goals, and accessibility goals. For each node, there are sub-nodes that represent specific considerations, e.g.,

TABLE 7.3 List of nodes, sub-nodes, and mention frequencies

Node	Sub-Node	Sub-Nodes Mentioned in Policy Documents	Frequency of Sub-Node	Total Frequency Sub-Node
Acceptability Goals	Equitability considerations	20	20	51
	Ethics considerations	18	24	
	Harmlessness considerations	7	7	
Accessibility Goals	Adaptability considerations	11	14	31
	Affordability considerations	2	2	
	Inclusiveness considerations	13	15	
Alignment Goals	Deliberateness considerations	8	9	33
	Meaningfulness considerations	13	18	
	Sustainability considerations	6	6	
Trustworthiness Goals	Explainability considerations	15	15	65
	Security considerations	22	33	
	Transparency considerations	15	17	
Well-governance Goals	Accountability considerations	13	16	33
	Participation considerations	8	8	
	Regulatory considerations	9	9	

security, transparency, and explainability within trustworthiness goals. Of the five nodes, the two with the highest total sub-node frequencies were 'trustworthiness goals' ($n = 65$) and 'acceptability goals' ($n = 51$). The other three nodes had nearly identical levels of total sub-node frequencies: 'alignment goals' ($n = 33$), 'well-governance goals' ($n = 33$), and 'accessibility goals' ($n = 31$). Among the 15 sub-nodes, 'security considerations' ($n = 33$) were most frequently mentioned, followed by 'ethics considerations' ($n = 24$). These findings suggest that these technology companies may

TABLE 7.4   Objectives of equitability considerations

Equitability Considerations	
Objectives	**Statements**
OBJ 1 Avoid bias	Does not create, reinforce, or propagate harmful or unfair biases in all stages of innovation and technology practice, from design to deployment and beyond.
OBJ 2 Guard against discrimination	Upholds the rights of all individuals and groups, embraces the full spectrum of social diversity, and actively prevents any form of discrimination.
OBJ 3 Strive for fairness	Proactively identifies and eliminates obstacles to ensure fair treatment for all and empower every individual equally through innovation and technology.

FIGURE 7.2   Word frequency in document content.

FIGURE 7.3   Word frequency in the coding.

place greater emphasis on security and ethics matters in their RIT guidance formulation. The reason might be that these companies want their products and solutions to be trustworthy so as not to jeopardise their own business while gaining more public trust and acceptance.

Additionally, they may be willing to adhere to ethical practices within the nature of the company [78–80]. In addition, external factors might also be one of the reasons why technology companies emphasise these aspects when formulating guidance documents, e.g., public concerns and sentiment [81], regulatory pressure [82], and ethical considerations [83]. Significantly, 'affordability considerations' ($n = 2$) are mentioned notably less than other individual sub-nodes (average $n = 14$). This reflects that there may be differing priorities among the policy community, academia, and industry in developing RIT guidelines based on their respective perspectives [10,23].

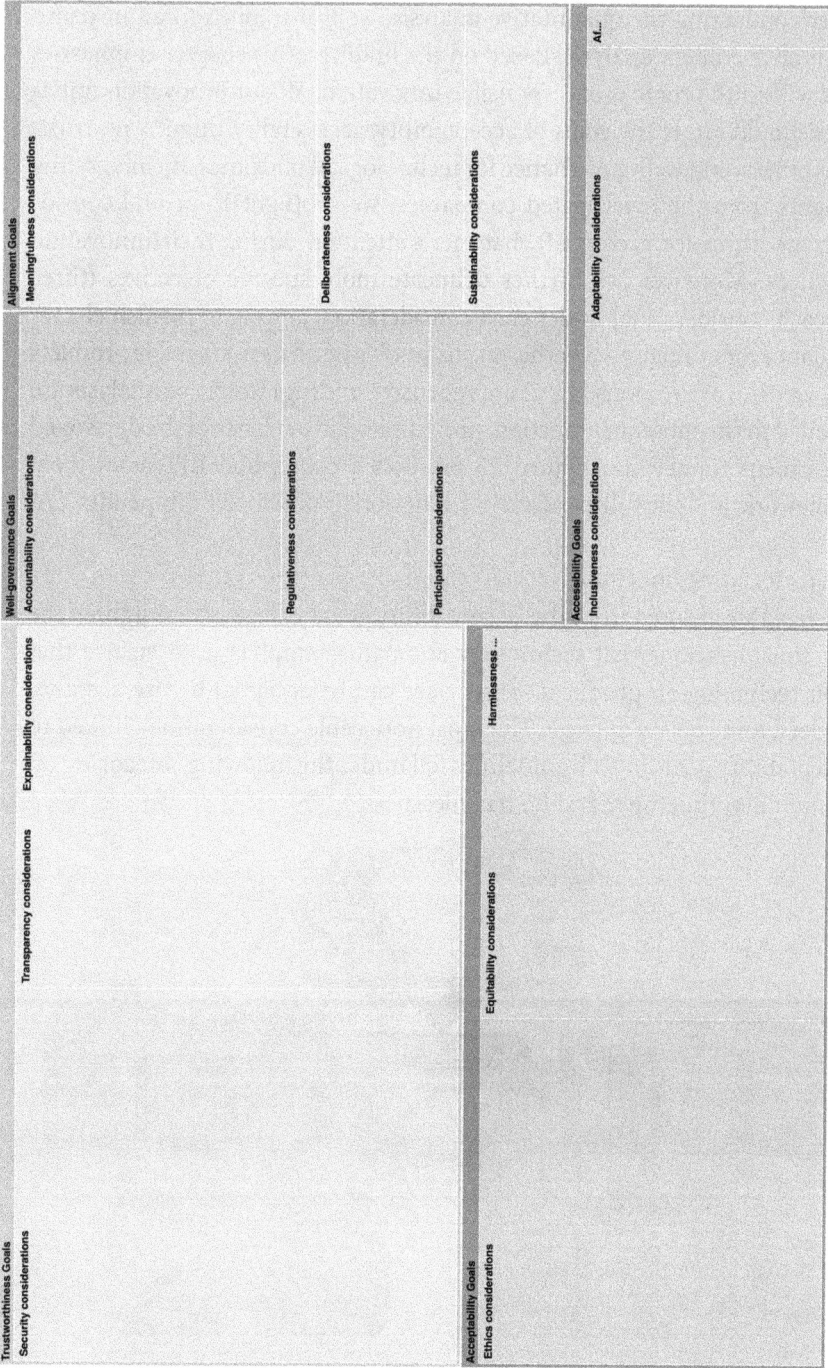

FIGURE 7.4 Hierarchy of nodes and sub-nodes.

4.2 Qualitative Content Analysis

After conducting the quantitative analysis, we transitioned to an in-depth qualitative content analysis. Based on the findings, this chapter emphasises that within the context of responsible innovation, all our innovation initiatives should target the goals of acceptability, accessibility, alignment, trustworthiness, and well-governance for technological outcomes. By integrating insights from the investigated companies, we spotlight 15 pivotal considerations (three for each goal) that merit attention during their innovation practices. Moreover, we further delineate more specific objectives (three for each consideration) under each consideration, aiming to furnish clearer guidance for evaluating whether the technology and its marketable products and services are responsible. Comprehensive findings from this analysis are detailed in the subsequent section, providing solid evidence of the developed assessment framework. Figure 7.5 presents a conceptual RIT assessment framework, and the full assessment framework is provided in Appendix 7.A.

4.2.1 Acceptability Goals

The insights gleaned from the 'acceptability goals' category highlighted the key considerations that technology companies emphasise to ensure that their technologies, products, or services can be accepted by users, stakeholders, and society at large. The most noticeable considerations linked to acceptability goals in RIT guidelines fell under the following categories: (a) equitability; (b) ethics; and (c) harmlessness.

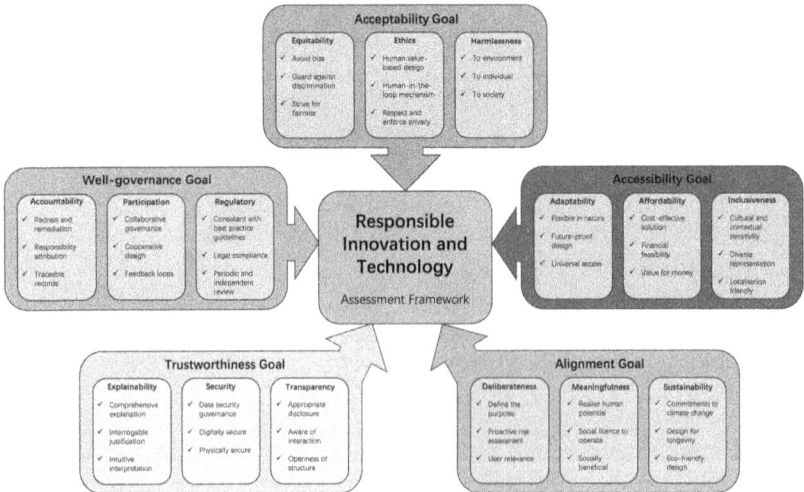

FIGURE 7.5   Conceptual RIT assessment framework.

4.2.1.1 Equitability Considerations

Oracle [68] indicated that AI systems can potentially reinforce or even exacerbate existing biases inherent in their training data. Such a magnification of biases can result in discriminatory effects, potentially restricting some individuals from accessing specific opportunities or services:

> AI systems can perpetuate and even amplify biases present in the data used to train them. This bias can lead to discriminatory outcomes, such as denying certain individuals access to opportunities or services.

Adobe [84] highlighted the necessity for meticulous attention to bias, especially when a product or service could profoundly influence facets of an individual's life:

> We understand that special care must be taken to address bias if a product or service will have a significant impact on an individual's life, such as with employment, housing, credit, and health.

Google [74] emphasised the importance of preventing adverse effects on individuals from AI applications, especially those tied to sensitive attributes:

> Avoiding unjust impacts on people, particularly those related to sensitive characteristics such as race, ethnicity, gender, nationality, income, sexual orientation, ability and political or religious belief.

Moreover, Sony [85] stated that when using AI, the diversity and human rights of all stakeholders should be upheld while ensuring non-discrimination:

> In its utilization of AI, Sony will respect diversity and human rights of its customers and other stakeholders without any discrimination while striving to contribute to the resolution of social problems through its activities in its own and related industries.

And VMware [86] advocated for the fair treatment of all individuals by AI, irrespective of race, gender, disabilities, income, or any other diversity markers:

> It is critical to invest stringent effort to identify such bias to avoid unfair and improper behavior from AI systems. Regardless of race,

gender, disabilities, income, and any other indicator of diversity, all people should be treated fairly by AI systems.

Atlassian [87] espoused similar views and expressed a desire for organisations and technologies to embody openness, inclusivity, fairness, and justice, mirroring human-centric values and inherent basic human rights:

> We want our company and our technologies to be open, inclusive, fair and just: to reflect the human-centric values and fundamental human rights that we all share.

Thus, this chapter emphasises that 'technology and its marketable products and services' (collectively referred to as 'TMPSs' in this chapter) should neither introduce nor amplify harmful biases, from design to deployment. TMPSs should uphold the rights of all individuals, respect social diversity, counter discrimination, and proactively eliminate barriers to guarantee fair treatment and empowerment for everyone. Table 7.4 presents our recommendations for specific objectives and statements within the context of equitability considerations.

4.2.1.2 Ethics Considerations

Oracle [68] stated that with the progression of AI technology, there is a possibility these systems could function autonomously and make independent decisions. Such advancements usher in concerns about accountability for these systems' actions and the need to guarantee their alignment with human ethics and values:

> As AI systems become more advanced, they may be able to operate independently and make decisions on their own. This potential development raises questions about who is responsible for the actions of these systems and how to ensure they align with human values.

Moreover, Cisco [73] highlighted that AI applications frequently utilise personal data, which could pose threats to individual privacy and civil rights if not handled appropriately:

> Applications of AI often use personal data that could impact individual privacy and civil liberties if not managed properly.

Further, NXP Semiconductors [70] emphasised the crucial role of ethics in AI system design and development. They pointed out that an approach driven solely by technological advancement might overlook human needs:

As designers and developers of AI systems, it is an imperative to understand the ethical considerations of our work. A technology-centric focus that solely revolves around improving the capabilities of an intelligent system doesn't sufficiently consider human needs. By empowering our designers and developers to make ethical decisions throughout all development stages, we ensure that they never work in a vacuum and always stay in tune with users' needs and concerns.

Schneider Electric [88] denoted that adhering to ethics and compliance is the foremost principle in the AI and data science development process. This can be achieved by adopting a human-centred approach and a vigilant oversight mechanism:

> The number one rule we apply when developing AI and data science is ethics and compliance in line with our Trust Charter. We leverage digital technologies for a sustainable future based on human-centered design with a 'do no harm' oversight.

IBM [66] emphasised that users should maintain control over their data and their applications. They explain that it is the duty of innovation teams to ensure users remain in control of their data and interactions. As AI technology advances, organisations must uphold ethical standards, using AI to enhance, not diminish, individual privacy:

> "Preserve and fortify users' power over their own data and its uses". "It's your team's responsibility to keep users empowered with control over their interactions and data". "Organizations have a responsibility to use AI ethically as the technology matures. AI should be used to amplify our privacy, rather than undermine it".

NXP semiconductors [70] summarised their stance on AI, stating that AI systems must respect human autonomy and ensure individuals are not subjugated or coerced by them. While there is an acknowledgement of humans ceding some decision-making to machines, the balance of power should always favour humans. It is essential to retain control and ensure AI aligns with privacy standards:

> AI systems should preserve the autonomy of human beings and warrant freedom from subordination to—or coercion by—AI systems. The conscious act to employ AI and its smart agency, while ceding some of our decision-making power to machines, should

TABLE 7.5   Objectives of ethics considerations

Ethics Considerations		
	Objectives	**Statements**
OBJ 4	Human value-based design	Prioritises human values and morals in the innovation process, ensuring that technological outcomes meet functional requirements and align with broader ethical and societal norms.
OBJ 5	Human-in-the-loop mechanism	Embeds appropriate human oversight and intervention into decision-making processes to balance efficiency with ethical considerations.
OBJ 6	Respect and enforce privacy	Incorporates privacy principles at every stage of the innovation lifecycle, ensuring that innovation and technological outcomes consistently prioritise and protect user privacy.

always remain under human control, so as to achieve a balance between the decision-making power we retain for ourselves and that which we delegate to artificial agents as well as ensure compliance with privacy principles.

Workday [71] suggested the adoption of a human-in-the-loop strategy, ensuring that end-users have control over the final decisions:

> We focus on improving and developing people's capabilities and experiences and leverage a 'human-in-the-loop' approach to enable end-user control over ultimate decisions.

Thus, this chapter emphasises that TMPSs should prioritise human values and morals during their design process, embed human oversight in decision-making to balance efficiency with ethics, and consistently incorporate privacy principles throughout their entire lifecycle to safeguard user privacy. Table 7.5 presents our recommendations for specific objectives and statements within the context of ethics considerations.

4.2.1.3 Harmlessness Considerations

NVIDIA [89] highlighted the importance of minimising potential harm arising from the deployment of AI models or systems. For example, the proliferation of AI in various sectors can inadvertently erode essential human skills due to increased automation, potentially leaving individuals ill-equipped without AI assistance. This surge in AI-driven changes can also catalyse economic and social disruptions, with industries transforming rapidly and possibly igniting social unrest if not navigated judiciously:

NVIDIA aims to reduce the risk of harm from deployment of AI models or systems.

Google [74] espoused similar views and has committed to actions based on their core principle of being harmless to limit applications that might be harmful or abusive:

> Be made available for uses that accord with these principles: We will work to limit potentially harmful or abusive applications.

From an environmental perspective, Qualcomm [90] places a premium on environmental concerns. Not only do they have specific objectives for reducing water withdrawal, but they also emphasise sustainable product design. Their aim is to mitigate risks to the environment and fortify the resilience of their supply chains:

> We want our products to be distinguished not only by their capabilities but also by the care and attention we put into producing them.

From a societal perspective, Oracle [68] stated that AI has the potential to automate numerous tasks, leading to job loss issues. Consequently, there are rising concerns about the support and measures required for workers and communities impacted by these technological shifts:

> AI can automate many tasks and processes, which can lead to job displacement. This displacement raises concerns about how to support workers and communities affected by these changes.

Hence, Xiaomi [72] emphasised that their goal is to offer users AI products and services that are both safe and dependable, ensuring no harm comes to society:

> Xiaomi is firmly dedicated to ensuring security and safety throughout development and application of trustworthy AI technologies, providing users with safe and trustworthy AI products and services and making sure that our trustworthy AI will not do any harm to society.

In addition, NXP Semiconductors [70] indicated that AI systems must be designed to prioritise the well-being of humans, ensuring they do not jeopardise individuals, whether in societal or work environments:

> AI systems should not harm human beings. By design, AI building blocks should protect the dignity, integrity, liberty, privacy, safety,

TABLE 7.6   Objectives of harmlessness considerations

Harmlessness Considerations	
Objectives	**Statements**
OBJ 7 Harmless to environment	Ensures respect and protection of the environment, avoiding practices that lead to environmental degradation or the corruption of natural resources.
OBJ 8 Harmless to individual	Does not lead to any harm against any individuals, including physical and psychological harm.
OBJ 9 Harmless to society	Minimises adverse impacts on society and commits to the harmonious progress of technology and society.

and security of human beings in society and at work. Human well-being should be the desired outcome in all system designs.

Thus, this chapter emphasises that TMPSs should ensure respect and protection of the environment, prevent any harm to individuals, both physical and psychological, and guarantee that negative impacts on society are minimised while fostering harmony between technology and societal progress. Table 7.6 presents our recommendations for specific objectives and statements within the context of harmlessness considerations.

4.2.2 Accessibility Goals
The insights gleaned from the 'accessibility goals' category highlight the key considerations that technology companies emphasise to ensure that their technologies, products, or services are accessible to as many people as possible, making sure no one is left behind in the digital age. The most noticeable considerations linked to accessibility goals in RIT guidelines fell under the following categories: (a) adaptability; (b) affordability; and (c) inclusiveness.

4.2.2.1 Adaptability Considerations
Microsoft [69] indicated that it is necessary to ensure that systems operate consistently and as designed, not just in controlled lab settings but also in real-world scenarios, especially when faced with evolving challenges:

> A reliable system functions consistently and as intended, not only in the lab conditions in which it is trained, but also in the open world or when they are under attack from adversaries.

VMware [86] underscored the necessity of a component-based design of AI-driven systems in response to the contingencies of use:

> Where appropriate, AI-powered systems should have control mechanisms to allow human operators to deactivate the AI component without affecting business continuity.

In addition, NVIDIA [89] highlighted the technical flexibility of their products and provided various programmes and tools to allow developers to create and accelerate applications for specific purposes and industries:

> Our products are programmable and general purpose in nature.

Palantir [91] also highlighted this point and presented the benefits of their product's interoperable and extensible architecture:

> Foundry's interoperable and extensible architecture has enabled data science teams worldwide to readily collaborate with their business and operational teams, enabling all stakeholders to create data-driven impact.

Furthermore, Samsung [92] put forward that their pursuit of technological innovation aims to provide equal and easy access for all users:

> We seek technological innovation to allow equal and convenient access to our products and services by all consumers. We apply the 4C Accessibility Design Principles when developing our products and services.

Thus, we underline that TMPSs should allow operating seamlessly with other systems for compatibility and integration, remain adaptable to evolving socio-technical challenges while serving their intended purposes throughout their lifecycle, and ensure accessibility for everyone. Table 7.7 presents our recommendations for specific objectives and statements within the context of adaptability considerations.

4.2.2.2 Affordability Considerations

The investigated technology companies seldom discussed the affordability of technology and provided only a few broad and ambiguous statements on this consideration. For example, Samsung [92] contended:

> We seek technological innovation to allow equal and convenient access to our products and services by all consumers.

TABLE 7.7   Objectives of adaptability considerations

Adaptability Considerations		
Objectives		**Statements**
OBJ 10	Flexible in nature	Allows individual modification or replacement and can seamlessly operate with other systems to ensure compatibility and ease of integration.
OBJ 11	Future-proof design	Remains adaptable in the face of evolving socio-technical challenges and paradigms and continues to serve intended purposes throughout their lifecycle.
OBJ 12	Universal access	Ensures accessibility for all individuals, irrespective of differences in physical ability, technological, cognitive, or actual usage context.

And Atlassian [87] claimed:

> We work for social and environmental progress in whatever we do, which includes a commitment to respect human rights; to invest in the diversity, equity, and inclusion of our teams and larger ecosystem; and to make Atlassian products and experiences fully accessible and usable for everyone.

Nevertheless, we contend that thoroughly considering affordability is crucial to broaden the notions of technology accessibility, especially from a user-friendly perspective [93–95]. Therefore, this chapter underscores that TMPSs should provide cost-effective solutions to diminish economic disparities in adoption. Technology outcome providers should attempt to decrease costs to facilitate access for wider population segments without financial setbacks and offer genuine value. This ensures that users receive benefits that justify the expenses, thus promoting lasting adoption and utility. Table 7.8 presents our recommendations for specific objectives and statements within the context of affordability considerations.

4.2.2.3 Inclusiveness Considerations

Intel [67] signifies the importance of equity, inclusion, and cultural awareness in AI development and deployment:

> We believe there is a need for equity, inclusion, and cultural sensitivity in the development and deployment of AI. We strive to ensure that the teams working on these technologies are diverse and inclusive. We believe that the AI technology domain should be developed and informed by diverse populations, perspectives, voices, and experiences.

TABLE 7.8   Objectives of affordability considerations

Affordability Considerations	
Objectives	Statements
OBJ 13 Cost-effective solution	Provides cost-effective solutions or alternatives to reduce economic disparities in technology adoption.
OBJ 14 Financial feasibility	Reduces costs to allow broad population segments to access and benefit from innovation and technological advances without negative financial implications.
OBJ 15 Value for money	Offers genuine value, ensuring that users receive meaningful benefits or solutions that justify the costs, promoting long-term adoption and utility.

IBM [66] indicates that unintended outcomes could manifest in both AI system algorithms and the data used for their training and testing. For instance, algorithmic bias might stem from the influences of cultural, social, or institutional norms. While it is challenging to eradicate all these consequences entirely, IBM believes it is imperative for a responsible team to embrace diverse perspectives. By ensuring a comprehensive collection of diverse and representative data and experiences, these influences on AI systems can be minimised:

> Diverse teams help to represent a wider variation of experiences.
> Embrace team members of different ages, ethnicities, genders, educational disciplines, and cultural perspectives.
> Although bias can never be fully eliminated, it is the role of a responsible team to minimize algorithmic bias through ongoing research and responsible data collection representative of a diverse population.

Salesforce [96] believes that AI ought to enhance the human condition and embody the values of everyone affected, not just its developers:

> AI should improve the human condition and represent the values of all those impacted, not just the creators. We will advance diversity, promote equality, and foster equity through AI.

VMware [86] states that integrating social diversity and inclusiveness into the innovation process can lead to better and more responsible results, particularly in AI practices:

> Diversity and inclusiveness in society result in teams that generate better outcomes—including in the practice of AI.

TABLE 7.9 Objectives of inclusiveness considerations

Inclusiveness Considerations	
Objectives	**Statements**
OBJ 16 Cultural and contextual sensitivity	Respects shared values while remaining sensitive and attuned to the nuances of diverse cultural norms and contexts.
OBJ 17 Diverse representation	Ensures marginalised or underrepresented groups are included and have representation in innovation and technology practices.
OBJ 18 Localisation-friendly	Localises to match local customs, languages, and preferences, ensuring relevance and acceptance in different geo-cultural contexts.

Automatic data processing [97] espouses similar views and believes that diverse teams are crucial for designing and developing AI applications. This diversity guarantees the inclusion of a broad spectrum of perspectives and experiences. Given that AI applications influence humans, it is vital that human experiences shape their impacts:

> We are committed to having diverse teams design and develop our ML models, to ensure a wide variety of perspectives and experience are considered. After all, ML models impact humans, and human experience should inform that impact.

Thus, this chapter emphasises that TMPSs should respect shared values and be attuned to diverse cultural nuances, ensuring the inclusion and representation of marginalised or underrepresented groups in innovation practices. Additionally, we consider that technology localisation is essential to ensure that TMPSs respect and align with local customs, languages, and preferences, maintaining relevance in various geo-cultural contexts. Table 7.9 presents our recommendations for specific objectives and statements within the context of inclusiveness considerations.

4.2.3 Alignment Goals

The insights gleaned from the 'alignment goals' category highlight as a key consideration that technology companies aim to ensure that their policies include the objective that technologies, products, or services are aligned with societal desirability and preferences as well as common human values to deliver positive outcomes. The most noticeable considerations linked to alignment goals in RIT guidelines fell under the following categories: (a) deliberateness; (b) meaningfulness; and (c) sustainability.

4.2.3.1 Deliberateness Considerations

Amazon [98] indicated that AI applications span a vast range of use cases, each with its unique goals, user demographics, and potential outcomes. While certain applications might pose minimal risks, others could have profound implications, especially when they affect human rights or safety. It is crucial for developers to weigh both the benefits and potential hazards specific to their AI application:

> There are a wide variety of use cases that may incorporate ML, with different goals, characteristics, user bases, and potential impacts. Developers should consider the benefits and potential risks of their specific use case. Given the broad nature and applicability of ML, many applications may pose limited or no risk (e.g., movie recommendation systems), while others could involve significant risk, especially if used in a way that impacts human rights or safety.

Further, Google [74] signified that their design approach for AI applications is rooted in caution and aligned with the best practices in AI safety research:

> Designed to be appropriately cautious and in accordance with best practices in AI safety research, including testing in constrained environments and monitoring as appropriate.

In addition, Adobe [84] declared they would commit to designing and sustaining their AI applications through a thorough evaluation process, always being mindful of the potential impacts and consequences that arise from their deployment:

> We will approach designing and maintaining our AI technology with thoughtful evaluation and careful consideration of the impact and consequences of its deployment.

Vmware [86] espoused similar views and further indicated the importance of ensuring that AI systems operate as intended. They advocated that meticulous planning and deliberation are paramount in developing AI systems to ensure they function accurately and consistently in line with their designers' expectations:

> "It is critical to take steps to ensure that AI systems function according to their design purpose". "Careful forethought is needed

TABLE 7.10   Objectives of deliberateness considerations

Deliberateness Considerations	
Objectives	**Statements**
OBJ 19	Define the purpose	Before taking any initiative or action, clarifies and articulates the intended purpose to ensure alignment with overall goals and values.
OBJ 20	Proactive risk assessment	Anticipates potential negative outcomes, challenges, or pitfalls, and designs strategies to mitigate or avoid them.
OBJ 21	User relevance	Gains a deep understanding of the users' needs, preferences, and challenges to ensure the solutions provided are directly relevant to and satisfy the end-user's needs.

to develop AI systems that accurately and consistently operate in accordance with their designers' expectations".

Workday [71] promised that they are both thoughtful and deliberate in their AI development, committing to producing AI solutions strictly in alignment with the core human values they recognise and respect:

> We're thoughtful and deliberate in our approach at Workday, and we only develop AI solutions that align with our values.

Thus, this chapter underlines that TMPSs should be designed with a clearly defined intended purpose consistent with overall goals and values. Developers should anticipate and strategise against potential adverse outcomes while deeply understanding users' needs and preferences to ensure relevant and effective solutions. Table 7.10 presents our recommendations for specific objectives and statements within the context of deliberateness considerations.

4.2.3.2   Meaningfulness Considerations

Sony [85] is aware of the profound effects that AI-driven products and services can have on society. They are committed to actively contributing to AI development with the goal of fostering a better society:

> Sony will be cognizant of the effects and impact of products and services that utilize AI on society and will proactively work to contribute to developing AI to create a better society and foster human talent capable of shaping our collective bright future through R&D and/or utilization of AI.

Baidu [99] believes that the true essence of AI lies in its potential to enable human growth and learning, rather than to outdo or replace humanity.

Their ultimate vision for AI is to usher in greater freedom and myriad opportunities for humankind:

> The value of AI is to empower mankind to learn and grow instead of surpassing and replacing mankind; the ultimate ideal of AI is to bring more freedom and possibilities to humankind.

Salesforce [96] espouses similar views and indicates that AI technology is truly useful when combined with human capabilities:

> We believe AI is best utilized when paired with human ability, augmenting people, and enabling them to make better decisions. We aspire to create technology that empowers everyone to be more productive and drive greater impact within their organizations.

Workday [71] declares that their AI design primarily aims to assist customers and their employees in unlocking potential and concentrating on impactful tasks. Their offerings are committed to enhancing human decision-making and empowering users to choose whether to follow the suggestions given by their AI-driven solutions:

> We design AI to help our customers and their employees unlock opportunities and focus on meaningful work. Our solutions support human decision-making, improve experiences, and put users in control to decide whether to accept the recommendations provided by our AI-based solutions.

Sony [85] believes that promoting the harmonious integration of AI into society can not only bolster individual empowerment but also contribute to sustainable societal progress:

> Through advancing its AI-related R&D and promoting the utilization of AI in a manner harmonized with society, Sony aims to support the exploration of the potential for each individual to empower their lives, and to contribute to enrichment of our culture and push our civilization forward by providing novel and creative types of Kando. Sony will engage in sustainable social development and endeavor to utilize the power of AI for contributing to global problem-solving and for the development of a peaceful and sustainable society.

In addition, Airbnb [100] offers another perspective regarding how technology supports local communities:

> We help create new sources of income for Hosts sharing their existing spaces and skills, making it possible to empower them

TABLE 7.11   Objectives of meaningfulness considerations

	Meaningfulness Considerations	
Objectives		**Statements**
OBJ 22	Realise human potential	Unlocks humanity's potential, empowering individuals to address complex challenges effectively with increased capability through complementary collaboration with technology.
OBJ 23	Social license to operate	Strives for continuous community recognition, emphasising legitimacy, trust, and ethical alignment beyond mere regulatory compliance.
OBJ 24	Socially beneficial	Promotes human welfare for both current and future generations, fostering growth, prosperity, and positive societal outcomes.

financially while fostering connection with people from around the world and supporting local communities in the process.

Atlassian [87] indicates that by responsibly and purposefully employing these technologies, they can promote even more positive impact within our communities:

> We know that behind every great human achievement, there is a team. We also believe that new technologies can help empower those teams to achieve even more. If we use these technologies (like AI) responsibly and intentionally, then we can supercharge this vision and contribute to better outcomes across our communities.

Thus, this chapter advocates that TMPSs should aim to unlock humanity's potential to tackle complex challenges and champion human welfare across generations, ensuring growth, prosperity, and positive societal impact. Moreover, we contend that securing continuous community recognition is essential for the harmonious integration of TMPSs into society. Table 7.11 presents our recommendations for specific objectives and statements within the context of meaningfulness considerations.

4.2.3.3 Sustainability Considerations

Based on these results, we found that the theme of sustainability is not elaborated upon in the guidance documents focusing on RIT. However, it is thoroughly discussed in corporate annual reports such as the Environmental, Social, and Governance (ESG) Report, Corporate Responsibility Report, and Sustainability Report. This may be because technology companies believe that achieving sustainability goals should be an overall commitment or vision for the entire company, rather than being limited to specific technologies, products, or services. Nevertheless, some companies do include brief discussions on sustainability considerations related to specific technologies and products in their guidance documents.

For example, NVIDIA [89] promoted a central objective throughout their research, development, and design stages, which is to enhance performance while optimising energy efficiency. Their goal for each successive generation of products is to outperform and be more energy-efficient than the last. Additionally, they claimed that their technology plays a pivotal role in pioneering advancements in climate modelling, carbon emission reductions, and the development of strategies to adapt to and mitigate the effects of global changes:

> Improving performance and energy efficiency is a principal goal in each step of our research, development, and design processes. We aim to make every new generation of GPUs faster and more energy efficient than its predecessor. And our technology is driving some of the most important advances for modelling our climate, reducing carbon emissions, and designing mitigation and adaptation strategies in a changing world.

Furthermore, Vmware [86] signified that the potential environmental impact of AI applications needs to be evaluated. The creation and use of AI technologies should not only be consistent with, but also reinforce, the company's ESG objectives:

> AI systems should be assessed regarding their impact on the environment. The development and consumption of AI technologies should align with and support the company's ESG goals.

Based on additional findings from the corporations' annual reporting documents, this chapter emphasises that TMPSs should target supporting global climate change adaptation and mitigation, adopt an eco-friendly design approach, and prioritise the longevity of technological outcomes in the design and production process. Table 7.12 presents our recommendations for specific objectives and statements within the context of sustainability considerations.

4.2.4 Trustworthiness Goals

The insights gleaned from the 'trustworthiness goals' category highlighted as key considerations that technology companies aim to ensure that their policies include the objective that their products or services can establish

TABLE 7.12   Objectives of sustainability considerations

Sustainability Considerations		
	Objectives	Statements
OBJ 25	Commitments to climate change	Targets a reduction in greenhouse gas emissions and prevention of waste generation, supporting global efforts to adapt to and mitigate climate change.
OBJ 26	Design for longevity	Ensures resources invested in creating products or solutions provide value over the long term and contribute to more sustainable and resilient societies.
OBJ 27	Eco-friendly design	Incorporates environmental considerations starting from the design phase to ensure products or solutions are eco-friendly throughout their lifecycle.

trust among stakeholders, end-users, and the wider public. The most noticeable considerations linked to trustworthiness goals in RIT guidelines fell under the following categories: (a) explainability; (b) security; and (c) transparency.

4.2.4.1 Explainability Considerations

Oracle [68] stated the complexity of AI can hinder explainability, posing challenges in sectors where accountability is crucial:

> AI systems can be difficult to understand, which can make it challenging to explain their decisions and assess their performance, for example, a medical-diagnosis AI system that can't explain its decision-making process, or a criminal-risk-assessment AI system that has a high rate of false positives for certain demographic groups.

Microsoft [69] signified that improving the explainability of AI helps to meet the challenges and also enhance user trust and product usability:

> Intelligibility can uncover potential sources of unfairness, help users decide how much trust to place in a system, and generally lead to more usable products.

NXP Semiconductors [70] espoused similar views and provided clear statements to describe the goal of explainability:

> "We encourage explainability and transparency of AI-decision-making processes in order to build and maintain trust in AI systems". "The goal of interpretability is to describe the internals of the system in a way that is understandable to humans. The system should be capable of producing descriptions that are simple

TABLE 7.13   Objectives of explainability considerations

Explainability Considerations		
	Objectives	Statements
OBJ 28	Comprehensive explanation	Consistently provides clear and understandable interpretations across a range of inputs and scenarios, rather than being limited to specific instances or datasets.
OBJ 29	Interrogable justification	Provides valid and clear reasons for decisions, operations, or predictions, aligned with pre-defined objectives and standards.
OBJ 30	Intuitive interpretation	Presents operations and outcomes in a manner that is immediately comprehensible to users, irrespective of their technical expertise.

enough for a person to understand. It should also use a vocabulary that is meaningful for the user and will enable the user to understand how a decision is made".

Moreover, Sony [85] stated that they would capture AI decision reasoning in product design and provide clear explanations about potential impacts to customers using AI-integrated products and services:

> During the planning and design stages for its products and services that utilize AI, Sony will strive to introduce methods of capturing the reasoning behind the decisions made by AI utilized in said products and services. Additionally, it will endeavor to provide intelligible explanations and information to customers about the possible impact of using these products and services.

Thus, this chapter highlights that TMPSs should provide valid reasons for any decisions while ensuring the decisions are aligned with established objectives. Systems should offer consistent and clear interpretations across scenarios and present results in a manner understandable to users irrespective of their technical expertise. Table 7.13 presents our recommendations for specific objectives and statements within the context of explainability considerations.

4.2.4.2 Security Considerations

Samsung [92] indicated that awareness of advanced cyberattacks and their potential damages is central to security considerations:

> Awareness about cybersecurity and the potential for damages caused by increasingly sophisticated cyberattacks remains at the forefront of our security considerations.

Dell [101] signified AI systems and their marketable products and services must ensure their safety and reliability:

> AI systems should be safe and reliable, guarding the wellbeing of users and yielding results consistent with our values.

Similarly, Oracle [68] explained that the security and safety of AI applications is crucial for end-users and the public:

> AI systems can be used in applications such as self-driving cars, military drones, and medical treatments. Ensuring that these systems are safe for their intended users and the public is crucial.

NXP Semiconductors [70] embedded top-tier security and data protection into every aspect, from design and functionality to operations and business models:

> We must adopt the highest appropriate level of security and data protection to all hardware and software, ensuring that it is pre-configured into the design, functionalities, processes, technologies, operations, architectures and business models.

Furthermore, Equinix [102] pointed out effective data governance is necessary to ensure the security of AI practices and applications:

> "In order to ensure the integrity of their AI outcomes, businesses must verify that none of these inputs have been corrupted and put rigorous checks in place to ensure data security and integrity". "In addition to protecting data integrity, data governance is also essential to providing the context that goes along with your AI outcomes".

Thus, this chapter advocates that TMPSs should incorporate systematic measures to protect digital assets and hardware against threats. Additionally, an ethical and compliant data governance approach should be adopted to maintain stakeholder trust. Table 7.14 presents our recommendations for specific objectives and statements within the context of security considerations.

4.2.4.3 Transparency Considerations

NVIDIA [89] acknowledged technology's profound effects on cities and societies. In their guidance document, NVIDIA explained that the

TABLE 7.14   Objectives of security considerations

Security Considerations	
Objectives	Statements
OBJ 31 Data security governance	Adopts ethical, safe, and regulatory-compliant data management mechanisms, ensuring security of organisational data and stakeholders' privacy, and upholding trust amidst innovation practices.
OBJ 32 Digitally secure	Establishes systematic measures to protect digital assets, data, and user privacy, ensuring user trust and building a resilient digital ecosystem.
OBJ 33 Physically secure	Implements tangible measures to protect hardware and data storage, ensuring operational continuity and guarding against physical threats to innovation and digital ecosystems.

company aims to promote positive change and ensure trust and transparency in AI development:

> Recognizing that technology can have a profound impact on people and the world, we've set priorities that are rooted in fostering positive change and enabling trust and transparency in AI development.

Cisco [73] prioritised transparency in their AI development process as well. They emphasised the importance of informing customers when AI influences their decisions and of responding to feedback, with the goal of nurturing and enhancing trust in their AI offerings among all stakeholders:

> As transparency is one of our Trust Principles and core to this framework, we inform customers when AI is being used to make decisions that affect them in material and consequential ways. Customers and users can then inform us of their concerns or let us know when they disagree with decisions. By keeping communications channels open, we intend to build, maintain, and grow the trust that our customers, users, employees, and other stakeholders place in our AI offerings.

Moreover, NXP semiconductors [70] indicated that users' awareness of their interactions with intelligent systems is critical for transparency reasons:

> Users need to be aware that they are interacting with an AI system, and they need the ability to retrace that AI system's decisions.

TABLE 7.15   Objectives of transparency considerations

Transparency Considerations	
Objectives	**Statements**
OBJ 34 Appropriate disclosure	Transparently discloses issues and matters that substantially affect stakeholders, equipping those engaging with innovation and new technologies with comprehensive information.
OBJ 35 Aware of interaction	Ensures users are consistently informed about their interactions with intelligent systems to promote user autonomy and prevent misuse or unintended consequences.
OBJ 36 Openness of structure	Existence of architecture, training data, algorithms, and operational works that are open, clear, and available for review.

Dell [101] emphasised that users should have disclosure and control over AI interactions and data use:

> Users should be provided appropriate disclosures and control over their interactions with AI and its use of their data.

Additionally, Palantir [91] stated that transparency should permeate entire AI solutions, from model development to post-deployment, enabling users to utilise AI responsibly and efficiently for organisational challenges:

> These objectives also transparently communicate state about a particular AI/ML solution—from model development to testing, to deployment and further post-deployment actions like monitoring and upgrades. This enables users to be more intentional, responsible, and effective in how they use AI to address their organization's operational challenges.

Thus, this chapter underscores that TMPSs should disclose substantial stakeholder-affecting issues transparently and consistently inform users about their interactions with intelligent systems to foster autonomy and prevent misuse. Additionally, there is a need to ensure that the architecture, training data, algorithms, and operational works are open and available for review to ensure transparency. Table 7.15 presents our recommendations for specific objectives and statements within the context of transparency considerations.

4.2.5 Well-Governance Goals

The insights gleaned from the 'well-governance goals' category highlighted the key considerations that technology companies emphasise to ensure that their technologies, products, or services are managed well to deliver the desired outcomes for cities and societies. The most noticeable considerations linked to well-governance goals in RIT guidelines fell under the following categories: (a) accountability; (b) participation; and (c) regulatory.

4.2.5.1 Accountability Considerations

Palantir [91] indicated that for AI solutions to be used both responsibly and effectively, they must possess the qualities of traceability, auditability, and governability:

> AI solutions must be traceable, auditable, and governable in order to be used effectively and responsibly.

Cisco [73] believes that accountability throughout the AI practice lifecycle is paramount for responsible development and operation. Given that AI tools can serve multiple purposes, including unforeseen or unintended applications, providers must take on the responsibility to ensure AI solutions function as intended and to prevent misuse:

> Accountability for AI solutions and the teams that develop them is essential to responsible development and operations throughout the AI lifecycle. AI tools often have more than one application, including unintended use cases and uses that might not have been foreseeable at the time of development. Companies that develop, deploy, and use AI solutions must take responsibility for their work in this area by implementing appropriate governance and controls to ensure that their AI solutions operate as intended and to help prevent inappropriate use.

Vmware [86] pointed out that individuals within the development organisation must be held accountable for every stage of an AI-powered system, from its conception to deployment. This includes taking responsibility for the system's outcomes, results, and any subsequent consequences of its utilisation:

> Individuals in your organization should be accountable for the ideation, design, implementation, and deployment of each AI-powered system they create and/or use—including the outcomes, results, and consequences of its use.

Automatic data processing [97] expressed that they have established audit procedures and risk assessments as foundational management methods for their AI applications. They remain committed to constantly overseeing and refining their models and systems, ensuring that variations in data or model conditions do not adversely influence the anticipated outcomes:

> We have implemented audit and risk assessments to test our ML models as the baseline of our oversight methodologies. We continue to actively monitor and improve our models and systems to ensure that changes in the underlying data or model conditions do not inappropriately affect the desired results. And we apply our existing compliance, business ethics, and risk management governance structures to our ML development activities.

Amazon [98] proposed the need for mechanisms that monitor and review processes during the AI system's development and operation. By setting up a traceable record, both internal and external groups can effectively assess the AI system's development and operation:

> Consider the need for implementing mechanisms to track and review steps taken during development and operation of the ML system, e.g., to trace root causes for problems or meet governance requirements. Evaluate the need to document relevant design decisions and inputs to assist in such reviews. Establishing a traceable record can help internal or external teams evaluate the development and functioning of the ML system.

Furthermore, Adobe [84] signified their commitment to taking responsibility for the results of their AI-enhanced tools as well as dedicating processes and resources in place to address any concerns related to their AI and are prepared to make necessary adjustments if needed:

> We take ownership over the outcomes of our AI-assisted tools. We will have processes and resources dedicated to receiving and responding to concerns about our AI and taking corrective action as appropriate.

Thus, this chapter advocates that TMPS should contain accountability procedures to identify responsible parties and rectify harm or errors clearly. Additionally, the comprehensive records of decisions and processes should be maintained to allow for open audits. Table 7.16 presents our recommendations for specific objectives and statements within the context of accountability considerations.

TABLE 7.16   Objectives of accountability considerations

Accountability Considerations		
Objectives		Statements
OBJ 37	Redress and remediation	Establishes procedures to rectify any harm or mistakes, compensating affected parties when necessary.
OBJ 38	Responsibility attribution	Clearly identifies parties responsible for decisions, outcomes, or errors arising from innovation and technology practices.
OBJ 39	Traceable records	Maintains detailed records of decisions, justifications, processes, and outcomes and ensures the records are accessible to relevant stakeholders.

4.2.5.2 Participation Considerations

Sony [85] highlighted the importance of participatory discussion with various stakeholders to address the challenges presented by AI applications. They indicated that by fostering dialogues with industry peers, organisations, and academic communities, the interests and concerns of stakeholders would get better attention and consideration in AI development:

> … Sony will seriously consider the interests and concerns of various stakeholders including its customers and creators, and proactively advance a dialogue with related industries, organizations, academic communities and more… Sony will construct the appropriate channels for ensuring that the content and results of these discussions are provided to officers and employees, including researchers and developers, who are involved in the corresponding businesses, as well as for ensuring further engagement with its various stakeholders.

Schneider Electric [88] promoted that collaborating through co-innovation and forming strategic partnerships are essential to fully leverage the capabilities of AI and expedite its developmental journey:

> Co-innovation and partnerships are key to harness the power of AI and accelerate the AI journey.

NXP semiconductors [70] expressed that, through cross-disciplinary scientific methods, they advocate for thought leadership in AI, aiming to further enhance AI technologies and their associated practices:

> Drawing on rigorous and multidisciplinary scientific approaches, we promote thought leadership in this area in close cooperation with a wide range of stakeholders. We will continue to share what

we've learned to improve AI technologies and practices. Thus, in order to promote cross-industrial approaches to AI risk mitigation, we foster multi-stakeholder networks to share new insights, best practices and information about incidents.

Meta Platforms [103] denoted the necessity for a collective effort involving not just themselves but also the broader tech industry, AI researchers, policymakers, and advocacy groups to establish clear standards for AI impact assessment. This collaborative approach helps to pinpoint and mitigate potential negative effects of AI while continuing to develop AI-driven products for the greater good:

> We—not just Facebook but also the tech industry, the AI research community, policymakers, advocacy groups, and others—need to collaborate on figuring out how to make AI impact assessment work at scale, based on clear and reasonable standards, so that we can identify and address potential negative AI-related impacts while still creating new AI-powered products that will benefit us all.

In addition, Atlassian [87] claimed they are committed to establishing feedback procedures that allow them to gather insights from their stakeholders and draw guidance both from within and outside the company:

> We are committed to putting in place processes that help us to obtain feedback from our stakeholders and take guidance from experts, internally and externally. We encourage our customers to tell us if something has gone wrong. In those cases, we will investigate and work to fix it.

Dell [101] has views that support this point and states that AI applications' development and deployment necessitate periodic evaluations by diverse professionals, spanning legal, ethics, technical, and business fields, to guarantee continuous compliance and transparency:

> The development and implementation of AI applications should be periodically reviewed by both internal and external legal, ethics, technical and business professionals to ensure ongoing compliance and transparency.

Thus, this chapter emphasises that TMPS should adopt the co-design approach for direct stakeholder contributions and provide ongoing feed-

TABLE 7.17   Objectives of participation considerations

Participation Considerations		
	Objectives	Statements
OBJ 40	Collaborative governance	Embraces a collaborative governance model which invites diverse stakeholders to jointly shape and monitor innovation practices, ensuring that technological outcomes resonate with and serve the wider public interest.
OBJ 41	Cooperative design	Facilitates co-design sessions to allow potential users and other stakeholders to directly contribute to the innovation's design or refinement.
OBJ 42	Feedback loops	Establishes mechanisms that allow for continuous feedback from users and stakeholders to adapt the innovation accordingly.

back procedures to refine innovations. A collaborative governance method can be applied to ensure technological outcomes align with the broader public interest by inviting diverse stakeholder participation. Table 7.17 presents our recommendations for specific objectives and statements within the context of participation considerations.

4.2.5.3 Regulatory Considerations

Schneider Electric [88] highlighted that in AI and data science development, the first principle is to ensure ethics and compliance:

> The number one rule we apply when developing AI and data science is ethics and compliance in line with our Trust Charter.

Sony [85] indicated they are committed to ensuring that their AI-integrated products and services adhere to both legal standards and their own internal guidelines. This commitment is based on their vision of respecting the intentions and trust of their customers:

> Sony, in compliance with laws and regulations as well as applicable internal rules and policies, seeks to enhance the security and protection of customers' personal data acquired via products and services utilizing AI, and build an environment where said personal data is processed in ways that respect the intention and trust of customers.

Amazon [98] stated that they actively consult with legal experts when developing AI to ensure AI-driven applications meet all regulatory requirements. They also emphasise the importance of being aware of varying legal stipulations across different regions and of the emerging AI regulations globally, ensuring continuous legal assessment throughout the deployment and operational phases:

> Engage with legal advisors to assess requirements for and implications of building your ML system. This may include vetting legal rights to use data and models, and determining applicability of laws around privacy, biometrics, anti-discrimination, and other use-case specific regulations. Be mindful of differing legal requirements across states, provinces, and countries, as well as new AI/ML regulation being considered and proposed around the world. Re-visit legal requirements and considerations through future deployment and operations phases.

Workday [71] advocated for practical, risk-based regulatory strategies that foster trust in AI technology while promoting innovation:

> We engage U.S. federal, state, and local governments, the European Union, and other governments around the world to advocate for workable, risk-based regulatory approaches that build trust in AI technology and enable innovation. As our development process continues to evolve to account for new best practices and emerging regulatory frameworks, we remain committed to supporting the delivery of trustworthy AI solutions that provide value to our customers, the workforce, and society.

Furthermore, Dell [101] denoted that the implementation and utilisation of AI should not only adhere to global laws in both essence and letter but also remain consistent with corporate conduct codes and emerging ethical standards. They underscored the importance of periodic reviews of AI applications by a diverse group of experts, spanning legal, ethical, technical, and business sectors, to guarantee sustained compliance and transparency:

> The implementation and use of AI should comply with the letter and spirit of globally applicable laws, be consistent with corporate codes of conduct and align with an evolving consensus on ethical practices. The development and implementation of AI applications should be periodically reviewed by both internal and external legal, ethics, technical and business professionals to ensure ongoing compliance and transparency.

TABLE 7.18   Objectives of regulatory considerations

Regulatory Considerations		
	Objectives	Statements
OBJ 43	Consistent with best practice guidelines	Observes industry-specific best practices, standards, and codes of conduct that might be set by professional bodies or associations.
OBJ 44	Legal compliance	Adheres to both the letter and spirit of global laws and regulations, while ensuring that innovation and technological outcomes remain adaptable to evolving regulatory landscapes.
OBJ 45	Periodic and independent review	Regularly conducts independent reviews of innovation and technology systems and adjusts as needed to ensure they function as intended and to pre-emptively deter misuse.

Thus, this chapter emphasises that TMPSs should not only observe industry-specific best practices and codes but also adhere to global laws in both intent and implementation. Furthermore, they need to undergo independent reviews regularly to ensure proper functioning and proactive prevention of misuse, all while remaining adaptable to changing regulatory environments. Table 7.18 presents our recommendations for specific objectives and statements within the context of regulatory considerations.

The completed RIT assessment framework is provided in Appendix 7.A.

5 FINDINGS AND DISCUSSION

Our research findings indicate that the selected companies largely focus their RIT considerations on AI. This focus may be attributed to several interrelated factors. First, AI is at the forefront of technological advancements, often deemed a pivotal driver of the Fourth Industrial Revolution (Industry 4.0) and the emerging Fifth Industrial Revolution (Industry 5.0) [104–106]. The transformative potential of AI across various sectors necessitates a concentrated effort to ensure its development aligns with ethical, legal, and social norms—a common tenet of RIT. Secondly, AI systems raise unique challenges, such as opacity (the 'black box' problem), algorithmic bias, and questions of accountability, which may demand appropriate RIT guidance to manage [64,107]. Finally, as public concerns related to privacy, autonomy, and decision-making bias of AI-driven products and services are rising, these leading technology companies may strategically prioritise AI in their RIT agendas to mitigate risks, align with regulatory trends, and bolster public confidence in their commitment to responsible innovation.

Consistent with the above, our findings discovered that these companies' key areas of emphasis include the trustworthiness and acceptability of technology and its marketable products and services (TMPSs), particularly concerning privacy-related challenges. Notably, discourse surrounding the affordability of technological outcomes and their subsequent adoption is limited. Additionally, while in-depth discussions on sustainability within the RIT context are lacking, this consideration is notably presented in their annual corporate reporting documents, such as the ESG Report, Corporate Responsibility Report, and Sustainability Report.

After synthesising the perspectives of leading technology companies on the RIT concept, this chapter has formulated a multifaceted evaluation approach to ensure that TMPSs align with the ultimate goals of the concept. The framework encompasses five interconnected pillars guiding organisations towards responsible innovation processes and product design strategy: *acceptability, accessibility, alignment, trustworthiness*, and *well-governance*.

The *acceptability goal* stresses the importance of ensuring TMPSs are aligned with human and social values while ensuring they are ethical and equitable and avoid harm to individuals, society, and the environment. This is critical as technologies like AI become increasingly integrated into every aspect of life; the way they are designed and deployed may have profound effects on individuals, society, and the environment, e.g., the cultural and creative industry [108], healthcare field [109], and public sector [110]. For example, an AI recruitment tool must be free from biases that could lead to discrimination, adhering to fairness as a fundamental principle, and the algorithms should be audited for bias. Tubadji et al. [108] indicated existing AI technology still grapples with a fundamental challenge: the capacity to connect with human emotions and, notably, the capability to intuitively grasp moral attitudes and values that define the essence of humanity. Acceptability should be the threshold for responsible AI and appears to be a key element for overcoming this challenge [110]. When AI provides societally unacceptable results (for example violating fundamental human rights or common values), this form of AI application should not be accepted because there are no identifiable ethically acceptable trade-offs [111].

The *accessibility goal* ensures that TMPSs are within reach for all sections of society, breaking down barriers related to affordability, cultural context, and usability. The purpose is to ensure technological achievements are not just for the minority but serve a broader demographic. In essence, the accessibility goal calls for a more inclusive approach to the design and deployment of TMPSs, whether in technical, financial, or cultural terms. This goal aligns with the broader goals of social justice and equity to foster a more inclusive digital future [112–114]. A familiar example is providing multilingual and multiform software interfaces that cater to wider user groups, thus embodying the inclusivity principle.

The *alignment goal* demands that TMPSs align with the broader objectives of societal and environmental well-being [26,39]. It emphasises the need for deliberate, meaningful, and sustainable innovation that recognises the interconnectedness of technology, society, and the natural world, and strives to create a positive impact on all these fronts [115–117]. This goal focuses on ensuring that industries and their outputs harmonise with broader societal and environmental objectives. Organisations should change their focus from the classical 'market and technology-driven' innovation perspective to pursuing broader social and environmental values to fulfil social responsibility [117–119]. For instance, a company could commit to reducing its carbon footprint by designing energy-efficient products, thereby aligning its mission with sustainability goals [120].

As technology becomes more complex and pervasive in our cities and societies, ensuring the trustworthiness of TMPSs is paramount [121–123]. Existing information system studies highlighted that the relationship between humans and technology is influenced by trust. This is because trust plays a crucial role as a precursor to engaging in risk-taking or adoption behaviours [124–126]. The *trustworthiness goal* emphasises the need for robust measures to protect data and provide transparency so that users can understand and trust the products they use [127,128]. By focusing on these considerations, companies can foster a strong relationship of trust with their users, which is essential in a digital age characterised by frequent data breaches and concerns over privacy and misuse of technology [129,130]. A company could, for example, adopt transparent data practices and explain to users how their data are used and protected, thus enhancing trustworthiness.

Finally, contemporary organisations are seen as the main setting for responsible innovation. Due to the inherent uncertainty in predicting innovation outcomes, which often leads to previously unencountered results, it is essential to govern responsibly, especially starting from the early stages of research and development. This is crucial to achieving responsible innovation goals within a business context [115,131]. The *well-governance goal* is to ensure that TMPSs are developed and managed responsibly. It covers accountability, regulatory compliance, and participation, ensuring that companies are accountable for their actions, decisions, and products, especially those that may affect customers, society, and the environment [25,107]. A practical example could be setting up a governance board that includes customer representatives, industry experts, and other relevant parties. Their participation ensures that a wide range of views and interests are considered, thus operationalising the well-governance goal [115,131].

While the technology companies highlight a predominant focus on RIT considerations within the realm of AI, it is crucial to recognise the broader technological landscape, especially in the context of Industry 4.0 and the emerging Industry 5.0 paradigms. Technologies such as the Internet of Things (IoT) and the Industrial Internet of Things (IIoT) play pivotal roles in shaping a hyper-connected world. In addition, other closely related technologies, like blockchain, augmented reality, and robotics, underpin the digital transformation of industries and societies [132–136]. Each of these technologies may present ethical, societal, environmental, or regulatory challenges that warrant rigorous RIT considerations. Industries, especially global technology giants, need to broaden their RIT focus to encompass this wider range of technologies, ensuring that we do so responsibly as we move towards an even more interconnected future.

Given the distinct industry or business backgrounds of various companies, their priorities concerning the RIT concept will naturally differ slightly. For instance, companies centred around AI-driven technologies might place a heightened emphasis on ethical considerations, while semiconductor manufacturers could prioritise environmental sustainability. Hence, we suggest that companies should integrate the RIT concept into their corporate missions and values. Enduringly successful companies frequently ascribe their successes to deeply held core values and beliefs that invariably inform their decision-making processes [137]. By incorporating RIT concept, technology companies can harmonise their initiatives with their foundational visions, e.g., *"we do this to respect human rights, encourage innovation, and reflect Cisco's purpose to power an inclusive future for all"* [73], *"we live Atlassian's mission and values in everything we do... seek to uphold those values when it comes to understanding what it means to act responsibly in building, deploying and using new technologies"* [87]. Such an alignment not only facilitates coherent and consistent decision-making but also aids in attracting and retaining individuals who resonate with these intrinsic values.

Additionally, this study reveals that leading technology companies place a heightened emphasis on the trustworthiness and acceptability of their technological solutions and results. By striving to shape their products and services to be more 'responsible', these companies aim to garner greater consumer trust and societal approval. We contend that integrating the RIT concept into corporate missions and values not only transcends the consideration of 'doing the right thing', but is also pivotal for following compelling reasons, especially for the private sector:

- Assurance of user protection and building trust: Ensuring user protection has become paramount in the contemporary environment, especially given the heightened concerns surrounding data privacy and the potential misuse of AI and related technologies. By adopting RIT practices, companies can proactively address these concerns, thereby reducing risks and bolstering user trust. Companies emphasising security and trust in their technological pursuits may distinguish themselves, thus fostering deeper and more enduring customer relationships, e.g., "*Xiaomi firmly believes that respecting and protecting the security of user's information and user privacy is the only approach to build long-term trust in Xiaomi products and services*" [72], "*our heritage is built on providing trustworthy and innovative solutions to our customers*" [84], and "*we will continue to promote responsible AI in order to maintain the trust of products and services by stakeholders*" [85];

- Commitment to social responsibility: As companies broaden their influence, the magnitude of their societal impact inevitably intensifies. Beyond profitability, there is a growing recognition of the role companies play in societal well-being. Adopting a vision centred on providing responsible technology and products, in line with a commitment to social responsibility, ensures that this impact is positive for both cities and broader societies, e.g., "*ensuring that our technology and the use of our technology benefits society*" [84], "*to develop technology that supports ethical growth and social responsibility*" [92], "*ensure that AI has a positive impact on society and helps to create a better future for all*" [68]. Furthermore, discerning contemporary consumers frequently favour businesses that prioritise societal welfare over mere profit accrual [116,138];

- Compliance with ethical norms and regulations: As governments and international bodies become more stringent about data privacy, ethical technology deployment, and environmental considerations. Proactively aligning with the RIT concept can help companies stay ahead of regulatory curves, ensuring that they are not just reacting to laws but actively shaping and adhering to best practices, e.g., *"understanding the important need for public trust, we work closely with policymakers across the country and around the world as they assess whether existing consumer protections remain fit-for-purpose in an AI era"* [98], *"take appropriate action to mitigate that abuse…unless we have high confidence that Intel's products are not being used to violate human rights"* [67], *"evaluate and mitigate potential legal, reputational, and ethical risks associated with AI use… be proactive rather than reactive in addressing AI ethics concerns"* [85];

- Obtaining a competitive advantage: The technology industry is fiercely competitive. Companies that embed the RIT concept into their products and services can leverage it as a 'unique' selling point, attracting ethically conscious consumers and partners. It also positions these companies as forward-thinking leaders in the industry, e.g., *"not just because we believe it gives us a competitive advantage, but because it is the right thing to do"* [97], *"to gain a key competitive advantage… organization that is committed to ethical AI… likely to see better returns from products and services that rely on AI"* [86], *"we are also leading technology industry initiatives to further advance responsible practices in minerals sourcing, mobility, and AI"* [67]. As consumers become more ethically and environmentally conscious, companies that prioritise RIT may differentiate themselves on the market, offering products and services that cater to this growing demographic of conscious consumers [139].

The growth of RIT practices among the leading technology companies reflects an interplay of increasing societal, regulatory, innovative, and ethical pressures. For the private sector, RIT is not just a moral or ethical consideration—it is a strategic imperative. Integrating the RIT concept into corporate missions and values ensures that companies remain resilient, adaptable, and successful in a rapidly evolving technological landscape, all while contributing positively to society and the environment [63].

6 CONCLUSION

Compared with previous studies, this chapter pioneers the exploration of the emerging topic of RIT—that is an umbrella term inclusive of 'Responsible Artificial Intelligence'—from an enterprise perspective, especially drawing more practical insights from leading technology companies worldwide. This chapter reveals that leading high-tech companies, so-called tech giants, have shifted their attention towards RIT, and their main RIT policy focus is AI technology. The most common policy areas are trustworthiness and acceptability of technology, and the most absent area related to the technology is affordability. Nevertheless, sustainability considerations are rarely part of RIT policy. The findings of this chapter illuminate these technology giants' understanding of weaving the RIT concept into their innovation practices, thereby aligning their marketable products and services with this concept. The main contribution of this chapter is to construct a holistic and actionable RIT assessment framework by merging these corporate insights with established viewpoints from the policy community and academia.

This framework (Figure 7.5 and Appendix 7.A) provides a systematic approach to holistically evaluate technological outcomes against a broad spectrum of responsible innovation principles. By transforming an abstract concept into specific objectives and statements, the framework operationalises the goals of RIT, making them more actionable and measurable within organisations. Moreover, this chapter advocates that by integrating the RIT concept into their corporate missions, values, and overarching strategic visions, companies can enhance their resilience, adaptability, and success amidst the dynamic technological landscape, ultimately fostering positive impacts on our cities and broader societies [4,14].

While the insights from this research offer some value, it is essential to acknowledge certain inherent limitations. This study applied content analysis, which is inherently exploratory but somewhat contingent on the authors' interpretative judgements. Its scope was confined to publicly available corporate documents and focused solely on the technosphere, which may limit its broader applicability. Some work-in-progress RIT guidance or some insights from other fields may have been omitted. Additionally, this study may have inadvertently excluded insights from diverse or non-listed companies by centring on the top listed global technology companies by market capitalisation. Nonetheless, given that the chosen technology companies are the most influential market leaders globally, they offer a representative insight into prevailing trends and dominant perspectives on the RIT discourse.

Despite these limitations, this research study establishes a foundation for the discourse on 'being responsible' within the technosphere. In particular, we developed a more practical and detailed framework, which can serve as a tool for governments, industries, practitioners, researchers, and other stakeholders to evaluate whether the technological outcomes are responsible. The topic of responsible research in the technosphere is still growing, and significant gaps need to be bridged. This chapter suggests that future research should delve deeper from an enterprise perspective because the private sector plays a pivotal role in innovation practices. Additionally, further investigations from a user/customer standpoint are necessary to ensure that enterprise propositions align with genuine societal expectations. The following issues/questions are important for prospective research to focus on and address:

- Discrepancies in RIT considerations: Why or what are some key considerations of RIT, which the academic community and user groups have emphasised, rarely mentioned, and discussed by high-tech companies?

- RIT and CSR: How do high-tech companies view the relationship between responsible innovation and CSR?

- Influences on RIT guidance: Do internal and external pressures directly influence how high-tech companies shape their RIT guidance, and if so, how? How do collaborations with various stakeholders and vendors in a business network (e.g., in the IoT and IIoT contexts) affect the focal organisation's RIT policies?

- Comparative perspectives on RIT: What are the differences or priorities between users' expectations for RIT and the viewpoints high-tech companies, academia, and the policy community advocate?

ACKNOWLEDGEMENTS

This chapter, with permission from the copyright holder, is a reproduced version of the following journal article: Li, W., Yigitcanlar, T., Nili, A., & Browne, W. (2023). Tech giants' responsible innovation and technology strategy: an international policy review. *Smart Cities*, 6(6), 3454–3492.

APPENDIX 7.A Responsible innovation and technology assessment framework

Acceptability Goals

Equitability Considerations

	Objectives	Statements	Scale				
			Low	Medium-Low	Medium	Medium-High	High
OBJ 1	Avoid bias	Does not create, reinforce, or propagate harmful or unfair biases in all stages of innovation and technology practice, from design to deployment and beyond.					
OBJ 2	Guard against discrimination	Upholds the rights of all individuals and groups, embraces the full spectrum of social diversity, and actively prevents any form of discrimination.					
OBJ 3	Strive for fairness	Proactively identifies and eliminates obstacles to ensure fair treatment for all and empower every individual equally through innovation and technology.					

Ethics Considerations

	Objectives	Statements	Scale				
			Low	Medium-Low	Medium	Medium-High	High
OBJ 4	Human value-based design	Prioritises human values and morals in the innovation process, ensuring that technological outcomes meet functional requirements and align with broader ethical and societal norms.					
OBJ 5	Human-in-the-loop mechanism	Embeds appropriate human oversight and intervention into decision-making processes to balance efficiency with ethical considerations.					
OBJ 6	Respect and enforce privacy	Incorporates privacy principles at every stage of the innovation lifecycle, ensuring that innovation and technological outcomes consistently prioritise and protect user privacy.					

(Continued)

APPENDIX 7.A Continued

Harmlessness Considerations

		Scale				
Objectives	**Statements**	**Low**	**Medium-Low**	**Medium**	**Medium-High**	**High**
OBJ 7 Harmless to environment	Ensures respect and protection of the environment, avoiding practices that lead to environmental degradation or the corruption of natural resources.					
OBJ 8 Harmless to individual	Does not lead to any harm against any individuals, including physical and psychological harm.					
OBJ 9 Harmless to society	Minimises adverse impacts on society and commits to the harmonious progress of technology and society.					

Accessibility Goals

Adaptability Considerations

		Scale				
Objectives	**Statements**	**Low**	**Medium-Low**	**Medium**	**Medium-High**	**High**
OBJ 10 Flexible in nature	Allows individual modification or replacement and can seamlessly operate with other systems to ensure compatibility and ease of integration.					
OBJ 11 Future-proof design	Remains adaptable in the face of evolving socio-technical challenges and paradigms and continues to serve intended purposes throughout their lifecycle.					
OBJ 12 Universal access	Ensures accessibility for all individuals, irrespective of differences in physical ability, technological, cognitive, or actual usage context.					

Affordability Considerations

Objectives	Statements	Scale				
		Low	Medium-Low	Medium	Medium-High	High
OBJ 13 Cost-effective solution	Provides cost-effective solutions or alternatives to reduce economic disparities in technology adoption.					
OBJ 14 Financial feasibility	Reduces costs to allow broad population segments to access and benefit from innovation and technological advances without negative financial implications.					
OBJ 15 Value for money	Offers genuine value, ensuring that users receive meaningful benefits or solutions that justify the costs, promoting long-term adoption and utility.					

Inclusiveness Considerations

Objectives	Statements	Scale				
		Low	Medium-Low	Medium	Medium-High	High
OBJ 16 Cultural and contextual sensitivity	Respects shared values while remaining sensitive and attuned to the nuances of diverse cultural norms and contexts.					
OBJ 17 Diverse representation	Ensures marginalised or underrepresented groups are included and have representation in innovation and technology practices.					
OBJ 18 Localisation-friendly	Localises to match local customs, languages, and preferences, ensuring relevance and acceptance in different geo-cultural contexts.					

(Continued)

APPENDIX 7.A Continued

Alignment Goals

Deliberateness Considerations

Objectives	Statements	Scale				
		Low	Medium-Low	Medium	Medium-High	High
OBJ 19 Define the purpose	Before taking any initiative or action, clarifies and articulates the intended purpose to ensure alignment with overall goals and values.					
OBJ 20 Proactive risk assessment	Anticipates potential negative outcomes, challenges, or pitfalls, and designs strategies to mitigate or avoid them.					
OBJ 21 User relevance	Gains a deep understanding of the users' needs, preferences, and challenges to ensure the solutions provided are directly relevant to and satisfy the end-user's needs.					

Meaningfulness Considerations

Objectives	Statements	Scale				
		Low	Medium-Low	Medium	Medium-High	High
OBJ 22 Realise human potential	Unlocks humanity's potential, empowering individuals to address complex challenges effectively with increased capability through complementary collaboration with technology.					
OBJ 23 Social license to operate	Strives for continuous community recognition, emphasising legitimacy, trust, and ethical alignment beyond mere regulatory compliance.					
OBJ 24 Socially beneficial	Promotes human welfare for both current and future generations, fostering growth, prosperity, and positive societal outcomes.					

Sustainability Considerations

		Scale				
Objectives	Statements	Low	Medium-Low	Medium	Medium-High	High
OBJ 25 Commitments to climate change	Targets a reduction in greenhouse gas emissions and prevention of waste generation, supporting global efforts to adapt to and mitigate climate change.					
OBJ 26 Design for longevity	Ensures resources invested in creating products or solutions provide value over the long term and contribute to more sustainable and resilient societies.					
OBJ 27 Eco-friendly design	Incorporates environmental considerations starting from the design phase to ensure products or solutions are eco-friendly throughout their lifecycle.					

Trustworthiness Goals

Explainability Considerations

		Scale				
Objectives	Statements	Low	Medium-Low	Medium	Medium-High	High
OBJ 28 Comprehensive explanation	Consistently provides clear and understandable interpretations across a range of inputs and scenarios, rather than being limited to specific instances or datasets.					
OBJ 29 Interrogable justification	Provides valid and clear reasons for decisions, operations, or predictions, aligned with pre-defined objectives and standards.					
OBJ 30 Intuitive interpretation	Presents operations and outcomes in a manner that is immediately comprehensible to users, irrespective of their technical expertise.					

(Continued)

APPENDIX 7.A   Continued

Security Considerations

Objectives	Statements	Scale				
		Low	Medium-Low	Medium	Medium-High	High
OBJ 31 Data security governance	Adopts ethical, safe, and regulatory-compliant data management mechanisms, ensuring security of organisational data and stakeholders' privacy, and upholding trust amidst innovation practices.					
OBJ 32 Digitally secure	Establishes systematic measures to protect digital assets, data, and user privacy, ensuring user trust and building a resilient digital ecosystem.					
OBJ 33 Physically secure	Implements tangible measures to protect hardware and data storage, ensuring operational continuity and guarding against physical threats to innovation and digital ecosystems.					

Transparency Considerations

Objectives	Statements	Scale				
		Low	Medium-Low	Medium	Medium-High	High
OBJ 34 Appropriate disclosure	Transparently discloses issues and matters that substantially affect stakeholders, equipping those engaging with innovation and new technologies with comprehensive information.					
OBJ 35 Aware of interaction	Ensures users are consistently informed about their interactions with intelligent systems to promote user autonomy and prevent misuse or unintended consequences.					
OBJ 36 Openness of structure	Existence of architecture, training data, algorithms, and operational works that are open, clear, and available for review.					

Well-governance Goals

Accountability Considerations

Objectives	Statements	Scale				
		Low	Medium-Low	Medium	Medium-High	High
OBJ 37 Redress and remediation	Establishes procedures to rectify any harm or mistakes, compensating affected parties when necessary.					
OBJ 38 Responsibility attribution	Clearly identifies parties responsible for decisions, outcomes, or errors arising from the innovation and technology practices.					
OBJ 39 Traceable records	Maintains detailed records of decisions, justifications, processes, and outcomes, and ensures the records are accessible to relevant stakeholders.					

Participation Considerations

Objectives	Statements	Scale				
		Low	Medium-Low	Medium	Medium-High	High
OBJ 40 Collaborative governance	Embraces a collaborative governance model, which invites diverse stakeholders to jointly shape and monitor innovation practices, ensuring that technological outcomes resonate with and serve the wider public interest.					
OBJ 41 Cooperative design	Facilitates co-design sessions to allow potential users and other stakeholders to directly contribute to the innovation's design or refinement.					
OBJ 42 Feedback loops	Establishes mechanisms that allow continuous feedback from users and stakeholders to adapt the innovation accordingly.					

(Continued)

APPENDIX 7.A   Continued

| | | Regulatory Considerations | Scale | | | | |
	Objectives	Statements	Low	Medium-Low	Medium	Medium-High	High
OBJ 43	Consistent with best practice guidelines	Observes industry-specific best practices, standards, and codes of conduct that might be set by professional bodies or associations.					
OBJ 44	Legal compliance	Adheres to both the letter and spirit of global laws and regulations, while ensuring that innovation and technological outcomes remain adaptable to evolving regulatory landscapes.					
OBJ 45	Periodic and independent review	Regularly conducts independent reviews of innovation and technology systems and adjusts as needed to ensure they function as intended and to pre-emptively deter misuse.					

REFERENCES

1. David, A., Yigitcanlar, T., Li, R., Corchado, J., Cheong, P., Mossberger, K., & Mehmood, R. (2023). Understanding local government digital technology adoption strategies: A PRISMA review. *Sustainability*, 15(12), 9645.
2. Son, T., Weedon, Z., Yigitcanlar, T., Sanchez, T., Corchado, J., & Mehmood, R. (2023). Algorithmic urban planning for smart and sustainable development: Systematic review of the literature. *Sustainable Cities and Society*, 94, 104562.
3. Li, W., Yigitcanlar, T., Liu, A., & Erol, I. (2022). Mapping two decades of smart home research: A systematic scientometric analysis. *Technological Forecasting and Social Change*, 179, 121676.
4. Yigitcanlar, T., Li, R., Beeramoole, P., & Paz, A. (2023). Artificial intelligence in local government services: Public perceptions from Australia and Hong Kong. *Government Information Quarterly*, 40(3), 101833.
5. Marasinghe, R., Yigitcanlar, T., Mayere, S., Washington, T., & Limb, M. (2024). Computer vision applications for urban planning: A systematic review of opportunities and constraints. *Sustainable Cities and Society*, 100, 105047.
6. Lewallen, J. (2021). Emerging technologies and problem definition uncertainty: The case of cybersecurity. *Regulation & Governance*, 15(4), 1035–1052.
7. Nili, A., Desouza, K., & Yigitcanlar, T. (2022). What can the public sector teach us about deploying artificial intelligence technologies? *IEEE Software*, 39(6), 58–63.
8. Regona, M., Yigitcanlar, T., Xia, B., & Li, R. (2022). Opportunities and adoption challenges of AI in the construction industry: A PRISMA review. *Journal of Open Innovation: Technology, Market, and Complexity*, 8(1), 45.
9. Moqaddamerad, S., & Tapinos, E. (2023). Managing business model innovation uncertainties in 5G technology: A future-oriented sensemaking perspective. *R&D Management*, 53(2), 244–259.
10. Lubberink, R., Blok, V., Van Ophem, J., & Omta, O. (2017). Lessons for responsible innovation in the business context: A systematic literature review of responsible, social and sustainable innovation practices. *Sustainability*, 9(5), 721.
11. Millar, C., Lockett, M., & Ladd, T. (2018). Disruption: Technology, innovation and society. *Technological Forecasting and Social Change*, 129, 254–260.
12. Floridi, L., Cowls, J., Beltrametti, M., Chatila, R., Chazerand, P., Dignum, V., ... & Vayena, E. (2021). An ethical framework for a good AI society: Opportunities, risks, principles, and recommendations. In Luciano Floridi. (Ed.) *Ethics, Governance, and Policies in Artificial Intelligence*, 19–39. Springer.
13. Jakobsen, S., Fløysand, A., & Overton, J. (2019). Expanding the field of responsible research and innovation (RRI): From responsible research to responsible innovation. *European Planning Studies*, 27(12), 2329–2343.
14. Yigitcanlar, T., Sabatini-Marques, J., da-Costa, E., Kamruzzaman, M., & Ioppolo, G. (2019). Stimulating technological innovation through incentives: Perceptions of Australian and Brazilian firms. *Technological Forecasting and Social Change*, 146, 403–412.

15. Boenink, M., & Kudina, O. (2020). Values in responsible research and innovation: From entities to practices. *Journal of Responsible Innovation*, 7(3), 450–470.

16. Stilgoe, J., Owen, R., & Macnaghten, P. (2013). Developing a framework for responsible innovation. *Research Policy*, 42(9), 1568–1580.

17. Ribeiro, B., Smith, R., & Millar, K. (2017). A mobilising concept? Unpacking academic representations of responsible research and innovation. *Science and Engineering Ethics*, 23, 81–103.

18. Thapa, R., Iakovleva, T., & Foss, L. (2019). Responsible research and innovation: A systematic review of the literature and its applications to regional studies. *European Planning Studies*, 27(12), 2470–2490.

19. Owen, R., Macnaghten, P., & Stilgoe, J. (2012). Responsible research and innovation: From science in society to science for society, with society. *Science & Public Policy*, 39(6), 751–760.

20. Von Schomberg, R. (2013). A vision of responsible research and innovation. In Richard Owen, John R. Bessant, Maggy Heintz (Eds.) *Responsible Innovation: Managing the Responsible Emergence of Science and Innovation in Society*, 51–74. https://doi.org/10.1002/9781118551424.ch3

21. Pavie, X., Carthy, D., & Scholten, V. (2014). *Responsible Innovation: From Concept to Practice*. Singapore: World Scientific.

22. Li, W., Yigitcanlar, T., Browne, W., & Nili, A. (2023). The making of responsible innovation and technology: An overview and framework. *Smart Cities*, 6(4), 1996–2034.

23. Gurzawska, A. (2021). Responsible innovation in business: Perceptions, evaluation practices and lessons learnt. *Sustainability*, 13(4), 1826.

24. Li, Y., Jiang, L., & Yang, P. (2023). How to drive corporate responsible innovation? A dual perspective from internal and external drivers of environmental protection enterprises. *Frontiers in Environmental Science*, 10, 1091859.

25. Hadj, T. (2020). Effects of corporate social responsibility towards stakeholders and environmental management on responsible innovation and competitiveness. *Journal of Cleaner Production*, 250, 119490.

26. Adomako, S., & Nguyen, N. (2023). Green creativity, responsible innovation, and product innovation performance: A study of entrepreneurial firms in an emerging economy. *Business Strategy and the Environment*. https://doi.org/10.1002/bse.3373

27. Xie, X., Wu, Y., & Tejerob, C. (2022). How responsible innovation builds business network resilience to achieve sustainable performance during global outbreaks: An extended resource-based view. *IEEE Transactions on Engineering Management*. https://doi.org/10.1109/TEM.2022.3186000

28. Jarmai, K., Tharani, A., & Nwafor, C. (2020). Responsible innovation in business. In: Jarmai, K. (eds) *Responsible Innovation*. SpringerBriefs in Research and Innovation Governance. pp. 7–17. Springer, Dordrecht.

29. Barros, A., Sindhgatta, R., & Nili, A. (2021). Scaling up chatbots for corporate service delivery systems. *Communications of the ACM*, 64(8), 88–97.

30. Adomako, S., & Tran, M. (2022). Environmental collaboration, responsible innovation, and firm performance: The moderating role of stakeholder pressure. *Business Strategy and the Environment*, 31(4), 1695–1704.
31. Makasi, T., Nili, A., Desouza, K., & Tate, M. (2022). Public service values and chatbots in the public sector: Reconciling designer efforts and user expectations. In *Proceedings of the 55th Hawaii International Conference on System Sciences* (pp. 2334–2343). University of Hawai'i at Manoa.
32. Lukovics, M., Nagy, B., Kwee, Z., & Yaghmaei, E. (2023). Facilitating adoption of responsible innovation in business through certification. *Journal of Responsible Innovation*, 1–19.
33. Tian, H., & Tian, J. (2021). The mediating role of responsible innovation in the relationship between stakeholder pressure and corporate sustainability performance in times of crisis: Evidence from selected regions in China. *International Journal of Environmental Research and Public Health*, 18(14), 7277.
34. Chin, T., Caputo, F., Shi, Y., Calabrese, M., Aouina-Mejri, C., & Papa, A. (2022). Depicting the role of cross-cultural legitimacy for responsible innovation in asian-pacific business models: A dialectical systems view of Yin-Yang harmony. *Corporate Social Responsibility and Environmental Management*, 29(6), 2083–2093.
35. Salzmann, O., Ionescu-Somers, A., & Steger, U. (2005). The business case for corporate sustainability: Literature review and research options. *European Management Journal*, 23(1), 27–36.
36. Carroll, A., & Shabana, K. (2010). The business case for corporate social responsibility: A review of concepts, research and practice. *International Journal of Management Reviews*, 12(1), 85–105.
37. Kolk, A., & Van Tulder, R. (2010). International business, corporate social responsibility and sustainable development. *International Business Review*, 19(2), 119–125.
38. Searcy, C. (2012). Corporate sustainability performance measurement systems: A review and research agenda. *Journal of Business Ethics*, 107, 239–253.
39. Novitzky, P., Bernstein, M. J., Blok, V., Braun, R., Chan, T. T., Lamers, W., ... & Griessler, E. (2020). Improve alignment of research policy and societal values. *Science*, 369(6499), 39–41.
40. Völker, T., Mazzonetto, M., Slaattelid, R., & Strand, R. (2023). Translating tools and indicators in territorial RRI. *Frontiers in Research Metrics and Analytics*, 7, 1038970.
41. Blok, V., & Lemmens, P. (2015). The emerging concept of responsible innovation. Three reasons why it is questionable and calls for a radical transformation of the concept of innovation. In: Koops, B. J. et al. (eds) *Responsible Innovation 2*. pp. 19–35. Springer, Cham.
42. Loureiro, P., & Conceição, C. (2019). Emerging patterns in the academic literature on responsible research and innovation. *Technology in Society*, 58, 101148.
43. Hellström, T. (2003). Systemic innovation and risk: Technology assessment and the challenge of responsible innovation. *Technology in Society*, 25(3), 369–384.

44. Von Schomberg, R. (2011). Towards responsible research and innovation in the information and communication technologies and security technologies fields (November 13, 2011). Available at SSRN: https://ssrn.com/abstract=2436399.

45. Baregheh, A., Rowley, J., & Sambrook, S. (2009). Towards a multidisciplinary definition of innovation. *Management Decision*, 47(8), 1323–1339.

46. Voegtlin, C., Scherer, A., Stahl, G., & Hawn, O. (2022). Grand societal challenges and responsible innovation. *Journal of Management Studies*, 59(1), 1–28.

47. Lehoux, P., Silva, H., Denis, J., Miller, F., Pozelli Sabio, R., & Mendell, M. (2021). Moving toward responsible value creation: Business model challenges faced by organizations producing responsible health innovations. *Journal of Product Innovation Management*, 38(5), 548–573.

48. Jarmai, K. (2020). Learning from sustainability-oriented innovation. In: Jarmai, K. (eds) *Responsible Innovation*. SpringerBriefs in Research and Innovation Governance. pp. 19–35. Springer, Dordrecht.

49. Auer, A., & Jarmai, K. (2017). Implementing responsible research and innovation practices in SMEs: Insights into drivers and barriers from the Austrian medical device sector. *Sustainability*, 10(1), 17.

50. Chatfield, K., Borsella, E., Mantovani, E., Porcari, A., & Stahl, B. (2017). An investigation into risk perception in the ICT industry as a core component of responsible research and innovation. *Sustainability*, 9(8), 1424.

51. Gurzawska, A., Mäkinen, M., & Brey, P. (2017). Implementation of Responsible Research and Innovation (RRI) practices in industry: Providing the right incentives. *Sustainability*, 9(10), 1759.

52. Centers for Disease Control and Prevention (CDC) (2023). Policy Analysis. https://www.cdc.gov/policy/polaris/policyprocess/policyanalysis/index.html (accessed on 28 November 2023).

53. Cook, L., LaVan, H., & Zilic, I. (2018). An exploratory analysis of corporate social responsibility reporting in US pharmaceutical companies. *Journal of Communication Management*, 22(2), 197–211.

54. Micozzi, N., & Yigitcanlar, T. (2022). Understanding smart city policy: Insights from the strategy documents of 52 local governments. *Sustainability*, 14(16), 10164.

55. Olivera, J., Ford, J., Sowden, S., & Bambra, C. (2022). Conceptualisation of health inequalities by local healthcare systems: A document analysis. *Health & Social Care in the Community*, 30(6), e3977–e3984.

56. CompaniesMarketCap (2023). Largest Tech Companies by Market Cap. https://companiesmarketcap.com/tech/largest-tech-companies-by-market-cap/ (accessed on 25 June 2023).

57. Federal Trade Commission (2021). Non-HSR Reported Acquisitions by Select Technology Platforms, 2010–2019: A Report of the FTC. https://www.ftc.gov/system/files/documents/reports/non-hsr-reported-acquisitions-select-technology-platforms-2010-2019-ftc-study/p201201technologyplatformstudy2021.pdf (accessed on 25 June 2023).

58. Capgemini and Efma (2021). Unprecedented Access to Capital Investment Fuels InsurTech and BigTech Maturity and Customer Adoption, World Insurtech Report. https://www.capgemini.com/in-en/wp-content/uploads/sites/18/2021/09/WORLD-INSURTECH-REPORT-2021.pdf (accessed on 25 June 2023).

59. Congressional Research Service (2022). Big Tech in Financial Services. https://crsreports.congress.gov/product/pdf/R/R47104 (accessed on 25 June 2023).

60. Kerzel, U. (2021). Enterprise AI canvas integrating artificial intelligence into business. *Applied Artificial Intelligence*, 35(1), 1–12.

61. Elliott, K., Price, R., Shaw, P., Spiliotopoulos, T., Ng, M., Coopamootoo, K., & van Moorsel, A. (2021). Towards an equitable digital society: Artificial intelligence (AI) and corporate digital responsibility (CDR). *Society*, 58(3), 179–188.

62. Brauner, P., Hick, A., Philipsen, R., & Ziefle, M. (2023). What does the public think about artificial intelligence? A criticality map to understand bias in the public perception of AI. *Frontiers in Computer Science*, 5, 1113903.

63. Zhdanov, D., Bhattacharjee, S., & Bragin, M. (2022). Incorporating FAT and privacy aware AI modeling approaches into business decision making frameworks. *Decision Support Systems*, 155, 113715.

64. Anagnostou, M., Karvounidou, O., Katritzidaki, C., Kechagia, C., Melidou, K., Mpeza, E., ... & Peristeras, V. (2022). Characteristics and challenges in the industries towards responsible AI: A systematic literature review. *Ethics and Information Technology*, 24(3), 37.

65. Kunz, W., & Wirtz, J. (2023). Corporate digital responsibility (CDR) in the age of AI: Implications for interactive marketing. *Journal of Research in Interactive Marketing.* https://doi.org/10.1108/JRIM-06-2023-0176

66. IBM (2022). Everyday Ethics for Artificial Intelligence. https://www.ibm.com/watson/assets/duo/pdf/everydayethics.pdf (accessed on 25 June 2023).

67. Intel (2023). 2022–2023 Corporate Responsibility Report. https://csrreportbuilder.intel.com/pdfbuilder/pdfs/CSR-2022-23-Full-Report.pdf (accessed on 25 June 2023).

68. Oracle (2023). Oracle's Guide to Ethical Considerations in AI Development and Deployment. https://blogs.oracle.com/ai-and-datascience/post/is-responsible-ai-synonymous-with-ai-ethics (accessed on 25 June 2023).

69. Microsoft (2020). Research Collection: Research Supporting Responsible AI. https://www.microsoft.com/en-us/research/blog/research-collection-research-supporting-responsible-ai/ (accessed on 25 June 2023).

70. NXP Semiconductors (2020). The Morals of Algorithms. https://www.nxp.com/docs/en/white-paper/AI-ETHICAL-FRAMEWORK-WP.pdf (accessed on 25 June 2023).

71. Workday (2022). Workday's Continued Diligence to Ethical AI and ML Trust. https://blog.workday.com/en-us/2022/workdays-continued-diligence-ethical-ai-and-ml-trust.html (accessed on 25 June 2023).

72. Xiaomi (2021). Xiaomi Trustworthy AI White Paper. https://www.studocu.com/vn/document/truong-dai-hoc-ha-noi/marketing/xiaomi-trustworthy-ai-white-paper-en-may-2021/76314130 (accessed on 25 June 2023).

73. Cisco (2022). Cisco Principles for Responsible Artificial Intelligence. https://www.cisco.com/c/dam/en_us/about/doing_business/trust-center/docs/cisco-responsible-artificial-intelligence-principles.pdf (accessed on 25 June 2023).

74. Google (2022). 2022 AI Principles Progress Update. https://ai.google/static/documents/ai-principles-2022-progress-update.pdf (accessed on 25 June 2023).

75. Nakao, Y., Stumpf, S., Ahmed, S., Naseer, A., & Strappelli, L. (2022). Toward involving end-users in interactive human-in-the-loop AI fairness. *ACM Transactions on Interactive Intelligent Systems*, 12(3), 1–30.

76. Balasubramaniam, N., Kauppinen, M., Rannisto, A., Hiekkanen, K., & Kujala, S. (2023). Transparency and explainability of AI systems: From ethical guidelines to requirements. *Information and Software Technology*, 159, 107197.

77. Sanderson, C., Douglas, D., Lu, Q., Schleiger, E., Whittle, J., Lacey, J., ... & Hansen, D. (2023). AI ethics principles in practice: Perspectives of designers and developers. *IEEE Transactions on Technology and Society*. 4(2), 171–187.

78. Malik, M., & Kanwal, L. (2018). Impact of corporate social responsibility disclosure on financial performance: Case study of listed pharmaceutical firms of Pakistan. *Journal of Business Ethics*, 150, 69–78.

79. Akbari, M., Rezvani, A., Shahriari, E., Zúñiga, M., & Pouladian, H. (2020). Acceptance of 5 G technology: Mediation role of Trust and Concentration. *Journal of Engineering and Technology Management*, 57, 101585.

80. Chouaibi, S., Rossi, M., Siggia, D., & Chouaibi, J. (2021). Exploring the moderating role of social and ethical practices in the relationship between environmental disclosure and financial performance: Evidence from ESG companies. *Sustainability*, 14(1), 209.

81. Kelly, S., Kaye, S., & Oviedo-Trespalacios, O. (2022). What factors contribute to acceptance of artificial intelligence? A systematic review. *Telematics and Informatics*, 77, 101925.

82. Webster, P. (2023). Tech companies criticise health AI regulations. *The Lancet*, 402(10401), 517–518.

83. McStay, A. (2020). Emotional AI and EdTech: Serving the public good? *Learning, Media and Technology*, 45(3), 270–283.

84. Adobe (2023). Adobe's Commitment to AI Ethics. https://www.adobe.com/content/dam/cc/en/ai-ethics/pdfs/Adobe-AI-Ethics-Principles.pdf (accessed on 25 June 2023).

85. Sony (2023). Sony Group's Initiatives for Responsible AI. 2023. https://www.sony.com/en/SonyInfo/sony_ai/responsible_ai.html (accessed on 25 June 2023).

86. VMware (2022). Why Your Organization Needs a Set of Ethical Principles for AI. https://octo.vmware.com/why-your-organization-needs-ethical-principles-for-ai/ (accessed on 25 June 2023).

87. Atlassian (2023). Atlassian's Responsible Technology Principles. https://www.atlassian.com/trust/responsible-tech-principles (accessed on 25 June 2023).

88. Schneider Electric (2023). AI Knowledge Base – Responsible and Ethical AI. 2023. https://www.se.com/ww/en/about-us/artificial-intelligence/knowledge-base.jsp (accessed on 25 June 2023).

89. NVIDIA (2022). Corporate Responsibility Report 2022. https://images.nvidia.com/aem-dam/en-zz/Solutions/csr/FY2022-NVIDIA-Corporate-Responsibility.pdf (accessed on 25 June 2023)

90. Qualcomm (2022). Qualcomm Corporate Responsibility Report. 2022. https://www.qualcomm.com/content/dam/qcomm-martech/dm-assets/documents/2022-qualcomm-corporate-responsibility-report.pdf (accessed on 25 June 2023).

91. Palantir (2023). Enabling Responsible AI in Palantir Foundry. 2023. https://blog.palantir.com/enabling-responsible-ai-in-palantir-foundry-ac23e3ad7500 (accessed on 25 June 2023).

92. Samsung (2022). Samsung Electronics Sustainability Report 2022. https://images.samsung.com/is/content/samsung/assets/uk/sustainability/overview/Samsung_Electronics_Sustainability_Report_2022.pdf (accessed on 25 June 2023).

93. Towse, A., & Mauskopf, J. (2018). Affordability of new technologies: The next frontier. *Value in Health*, 21(3), 249–251.

94. Li, W., Yigitcanlar, T., Erol, I., & Liu, A. (2021). Motivations, barriers and risks of smart home adoption: From systematic literature review to conceptual framework. *Energy Research & Social Science*, 80, 102211.

95. Tuncer, I. (2021). The relationship between IT affordance, flow experience, trust, and social commerce intention: An exploration using the SOR paradigm. *Technology in Society*, 65, 101567.

96. Salesforce. (2023). Meet Salesforce's Trusted AI Principles. https://blog.salesforceairesearch.com/meet-salesforces-trusted-ai-principles/ (accessed on 25 June 2023).

97. Automatic Data Processing (2022). ADP: Ethics in Artificial Intelligence. https://www.adp.com/-/media/adp/redesign2018/pdf/data-privacy/ai-ethics-statement.pdf?rev=934d7063975f402889c4ed8610324c36&hash=9FA7B34280D71654740CC51D14F74E79 (accessed on 25 June 2023).

98. Amazon (2022). Introducing AWS AI Service Cards: A New Resource to Enhance Transparency and Advance Responsible AI. https://aws.amazon.com/blogs/machine-learning/introducing-aws-ai-service-cards-a-new-resource-to-enhance-transparency-and-advance-responsible-ai/ (accessed on 25 June 2023).

99. Baidu (2023). Responsible AI. https://esg.baidu.com/detail/560.html (accessed on 25 June 2023).

100. Airbnb (2021). Airbnb's Work on Human Rights. https://news.airbnb.com/airbnbs-work-on-human-rights/ (accessed on 25 June 2023).

101. Dell (2022). Dell Technologies Principles for Ethical Artificial Intelligence. https://www.delltechnologies.com/asset/en-us/solutions/business-solutions/briefs-summaries/principles-for-ethical-ai.pdf (accessed on 25 June 2023).

102. Equinix (2023). 4 Factors That Define Responsible AI. https://blog.equinix.com/blog/2023/01/09/4-factors-that-define-responsible-ai/ (accessed on 25 June 2023).

103. Meta Platforms (2021). Facebook's Five Pillars of Responsible AI. https://ai.meta.com/blog/facebooks-five-pillars-of-responsible-ai/ (accessed on 25 June 2023).

104. Makridakis, S. (2017). The forthcoming Artificial Intelligence (AI) revolution: Its impact on society and firms. *Futures*, 90, 46–60.

105. French, A., Shim, J. P., Risius, M., Larsen, K. R., & Jain, H. (2021). The 4th Industrial Revolution powered by the integration of AI, blockchain, and 5G. *Communications of the Association for Information Systems*, 49(1), 6.

106. Ahmed, T., Karmaker, C. L., Nasir, S. B., Moktadir, M. A., & Paul, S. K. (2023). Modeling the artificial intelligence-based imperatives of industry 5.0 towards resilient supply chains: A post-COVID-19 pandemic perspective. *Computers & Industrial Engineering*, 177, 109055.

107. Buhmann, A., & Fieseler, C. (2021). Towards a deliberative framework for responsible innovation in artificial intelligence. *Technology in Society*, 64, 101475.

108. Tubadji, A., Huang, H., & Webber, D. J. (2021). Cultural proximity bias in AI-acceptability: The importance of being human. *Technological Forecasting and Social Change*, 173, 121100.

109. Laux, J., Wachter, S., & Mittelstadt, B. (2024). Trustworthy artificial intelligence and the European Union AI act: On the conflation of trustworthiness and acceptability of risk. *Regulation & Governance*, 18(1), 3–32.

110. Laux, J., Wachter, S., & Mittelstadt, B. (2023). Trustworthy artificial intelligence and the European Union AI act: On the conflation of trustworthiness and acceptability of risk. *Regulation & Governance*. https://doi.org/10.1111/rego.12512

111. AI, H. (2019). High-level expert group on artificial intelligence. *Ethics guidelines for trustworthy AI*, 6. https://www.aepd.es/sites/default/files/2019-09/ai-definition.pdf

112. Sovacool, B. K., Kester, J., Noel, L., & de Rubens, G. Z. (2019). Energy injustice and Nordic electric mobility: Inequality, elitism, and externalities in the electrification of vehicle-to-grid (V2G) transport. *Ecological Economics*, 157, 205–217.

113. Altinay, F., Ossiannilsson, E., Altinay, Z., & Dagli, G. (2020). Accessible services for smart societies in learning. *The International Journal of Information and Learning Technology*, 38(1), 75–89.

114. Early, J., & Hernandez, A. (2021). Digital disenfranchisement and COVID-19: Broadband internet access as a social determinant of health. *Health Promotion Practice*, 22(5), 605–610.

115. Brand, T., & Blok, V. (2019). Responsible innovation in business: A critical reflection on deliberative engagement as a central governance mechanism. *Journal of Responsible Innovation*, 6(1), 4–24.

116. Padilla-Lozano, C. P., & Collazzo, P. (2021). Corporate social responsibility, green innovation and competitiveness–causality in manufacturing. *Competitiveness Review: An International Business Journal*, 32(7), 21–39.

117. Wang, L., Qu, G., & Chen, J. (2022). Towards a meaningful innovation paradigm: Conceptual framework and practice of leading world-class enterprise. *Chinese Management Studies*, 16(4), 942–964.

118. Carayannis, E. G., Grigoroudis, E., Stamati, D., & Valvi, T. (2019). Social business model innovation: A quadruple/quintuple helix-based social innovation ecosystem. *IEEE Transactions on Engineering Management*, 68(1), 235–248.

119. Hagedoorn, J., Haugh, H., Robson, P., & Sugar, K. (2023). Social innovation, goal orientation, and openness: Insights from social enterprise hybrids. *Small Business Economics*, 60(1), 173–198.

120. Sáez-Martínez, F. J., Ferrari, G., & Mondéjar-Jiménez, J. (2015). Eco-innovation: Trends and approaches for a field of study. *Innovation*, 17(1), 1–5.

121. Nickel, P. J., Franssen, M., & Kroes, P. (2010). Can we make sense of the notion of trustworthy technology? *Knowledge, Technology & Policy*, 23, 429–444.

122. Liu, H., Wang, Y., Fan, W., Liu, X., Li, Y., Jain, S., ... & Tang, J. (2022). Trustworthy ai: A computational perspective. *ACM Transactions on Intelligent Systems and Technology*, 14(1), 1–59.

123. Petkovic, D. (2023). It is not "Accuracy vs. Explainability"—we need both for trustworthy AI systems. *IEEE Transactions on Technology and Society*, 4(1), 46–53.

124. Chi, O. H., Jia, S., Li, Y., & Gursoy, D. (2021). Developing a formative scale to measure consumers' trust toward interaction with artificially intelligent (AI) social robots in service delivery. *Computers in Human Behavior*, 118, 106700.

125. Jacovi, A., Marasović, A., Miller, T., & Goldberg, Y. (2021, March). Formalizing trust in artificial intelligence: Prerequisites, causes and goals of human trust in AI. In *Proceedings of the 2021 ACM Conference on Fairness, Accountability, and Transparency* (pp. 624–635). Toronto, Canada.

126. Keding, C., & Meissner, P. (2021). Managerial overreliance on AI-augmented decision-making processes: How the use of AI-based advisory systems shapes choice behavior in R&D investment decisions. *Technological Forecasting and Social Change*, 171, 120970.

127. Shneiderman, B. (2020). Bridging the gap between ethics and practice: Guidelines for reliable, safe, and trustworthy human-centered AI systems. *ACM Transactions on Interactive Intelligent Systems (TiiS)*, 10(4), 1–31.

128. Wells, L., & Bednarz, T. (2021). Explainable ai and reinforcement learning—A systematic review of current approaches and trends. *Frontiers in Artificial Intelligence*, 4, 550030.

129. Hickman, E., & Petrin, M. (2021). Trustworthy AI and corporate governance: The EU's ethics guidelines for trustworthy artificial intelligence from a company law perspective. *European Business Organization Law Review*, 22, 593–625.

130. Omrani, N., Rivieccio, G., Fiore, U., Schiavone, F., & Agreda, S. G. (2022). To trust or not to trust? An assessment of trust in AI-based systems: Concerns, ethics and contexts. *Technological Forecasting and Social Change*, 181, 121763.

131. Bacq, S., & Aguilera, R. V. (2022). Stakeholder governance for responsible innovation: A theory of value creation, appropriation, and distribution. *Journal of Management Studies*, 59(1), 29–60.

132. Hahn, G. (2020). Industry 4.0: A supply chain innovation perspective. *International Journal of Production Research*, 58(5), 1425–1441.

133. Huang, S., Wang, B., Li, X., Zheng, P., Mourtzis, D., & Wang, L. (2022). Industry 5.0 and Society 5.0—Comparison, complementation and co-evolution. *Journal of Manufacturing Systems*, 64, 424–428.

134. Kovacs, O. (2022). Inclusive industry 4.0 in Europe: Japanese lessons on socially responsible industry 4.0. *Social Sciences*, 11(1), 29.

135. Ivanov, D. (2023). The industry 5.0 framework: Viability-based integration of the resilience, sustainability, and human-centricity perspectives. *International Journal of Production Research*, 61(5), 1683–1695.

136. Saihi, A., Awad, M., & Ben-Daya, M. (2023). Quality 4.0: Leveraging Industry 4.0 technologies to improve quality management practices–a systematic review. *International Journal of Quality & Reliability Management*, 40(2), 628–650.

137. Sai Manohar, S., & Pandit, S. R. (2014). Core values and beliefs: A study of leading innovative organizations. *Journal of Business Ethics*, 125, 667–680.

138. Chen, J., Sun, C., & Liu, J. (2022). Corporate social responsibility, consumer sensitivity, and overcapacity. *Managerial and Decision Economics*, 43(2), 544–554.

139. Segarra-Oña, M., Peiró-Signes, Á., & Mondéjar-Jiménez, J. (2016). Twisting the twist: How manufacturing & knowledge-intensive firms excel over manufacturing & operational and all service sectors in their eco-innovative orientation. *Journal of Cleaner Production*, 138, 19–27.

CHAPTER 8

Artificial Intelligence and Sustainable Development Goals

1 INTRODUCTION

In recent years, the construction industry has experienced a transformative shift due to the progression of artificial intelligence (AI) technologies (Li et al., 2021). This shift has been characterised by a fundamental redirection in how construction processes are designed, planned, and executed (Wang et al., 2023). AI's integration has transformed current practices, introducing innovative methodologies and tools that optimise various facets of the construction lifecycle. For instance, AI-powered drones are being used for site surveying and progress monitoring. These drones can quickly and accurately capture data, allowing stakeholders to identify potential issues and make informed decisions in real time. It has empowered the industry with predictive analytics, automation, and data-driven insights that streamline workflows, enhance decision-making, and improve overall project outcomes (Yigitcanlar et al., 2023).

The 2015 introduction of the United Nations' 2030 Agenda prompted a strategic shift within the construction industry (General, 2015). This shift occurred because the construction industry impacts various dimensions of society, economy, and environment, rendering it a crucial contributor to the realisation of sustainable development goals (SDGs). In shaping the built

DOI: 10.1201/9781003521440-10

environment, the industry affects communities, making its role indispensable in advancing the global sustainability agenda outlined by the SDGs (Pan and Zhang, 2023). The industry is progressively embracing sustainability principles influenced by the SDGs by incorporating sustainable design, adopting green building standards, concentrating on sustainable materials and technologies, highlighting lifecycle assessment, forging collaborative partnerships, and investing in research and innovation. This alignment with the SDGs is driving the industry towards a more environmentally friendly and inclusive future, with stakeholders throughout the value chain collaborating to accomplish SDGs on a global scale (Bang and Andersen, 2022).

Moreover, the construction industry is undergoing a significant transformation, acknowledging the imperative to incorporate AI as an essential tool in daily operations (Regona et al., 2022a). AI's crucial role is evident in its ability to optimise building designs for enhanced energy efficiency and minimised environmental impact. Through extensive data analysis and scenario simulations, AI empowers architects and engineers to make well-informed decisions that comply with rigorous green building standards. In addition to promoting sustainability, AI has played a pivotal role in advancing construction materials and technologies, driving innovation and resilience in structural design (Omer and Noguchi, 2020). By harnessing AI algorithms and machine learning techniques, researchers and industry professionals have developed novel materials that enhance the durability, safety, and efficiency of structures. These advancements not only contribute to sustainable construction practices but also address evolving challenges in urbanisation, climate change, and infrastructure development (Yigitcanlar et al., 2008).

The construction industry is beginning to realise the advantages of sustainable construction practices and technological advancements into their traditional frameworks (Wang et al., 2021). This shift signifies a broader recognition of the imperative to mitigate environmental impact and enhance long-term sustainability across all facets of construction activity. As awareness grows regarding the potential advantages of sustainable approaches, stakeholders within the industry are increasingly exploring innovative solutions and integrating them into their established practices (Baduge et al., 2022). This evolution reflects a proactive response to emerging global challenges such as resource depletion, cost overruns, safety issues, and urbanisation pressures, signalling a pivotal moment in the industry's journey towards greater resilience and responsibility (Akinosho et al., 2020).

AI presents itself as a compelling pathway for the construction sector to attain sustainable goals by harnessing its capacity to process vast datasets, anticipate outcomes, and automate tasks. These capabilities serve as a catalyst for driving sustainable practices within construction. Evident by its ability to optimise energy usage in building designs, minimise waste through accurate resource allocation, enhance safety measures by pre-emptively identifying risks, streamline project management for efficiency, ensure real-time compliance with sustainability standards through monitoring, and enable life-cycle assessments for eco-friendly materials (Zhang, 2021). These AI-driven tools not only streamline operations but also form the bedrock for advancing the industry's sustainability journey.

Despite extensive research on SDGs, a significant research gap exists regarding the construction industry's role in achieving them, particularly concerning AI implementation (Sachs et al., 2019). While the industry is vital for sustainability, empirical studies on its contribution to SDGs are limited. This gap highlights the need for focused investigation into aligning construction practices and AI technologies with SDGs to advance global sustainable development (Yigitcanlar et al., 2022). The intersection of AI, sustainability, and construction remains underexplored, with minimal empirical research on AI's practical applications in enhancing sustainability outcomes in construction projects. Additionally, as AI evolves and becomes more prevalent in construction projects, understanding its coexistence with sustainability is crucial. Existing literature provides theoretical frameworks for addressing broader sustainability aspects of construction projects, but limited sources explore how AI can aid the construction industry in achieving relevant SDGs. This scarcity underscores the need for further exploration and analysis to fully harness AI's potential as a catalyst for sustainable advancements within construction practices (Saeed et al., 2022).

Bridging this gap requires interdisciplinary collaboration to develop holistic approaches for integrating AI into construction practices. Empirical studies are necessary to identify barriers and opportunities associated with AI adoption in construction, including data privacy, technical challenges, and workforce readiness. Addressing the research gap is crucial for informing policy decisions and industry practices to achieve SDGs effectively. Closing this gap can unlock the construction industry's full potential in advancing sustainability and contributing to SDGs.

This study aims to comprehensively explore the intersection of technological innovation, particularly AI, with sustainable construction practices.

The objective is to elucidate the role of AI in enhancing sustainability across various phases of construction projects, thereby contributing to the achievement of SDGs. Through an in-depth analysis, the study seeks to identify opportunities and challenges associated with integrating AI-driven solutions in construction processes. Additionally, it aims to provide actionable insights for industry stakeholders on optimal ways of leveraging AI to achieve sustainability outcomes, improve project efficiency, and foster responsible practices within the construction sector.

The chapter is organised into several sections to provide a structured analysis of the research findings. In Section 3, the research methodology employed in this study is outlined, detailing the systematic approach adopted to conduct the literature review and map AI applications in the construction industry. Section 4 presents the results and general observations regarding the relevant SDGs in construction, clarifying how AI can contribute to achieving these goals effectively. This section provides a comprehensive overview of the direct and indirect impact of AI on a construction project. In Section 5, a detailed discussion of the results is offered, providing deeper insights and interpretations into the implications of AI integration for sustainability in the construction sector. Finally, Section 6 concludes the chapter by summarising the key findings, discussing their implications for future research and practice, and suggesting potential avenues for further exploration.

2 LITERATURE BACKGROUND

2.1 Sustainable Construction

At the inaugural First International Conference on Sustainable Construction in November 1994, the concept of sustainability was defined as follows: "Building a healthy built environment through the application of resource-efficient, ecologically sound principles (Poprach et al., 2019)". Sustainable construction represents a commitment to three fundamental pillars:

- Economic sustainability involves enhancing profitability by optimising the utilisation of resources, including labour, materials, water, and energy (Oluleye et al., 2022).

- Environmental sustainability entails preventing harmful and potentially irreversible effects on the environment through prudent use of natural resources, minimising waste generation, and safeguarding (Barros and Rusche, 2021).

- Social sustainability encompasses meeting the needs of individuals throughout the construction process, from inception to demolition, by ensuring high customer satisfaction and fostering close collaboration with project-related stakeholders (Fnais et al., 2022).

In traditional design and construction, the focus is on cost, performance, and quality objectives (Opoku, 2019). However, sustainable design and construction bring in additional considerations, such as minimising resource depletion, reducing environmental degradation, and promoting a healthy built environment. This shift towards sustainability marks a new era within the building design and construction industry, where sustainable goals are embedded in decision-making processes throughout the lifecycle of the facility. Figure 8.1 outlines the evolution and challenges of the sustainable construction concept in a global context.

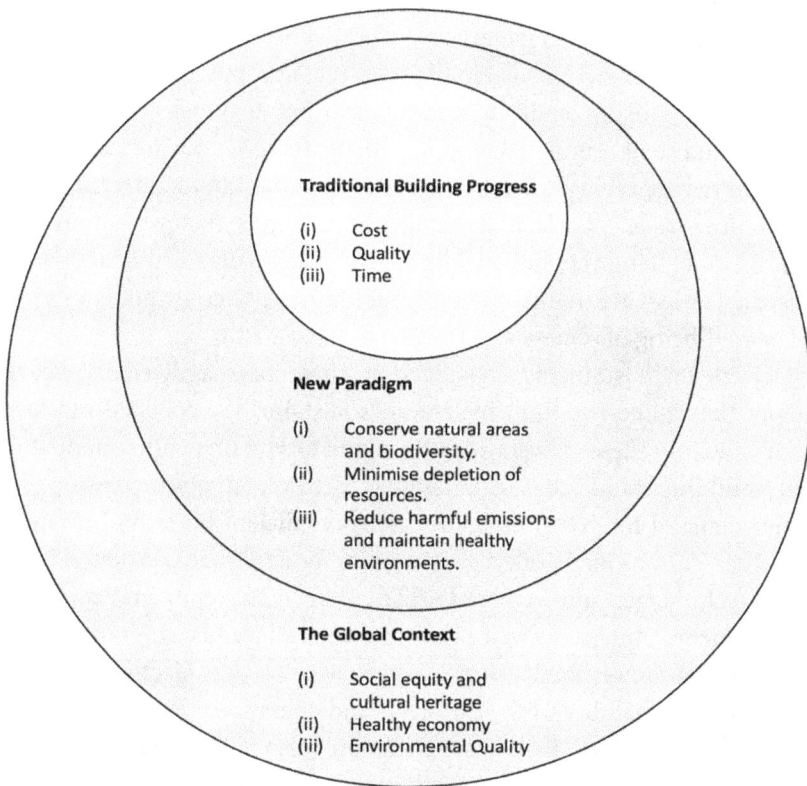

Traditional Building Progress

(i) Cost
(ii) Quality
(iii) Time

New Paradigm

(i) Conserve natural areas and biodiversity.
(ii) Minimise depletion of resources.
(iii) Reduce harmful emissions and maintain healthy environments.

The Global Context

(i) Social equity and cultural heritage
(ii) Healthy economy
(iii) Environmental Quality

FIGURE 8.1   Challenges of sustainable construction in a global context.

In contrast to many other industries, the construction sector in Australia holds a unique position due to its enduring nature and profound, long-lasting impact on society. Buildings and infrastructure developed within this industry can shape communities and landscapes for generations, emphasising the critical importance of thoughtful planning and design (Barbosa et al., 2023). The decisions made during the initial phases of a construction project have far-reaching implications for its sustainable performance over time.

For a structure to achieve exemplary performance while minimising its ecological footprint, it is imperative to embed sustainability principles from the outset of the project. This proactive approach ensures that considerations of environmental, economic, and social sustainability are integrated seamlessly into every stage of development and construction. As highlighted by Regona et al. (2022b), the early incorporation of sustainability principles is fundamental to realising a built environment that not only meets the needs of the present but also safeguards the well-being of future generations. Harmonising the diverse facets of sustainability is essential in achieving sustainable construction practices. This entails not only minimising environmental impact but also fostering economic viability and social equity (Boje et al., 2020). Table 8.1 illustrates the multifaceted nature of sustainability, emphasising the interconnectedness of environmental, economic, and social considerations in construction projects. By embracing this holistic approach, stakeholders can create buildings and infrastructure that not only endure but also contribute positively to the well-being of society and the planet (Gao 2020).

Drivers for sustainable construction encompass a diverse range of factors that guide the industry towards sustainable beneficial practices (Fayek, 2020). These drivers include regulatory frameworks mandating green building standards and emissions reduction targets, increasing consumer demand for eco-friendly and energy-efficient buildings, financial incentives such as tax breaks and subsidies for sustainable projects, innovative technologies and materials that enhance efficiency and minimise environmental impact, as well as broader societal trends emphasising corporate social responsibility, safety, and well-being of project stakeholders and long-term viability in their projects and operations (Pizzi et al., 2020).

Challenges in sustainable construction pose substantial barriers to the broad acceptance of sustainable practices due to several factors. These include the considerable upfront expenses linked to green building materials and technologies, a shortage of skilled labour and proficiency

TABLE 8.1 Principles of sustainable construction, derived from Chen et al. (2022)

Sustainability Pillar	Description
Environmental	• Decrease material intensity through substitution technologies • Improve material recyclability • Minimise and regulate the usage and spread of harmful materials • Lower the energy needed for processing goods and delivering services • Back international conventions and agreements • Optimise the sustainable utilisation of biological and renewable resources • Consider the effects of planned projects on air, soil, water, flora, and fauna.
Social	• Encourage community involvement • Foster the establishment of suitable institutional structures • Evaluate the influence on the current social context • Assess the effects on health and overall well-being
Economic	• Incorporate external costs • Explore alternative financing methods • Create suitable economic tools to encourage sustainable consumption • Evaluate the economic impact on local structures

in sustainable construction techniques, complex and inconsistent regulations that hinder adherence to green building criteria, perceived risks and uncertainties regarding the effectiveness and durability of sustainable designs and technologies, and entrenched industry attitudes and practices that favour traditional construction methods (Fnais et al., 2022).

Additionally, financing constraints, market barriers, and insufficient awareness or understanding of the benefits of sustainable construction among stakeholders further hinder progress towards a more sustainable built environment. Overcoming these barriers requires concerted efforts from governments, industry stakeholders, and the broader community to address policy gaps, promote innovation, provide financial incentives, and foster collaboration towards achieving SDGs.

Significant opportunities remain for projects to become more sustainable for the entirety of the building lifecycle (Regona et al., 2023). While there are variations from previous studies that investigated the environmental consequences of construction, there appears to be a broad consensus on the need for significant transformation in the industry to embrace sustainability goals and contribute to sustainable development (Wang et al., 2023).

2.2 Sustainable Development Goals in Construction

SDGs emerged as an outcome of the Rio United Nations Summit in 2013 (Yigitcanlar, 2010). The summit concluded by recognising the need to establish global development guidelines that consider human needs, environmental sustainability, human rights, and partnerships (Ma et al., 2019). The 17 goals and their 169 targets are structured around the three sustainability pillars—social, environmental, and economic, as seen in Figure 8.2. Unlike the Millennium Development Goals (MDG) predecessor, the SDGs offer better coverage and balance between economic, social, and environmental dimensions of sustainable development (CISL, 2017).

SDGs represent a comprehensive framework for addressing social, economic, and environmental challenges through predefined targets and indicators (Kroll et al., 2019). The construction industry has a direct impact on several crucial SDGs, namely, SDGs 6–9, SDGs 11–13, SDG 15, and SDG 17, as illustrated in Figure 8.3. SDG 6, which focuses on clean water and sanitation, is affected by construction projects that can alter water resources and contribute to the development of water infrastructure. SDG 7, concerning affordable and clean energy, intersects with construction through energy consumption and sustainable practices. SDG 8, addressing decent work and economic growth, is influenced by the construction sector's job

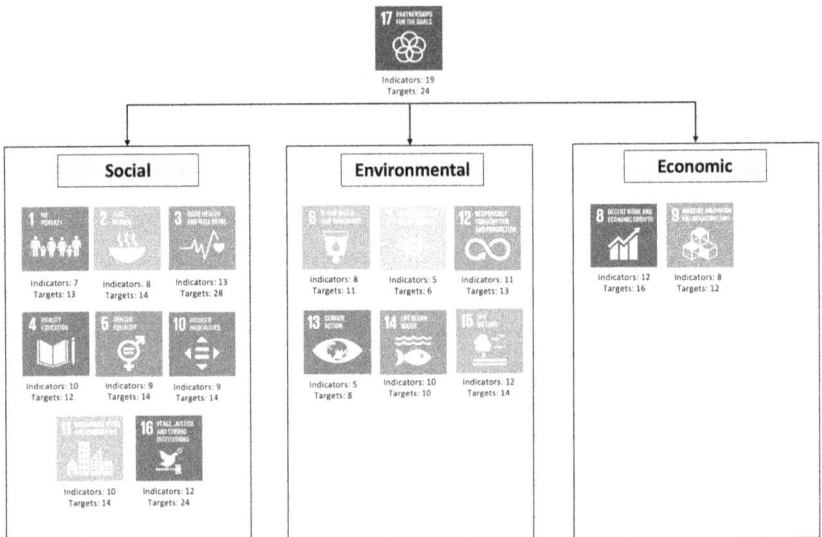

FIGURE 8.2 Sustainable development goals, targets, and indicators derived from United Nations (2015).

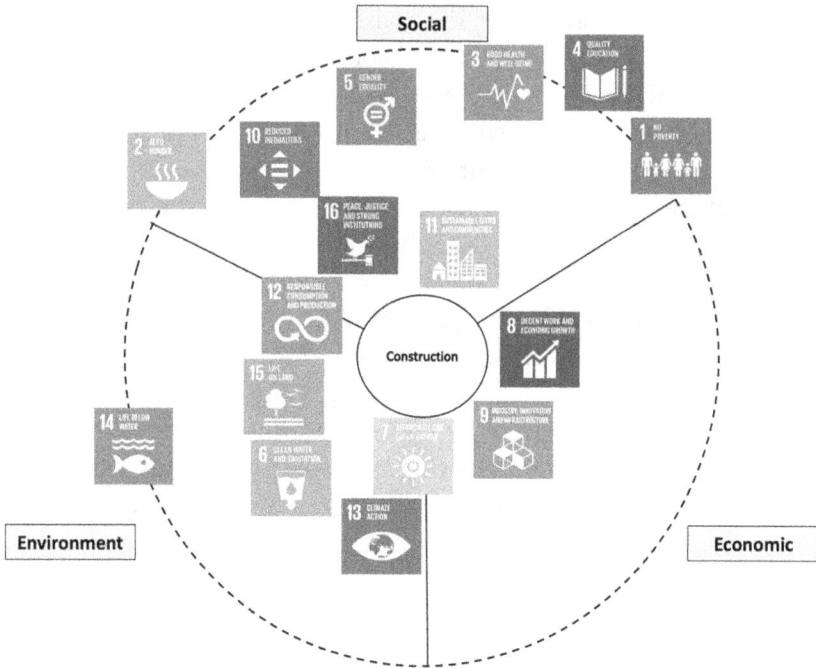

FIGURE 8.3   SDGs based on the three pillars of sustainability.

opportunities and labour practices. SDG 9, emphasising industry, innovation, and infrastructure, is closely tied to construction's role in developing infrastructure and driving economic growth. Additionally, SDGs 11–13, which focus on sustainable cities and communities, responsible consumption and production, and climate action, are significantly impacted by the construction sector. SDG 11 highlights the critical role of the construction industry in urban development and promoting resilient urban spaces.

Furthermore, SDG 12 emphasises resource efficiency and waste reduction through sustainable construction practices such as using recycled materials, minimising waste generation, and promoting circular economy principles. Additionally, SDG 13 underscores the urgent need to adopt low-carbon building materials, energy-efficient design strategies, and climate-resilient infrastructure. Moreover, SDG 15, which addresses life on land, is closely linked to construction practices, particularly concerning land use and biodiversity conservation. Lastly, SDG 17, which emphasises partnerships for all 16 goals, underscores the collaborative efforts required between various stakeholders within the construction industry and beyond to achieve sustainable development objectives. These findings highlight the essential role of the construction industry in advancing SDGs and actively contributing to the global sustainability agenda.

While the SDGs are often portrayed as distinct and separate objectives, they are intricately interconnected rather than isolated. This interdependence provides a route for crafting strategic policies and solutions that simultaneously address multiple goals (Nilsson et al., 2016). This integrated framework implies that advancements in one target can lead to diverse impacts that influence other targets. Consequently, this raises the potential for unintended consequences arising from trade-offs between goals if policies and actions lack careful planning.

Implementing SDGs' principles in a project presents the construction industry with an opportunity to broaden its focus beyond environmental sustainability. However, achieving these goals isn't solely the responsibility of governments and policymakers; it requires collaborative efforts from various sectors to embed sustainability principles into their practices (Waage et al., 2015). To meet the SDGs by the 2030 deadline, it's crucial for all levels of government and sectors of society to integrate these goals into local planning schemes. Local governments have a vital role here, as they can embrace the inclusive nature of the SDGs and identify sectors that may need extra support (Pizzi et al., 2020).

Therefore, successful implementation relies on effective communication between government, private, and public sectors to come to agreements. By fostering partnerships and facilitating dialogue among stakeholders, the construction industry can contribute significantly to advancing the SDGs and promoting sustainable development (Rakesh et al., 2023). This collaborative approach not only enhances our ability to address societal and environmental challenges but also fosters a culture of shared responsibility and innovation essential for lasting impact.

2.3 Artificial Intelligence in Construction

AI is rapidly becoming a cornerstone in the construction industry, reshaping the entire project lifecycle from the design, construction, to the repair and maintenance stage. This transformative evolution is marked by the integration of AI-powered solutions that leverage data analytics, automation, and sophisticated algorithms to optimise every aspect of construction projects (Yigitcanlar et al., 2024).

While many definitions exist for AI, it is widely acknowledged that AI within the built environment involves "the creation of intelligent machines and software that emulate cognitive systems for learning and problem-solving" (Baum et al., 2017). AI is a broad field of computer science that focuses on creating systems or machines capable of performing tasks that typically require human intelligence. The following are the three subsets of AI:

- Artificial Intelligence (AI): The broad field of computer science focused on creating intelligent systems capable of tasks requiring human-like intelligence (Baduge et al., 2022).

- Machine Learning (ML): A subset of AI that develops algorithms enabling computers to learn from data and improve performance without explicit programming (Bang et al., 2022).

- Deep Learning (DL): A subset of ML that automatically employs deep neural networks to learn and represent complex patterns, revolutionising AI applications like image recognition and natural language processing (Baduge et al., 2022).

These AI subcategories have garnered increased interest due to notable progress in data collection and consolidation, analytical capabilities, and the availability of adequate computing power. This is evident as the construction industry spent USD 26 billion on engineering and construction technologies from 2014 to 2019, an increase of USD 8 billion over the previous five years (Young et al., 2021).

In recent years, the construction industry has begun to embrace technology as firms recognise the benefits and operational advantages it offers (Regona et al., 2023). This shift reflects a growing awareness among industry stakeholders of the transformative potential of technological integration in improving project outcomes and enhancing overall productivity. The adoption of technology in construction holds promise for addressing key challenges faced by the urban built environment, including safety concerns, labour shortages, and cost and schedule overruns (Yun et al., 2016). By leveraging innovative technologies, construction companies can navigate these challenges more effectively while also unlocking new opportunities for efficiency and innovation.

Tables 8.2 and 8.3 provide comprehensive insights into the potential and future real-time and technical applications of AI within the construction industry. The two tables outline the diverse ways in which AI-driven solutions are being deployed to optimise various aspects of construction projects, from real-time monitoring and predictive analytics to advanced

TABLE 8.2 Real-time uses of artificial intelligence in construction

AI in Construction	Description
Predictive Maintenance	AI algorithms can analyse real-time data from sensors embedded in machinery and equipment on construction sites (Gasper et al., 2019). By monitoring variables such as temperature, vibration, and usage patterns, AI can predict potential equipment failures before they occur. This enables proactive maintenance, minimising downtime and reducing costly repairs (Yaseen et al., 2020).
Safety Monitoring	AI-powered cameras and sensors can continuously monitor construction sites in real-time to detect safety hazards such as falls, unauthorised personnel, or equipment malfunctions. These systems can issue alerts to supervisors or automatically shut down operations in hazardous situations, thus improving overall safety on the site (Winge et al., 2019).
Resource Optimisation	AI algorithms can optimise the use of resources such as materials, labour, and equipment based on real-time data inputs. AI can adjust construction schedules and resource allocations in response to changing weather conditions, material availability, or workforce productivity, thereby maximising efficiency and reducing waste (Ginzburg et al., 2018).
Quality Control	AI-driven computer vision systems can inspect construction materials and components in real-time to ensure compliance with quality standards and specifications. These systems can detect defects, deviations, or anomalies during the construction process, allowing for immediate corrective actions to be taken to maintain quality standards (Fang et al., 2021).

automation and decision support systems (Mocerino, 2018). By harnessing AI technologies, construction firms can achieve greater precision, accuracy, and efficiency in project execution, thereby improving project outcomes and enhancing overall competitiveness in the market. Moreover, the integration of AI fosters a culture of continuous improvement and innovation within the industry, driving progress towards a more sustainable, resilient, and technologically advanced built environment (Kankanamge et al., 2021).

These standard technologies allow construction companies to have increased accuracy in data analysis and formulate better strategies that benefit all stakeholders involved (Lee et al., 2018). In addition, as projects are temporary and multi-organisational and rely on planning and scheduling models, the construction industry would benefit more than other industries to incorporate technologies (Akinosho et al., 2020).

TABLE 8.3   Technical uses of artificial intelligence in construction

AI in Construction	Description
Data Integration:	AI applications in construction rely on integrating data from various sources such as sensors, drones, BIM models, and historical project data. Ensuring seamless data integration requires robust data management systems and interoperability standards to enable AI algorithms to access and analyse diverse data sources effectively (Rakesh et al., 2023).
Machine Learning Models	AI algorithms used in construction often employ machine learning techniques such as supervised learning, unsupervised learning, and reinforcement learning. These models require extensive training on labelled datasets to learn patterns, predict outcomes, and make decisions autonomously (Sacks et al., 2020). Ensuring the accuracy and reliability of machine learning models is crucial for their successful deployment in construction applications.
Edge Computing	Real-time AI applications in construction often require processing large volumes of data generated by sensors and cameras at the edge of the network, near the construction site. Edge computing technologies enable AI algorithms to run locally on edge devices, reducing latency and bandwidth requirements while ensuring timely decision-making and response in dynamic construction environments (Saeed et al., 2022).
Human-AI Collaboration	Effective deployment of AI in construction requires collaboration between AI systems and human workers. Human-AI collaboration involves integrating AI tools into existing workflows, providing training and support for workers to interact with AI systems, and leveraging human expertise to interpret AI-generated insights and make informed decisions on construction projects (Sachs et al., 2019).

Following this introduction and background, the chapter is organised into several sections to provide a structured analysis of the research findings. In Section 3, the research methodology employed in this study is outlined, detailing the systematic approach adopted to conduct the literature review and map AI applications in the construction industry. Section 4 presents the results and general observations regarding the relevant SDGs in construction, clarifying how AI can contribute to achieving these goals effectively. This section provides a comprehensive overview of the direct and indirect impact of AI on a construction project. In Section 5, a detailed discussion of the results is offered, providing deeper insights and interpretations into the implications of AI integration for sustainability in the construction sector. Finally, Section 6 concludes the chapter by summarising the key findings, discussing their implications for future research and practice, and suggesting potential avenues for further exploration.

3 RESEARCH DESIGN

This study aims to outline key AI technologies aiding the construction industry in achieving the UN's SDGs. Integrating AI can enhance construction efficiency, sustainability, and alignment with SDG principles. A systematic literature review was conducted which adheres to the PRISMA protocol and ensures research replicability (see http://prisma-statement. org). The chosen methodology was based on the structured and unbiased synthesis that systematic reviews offer. By providing comprehensive overviews of existing research on specific topics, they inform decision-making and guide future actions across various fields as seen in Table 8.4.

AI-related literature is exclusively sourced from reputable academic journals, covering established and emerging technologies. By examining current applications, maturity levels, and their potential to address

TABLE 8.4   Statistical information of the data

Domain	Specifics
Data Source	Scopus bibliographic repository
Covered Period	From 1 January 2000 to 15 November 2023
Number of Publications	91
Covered country contexts	64
Number of authors	314
Number of universities	53
Number of publishing sources	39
Number of papers in each SDG	SDG 6: 5
	SDG 7: 12
	SDG 8: 16
	SDG 9: 22
	SDG 11: 13
	SDG 12: 7
	SDG 13 12
	SDG 15: 5
	SDG 17: Interconnected within all the papers
Number of papers in each project state	Design: 14
	Planning: 19
	Construction:36
	Operation and maintenance: 22

sustainability challenges, this study provides insights crucial for advancing sustainable practices in construction.

Stage 1 (planning stage) focused on the included research objectives that answer the research questions, keywords, and a set of exclusion and inclusion criteria. The primary aim of this study is to identify and highlight the predominant AI technologies that are making significant contributions towards achieving the UN's sustainability goals. While some research papers examine effective construction methods that contribute to sustainability objectives, there needs to be more research analysing the prevalent AI technologies employed across the four primary phases of Construction. Based on the literature analysis conducted, it was found that nine SDGs emerged as frequently mentioned and emphasised across various studies and sources as seen in Table 8.5. These SDGs represent key areas of focus and priority for policymakers, organisations, and stakeholders involved in sustainable development efforts.

The following 26 keywords were used to form a defined a search criteria: "Artificial Intelligence" AND "Sustainability" AND "Construction Industry" OR "Buildings" OR "Construction Projects" OR "UN Sustainable Development Goals" OR "Development Goals" OR "Automation" OR "Robotics" OR "Urban Development" OR "AI Application" OR "Sustainable Building Design" OR "Sustainable Urban Planning" OR "Sustainable Architecture" OR "Renewable Energy" OR "Energy Efficiency" OR "Green Building" OR "Sustainable Materials" OR "Waste Reduction" OR "Climate Change" OR "Emissions" OR "Lifecycle" OR "Civil Construction".

Based on the findings from the literature review, keywords co-occurrences for AI in Construction concerning sustainability development goals were established. The results showed that 38 broad technology themes are co-occurrent with sustainability goals. Additional keywords were analysed but did not establish or identify other technology areas. Seven of the thirty-eight key AI themes were frequently mentioned in the journal articles, namely, Decision support systems ($n = 230$), Machine learning ($n = 159$), Internet of Things (IoT; $n = 128$), Automation ($n = 109$), Big data ($n = 76$), Deep Learning ($n = 76$), and Robotics ($n = 55$). The clustering of central keywords not only outlined the research landscape's boundaries but also a comprehensive overview of AI technologies that showcase a notable alignment with the sustainability goals of the United Nations.

The keyword search was conducted in July 2023 and obtained 1,490 results that met the stipulated search parameters (from January 2000 to July 2023). Following the removal of duplicated journal articles, the retained

TABLE 8.5   Construction-related sustainable development goals (Berawi, 2023)

Goals	Objectives
SDG 6: Water	– Ensure access to clean and safe water. – Improve water quality through effective treatment and pollution reduction. – Enhance water use efficiency and sustainable management of water resources.
SDG 7: Clean Energy	– Increase the share of renewable energy sources in the global energy mix. – Promote energy efficiency measures across various sectors. – Ensure universal access to affordable, reliable, and modern energy services.
SDG 8: Decent Work and Economic Growth	– Promote sustained, inclusive, and sustainable economic growth. – Focus on youth employment – Protect labour rights
SDG 9: Innovation and Infrastructure	– Develop and upgrade reliable and sustainable infrastructure. – Foster innovation and encourage the growth of sustainable industries. – Enhance research and development activities to support technological advancements.
SDG 11: Sustainable Cities	– Promote inclusive and sustainable urbanisation. – Provide adequate and affordable housing. – Ensure efficient and sustainable transportation systems within cities.
SDG 12: Consumption and Production	– Reduce waste generation and promote recycling and reuse. – Encourage sustainable practices in industries, businesses, and households. – Support the transition to circular economies.
SDG 13: Climate Action	– Strengthen resilience and adaptive capacity to climate-related impacts. – Integrate climate change measures into national policies, strategies, and planning. – Raise awareness and enhance capacity for climate change mitigation, adaptation, impact reduction, and early warning.
SDG 15: Life of Land (Biodiversity)	– Protect, restore, and sustainably manage ecosystems. – Combat land degradation and deforestation. – Promote the conservation of plant and animal species.
SDG 17: Partnership for the Goals	– Collaboration and partnerships among stakeholders – Increase financial resources to support the implementation of sustainable development initiatives – Importance of data collection, monitoring, and reporting mechanisms

TABLE 8.6   Exclusion and inclusion criteria derived from Yigitcanlar et al. (2020)

Primary Data		Secondary Data	
Inclusionary	**Exclusionary**	**Inclusionary**	**Exclusionary**
Journal articles Peer-reviewed Full-text available online Published in English Government reports Conferences	Duplicate records Books and chapter Industry reports	AI in construction Opportunities and Challenges in construction Relevant to the research objective	Not AI in construction-related Irrelevant research objectives

count settled at 1,358 papers, encompassing research contributions that extend beyond the confines of the university library database. The search mechanism encompassed over 400 bibliographic repositories, notably bibliographic repositories such as Scopus, ScienceDirect, Web of Science, the Directory of Open Access Journals, and Wiley Online Library. As demonstrated in Table 8.6, a set of criteria was systematically formulated to curtail the volume and intricacy of the forthcoming review process, thereby facilitating efficient article screening. In this context, primary data encompasses research outputs meticulously generated by individuals, serving as purposeful tools for comprehending and resolving intricate research questions. Conversely, secondary data assumes the form of research contributions from governmental bodies and construction institutions, collectively contributing to the knowledge landscape.

In stage 2 (review stage), the remaining 1,263 articles were assessed against category formulation, as seen in Table 8.7. As a result of this evaluation, the articles were further refined, which resulted in 578 remaining articles. The title, abstract, and keywords of the remaining 578 articles were screened according to the exclusion criteria, and the number of relevant articles was eventually reduced to 91.

In Stage 3 (reporting), the screening processes aimed to analyse the selected articles according to pre-defined categories to assess similarities and differences (Yigitcanlar et al., 2021a). The four-step method was then used to classify the reviewed literature into specific themes (Degirmenci et al., 2021; Yigitcanlar et al., 2021b). The first step highlighted significant technologies that play a part in reducing carbon emissions and helping meet sustainability goals. Second, the most important ranked themes were categorised and reviewed as aligned with the research aims. The third step was cross-checking the categories with other review studies

TABLE 8.7   Category formulation criteria derived from Butler et al. (2021) and
Yigitcanlar et al. (2020)

	Selection Criteria
Authors	– Using qualitative data, identify the key authors relevant to AI in the construction industry
	– Group AI technologies relating to a particular phase, sustainability goal and form categories
	– Check the consistency and AI categories against other literature
	– Shortlist categories and analyse recent literature reviews
	– The final categories are verified, classified, and finalised
	– Relevant categories are distributed and selected the most pertinent categories
Literature	– Relevant to sustainable construction
	– Quality from scholarly publications
	– Comprehensive coverage of AI role in construction
	– Priories literature with real-world case studies
	– Include literature that explore AI technologies and methodologies specifically addressing sustainability challenges and opportunity
	– Scalability of AI solutions
	– Consider literature that emphasises the involvement of industry stakeholders, policymakers, and community members
	– Discus emerging trends in the field of AI in construction to enhance sustainability

and identifying any additional relevant SDG in construction. Lastly, the themes were categorised and finalised in the planning, design, construction operation, and maintenance under common themes. Figure 8.4 displays the overview of the process of selecting papers.

Subsequently, the selected papers were examined and categorised based on the specific sustainability development goals (SDGs) they addressed and AI sub-categories that were frequently mentioned in the articles. Additionally, the papers were later organised based on the direct and indirect relationships between SDG targets and the construction industry, as seen in Figure 8.5. They were then assessed to determine which specific AI subcategory would exert the most significant influence on each respective target. Other relevant peer-reviewed studies further evaluated these overarching themes.

Additionally, the chosen articles were segmented into four distinct phases: the "planning phase" ($n = 21$), the "design phase" ($n = 25$), the actual "construction phase" ($n = 27$), and the subsequent phases of "operation and maintenance" ($n = 18$). Each phase involves common and distinct AI technologies, which vary in impact on the associated SDG

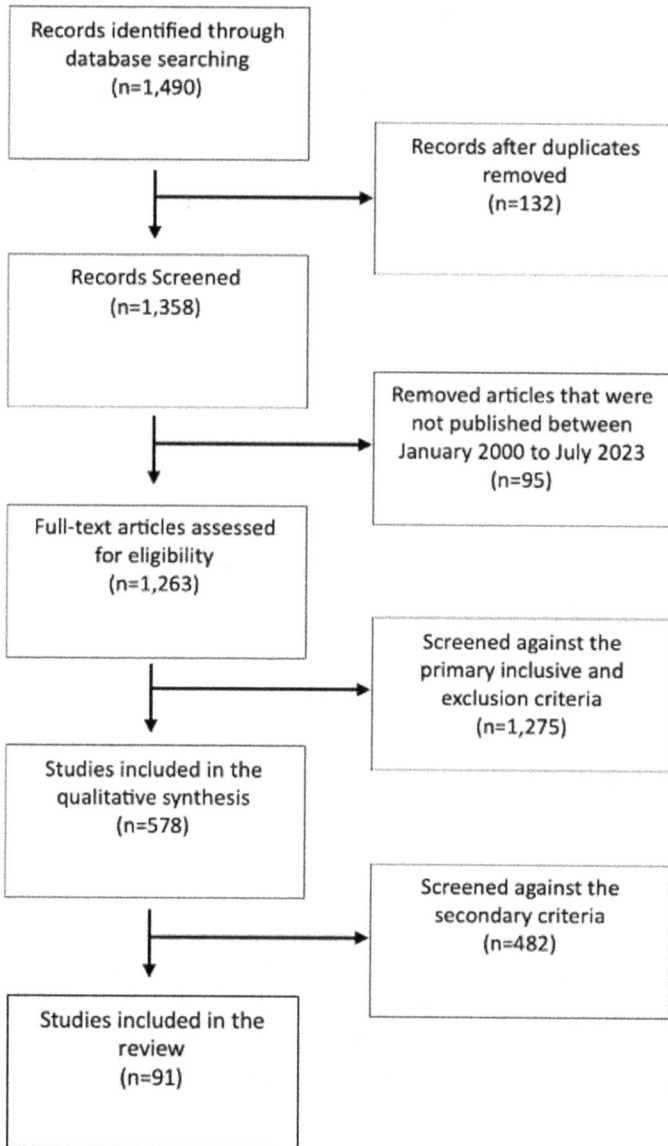

FIGURE 8.4 The PRISMA selection process of relevant literature.

4 ANALYSIS AND RESULTS

4.1 General Observations

The number of publications published in recent years has shown the increased interest in AI helping the built environment become more environmentally sustainable. This is evident as 78% of the 91 articles were

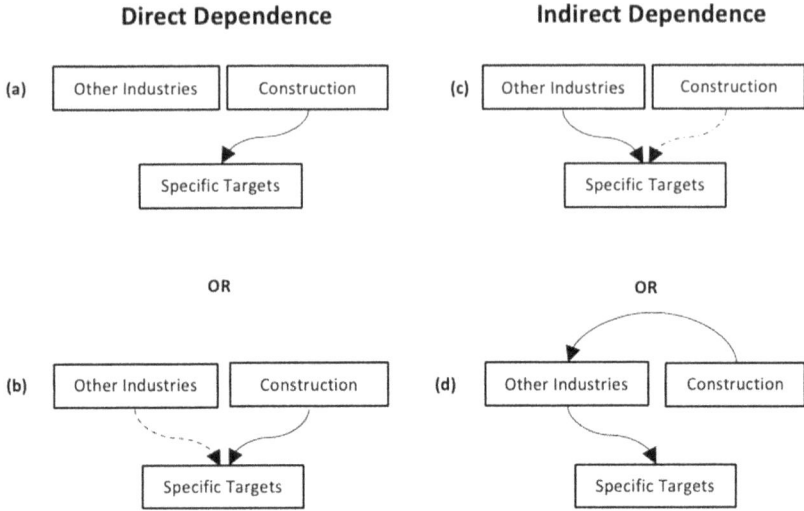

FIGURE 8.5 Process for defining direct and indirect dependencies of SDG targets in the construction industry.

published in the last three years (5 in 2020, 9 in 2021, 25 in 2022, and 22 in 2023). Many of the authors were affiliated with academic institutions in China ($n = 12$), Australia ($n = 9$) and the United States ($n = 8$). This global distribution signifies collaborative efforts in utilising AI to address ecological challenges in the industry. China's significant representation underscores its commitment to innovative solutions, while Australia and the United States demonstrate their dedication to sustainable construction practices. This collective international engagement highlights AI's potential for fostering sustainability in construction on a worldwide scale. As shown in Figure 8.6, the growth of AI literature concerning environmentally sustainable practices has drastically increased in recent years.

The articles were categorised and analysed based on how often different AI sub-categories were mentioned, as detailed in Table 8.8. These findings consistently highlight the importance of AI-related tasks across various project phases. The prevalence of AI tasks across these phases emphasises its significance, showcasing its broad applicability and central role in addressing diverse project needs. ML appears regularly as a contributor, with varying levels of involvement, indicating its supportive role in phases where data-driven approaches are crucial. While DL's consistently lower presence suggests a more specialised application within the project, emphasising its relevance in specific areas where its unique capabilities are essential.

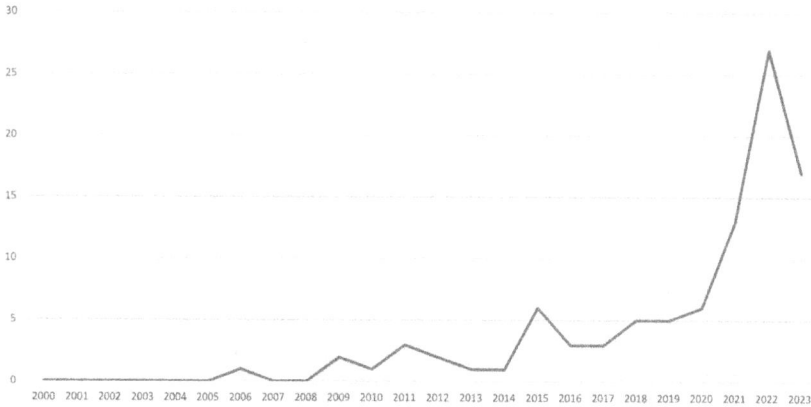

FIGURE 8.6   Distributions of publication by year.

TABLE 8.8   Sub-categories of AI mentioned in the analysed articles

Phases	Artificial Intelligence	Machine Learning	Deep Learning	Total
Planning	8	9	4	21
Design	15	7	3	25
Construction	14	9	4	27
Operation and maintenance (O&M)	15	1	2	18
Total	52	26	13	

In the context of leveraging AI to enhance sustainability within construction projects, an analysis of 91 articles revealed a distinct pattern: only two authors emerged as contributors to more than one paper: Frahzadil ($n = 2$) and Kiomarsi ($n = 2$). An assessment of the author's h-index scores highlighted three prominent researchers: Kovac ($n = 38$), Banins ($n = 33$), and Clements-Croom ($n = 28$). These scores recognise the influence of these authors in utilising AI to advance sustainability within construction contexts. Interestingly, no single author held a dominant position in the literature. The contributions were widespread across various authors, outlined the diverse and collaborative nature of the literature, and depicted a range of viewpoints and cumulative knowledge.

Given the specific focus on AI and sustainability in the literature review, it is unsurprising that a significant portion of the articles ($n = 10$) were published in the Sustainability journal. These papers addressed the entire lifecycle of construction. The remaining papers were published in

TABLE 8.9   Sustainable development goals relevant to the construction industry

Phases	SDG 6	SDG 7	SDG 8	SDG 9	SDG 11	SDG 12	SDG 13	SDG 15
Planning	11	5	3	6	3	5	4	9
Design	1	24	9	11	4	6	8	0
Construction	4	10	4	23	6	24	15	4
O&M	4	17	11	4	14	3	11	1
Total	20	56	27	44	27	38	38	14

TABLE 8.10   Sustainable development goals not relevant to the construction industry

Phases	SDG 1	SDG 2	SDG 3	SDG 4	SDG 5	SDG 10	SDG 14	SDG 16
Planning	0	0	0	0	0	0	0	0
Design	0	0	1	0	0	0	0	0
Construction	0	0	0	0	0	0	0	0
O&M	0	0	1	0	0	0	0	0
Total	0	0	2	0	0	0	0	0

38 journals, focusing on sustainability and how construction projects can utilise AI technologies to reduce carbon emissions, efficient energy consumption, and project waste. Table 8.9 presents the relevant SDG and which construction phases were frequently mentioned within the papers.

It is essential to highlight that several SDGs are not inherently aligned with the core functions of the construction industry, as seen in Table 8.10. While the construction sector plays a crucial role in shaping our built environment, fostering economic growth, and enhancing infrastructure, it does not directly address or encompass the entire spectrum of SDGs, as seen in Table 8.2. These goals, each addressing distinct global challenges, require specialised approaches and interventions outside the immediate purview of construction.

For instance, SDG 1 focuses on addressing poverty, which requires a broader set of economic and social policies that extend beyond the construction industry's primary functions. Similarly, SDG 2, which seeks to achieve zero hunger, is primarily concerned with food security, an area unrelated to the construction's primary function of building physical infrastructure. Further, although the construction sector can impact worker health and safety, it needs to encompass the comprehensive healthcare and well-being objectives of SDG 3.

While construction projects can contribute to educational initiatives, such as constructing schools, the industry itself does not serve as the primary driver of the educational goals central to SDG 4. Gender equality, a pivotal issue in SDG 5, remains a critical social concern but isn't the principal focus of the construction sector, despite ongoing efforts to foster gender diversity and equality within it. Income inequality, a central theme of SDG 10, is indirectly influenced by the construction industry through job creation and wage structures, but addressing income inequality is not the core mission of the sector. SDG 14, addressing life below water, is unrelated to construction, an industry predominantly on land, with no direct impact on marine ecosystems. Lastly, establishing peace, justice, and strong institutions, as mentioned in SDG 16, does not fall within the immediate purview of the construction industry, as this goal primarily involves governance, law enforcement, and related matters.

4.2 Sustainable Development Goals and Artificial Intelligence in Construction

The pivotal role of the construction sector in driving societal development and infrastructure creation underscores the urgent imperative to achieve nine directly related sustainable SDGs goals. Embracing AI within the construction industry has emerged as a critical strategy for projects to elevate sustainability efforts while pursuing SDGs. The seamless fusion of SDGs and AI technologies within construction underscores their transformative potential, catalysing a significant reshaping of the construction landscape.

However, the literature review highlights the imperative for the construction industry to expedite the integration of AI and SDGs, attributed to the growing complexity of construction projects and a tendency to prioritise economic gains over environmental and social considerations (Egwim et al., 2021). As such, there is a pressing need for concerted efforts from all relevant stakeholders to actively contribute to attaining these crucial goals. Collaboration among industry stakeholders is essential to promote sustainable practices that align with the SDGs (Pan and Zhang, 2023). Governments can implement supportive policies, local communities can ensure projects meet their needs and adhere to sustainability principles, construction companies can invest in green solutions, and architects and engineers can incorporate sustainable designs (Ioppolo et al., 2019). This collective effort advances the SDGs and establishes new industry standards, becoming a global model for sustainable construction practices.

4.2.1 SDG 6: Clean Water and Sanitation

Providing clean water and sanitation is considered in SDG 6, which consists of 8 targets and 11 indicators. The main objectives centre around water and sanitation availability and sustainability management that promotes responsible construction practices. Construction projects demand substantial water for concrete mixing, dust control, and site preparation. Construction organisations must protect freshwater bodies near construction sites from contamination (Opoku, 2022). The design of intelligent, innovative, and sustainable buildings offers improvement in water use and could contribute to minimising its impact on the natural environment. The construction industry can leverage the following, as shown in Table 8.11, vital technologies to attain this objective.

TABLE 8.11   Potential of AI for SDG6

Construction Activity	Stage	Sustainability Pillar	Potential of AI
Water Distributions	Design and Planning	Environmental	Predicting and preventing pipeline failures, optimising water flow, and ensuring a consistent and reliable water supply (Naser, 2019).
Monitoring Water	Planning and Construction	Environmental	Leverage satellite data to monitor water resources, identify water scarcity areas, and plan projects accordingly (Arsiwala et al., 2023).
Simulating Water Patterns	Construction	Environmental	Simulating and analysing the patterns of water infrastructure to model the impact of various scenarios and optimise water usage (Aparicio et al., 2020).
Smart Water Infrastructure	Construction	Environmental	Deploying sensors and Internet of Things (IoT) devices throughout water infrastructure to monitor water quality, detect leaks, and manage water distribution (Rodriguez-Gracia et al., 2023).
Water Quality Monitoring	Construction	Environmental	ML algorithms analyse real-time water quality data by identifying contaminants, assessing the impact of construction activities on water quality, and ensuring compliance with environmental standards (Arsiwala et al., 2023).

4.2.2 SDG 7: Clean Energy

SDG 7's emphasis on clean and sustainable energy aligns closely with the objectives of AI in construction through 5 targets and 6 indicators. Construction significantly contributes to greenhouse gas emissions, primarily through the energy used in building operations and construction processes. It ensures access to affordable, reliable, sustainable, and modern community energy (Fei et al., 2021). This goal recognises the pivotal role of energy in construction projects and highlights the need for responsible and sustainable energy practices. This alignment benefits the construction sector and contributes to global efforts to combat climate change and ensure access to clean energy (Hartenburger et al., 2018). The construction industry can leverage the following, as shown in Table 8.12, key technologies to attain this objective.

4.2.3 SDG 8: Decent Work and Economic Growth

SDG 8 promotes sustained, inclusive, and sustainable economic and work growth through 12 targets and 16 indicators. It emphasises the interconnectedness of sustainable policies, striving for inclusive prosperity and decent work opportunities (Aparicio et al., 2020). The construction industry enhances safety standards by identifying and mitigating potential hazards in real-time, aligning with a focus on decent work opportunities and workforce well-being. This diversification fosters innovation and bridges traditional practices with technology, ensuring the construction sector remains adaptable and competitive. Furthermore, these efficiencies benefit the construction sector by increasing profitability and competitiveness, creating jobs, and improving infrastructure (Hattacharya and Bose, 2023). The construction industry can leverage the following, as shown in Table 8.13, key technologies to attain this objective.

4.2.4 SDG 9: Innovation and Infrastructure

SDG 9 seeks to foster innovation, build resilient infrastructure, and promote sustainable industrialisation through 8 targets and 12 indicators. The construction industry can contribute to the goal by prioritising resilient infrastructure and circular economy practices, adopting innovative technologies, and fostering public-private partnerships. The construction industry should develop and align its strategies with the SDG at the organisation and the project level. It plays a crucial role in reducing the environmental footprint of buildings, aligning with sustainable urban development goals (Mocerino, 2018). This ensures that projects delivered (new build or refurbishment) demonstrate sustainable development. The

TABLE 8.12   Potential of AI for SDG7

Construction Activity	Stage	Sustainability Pillar	Potential of AI
Predictive Maintenance	Repair and Maintenance	Economic	Utilise AI algorithms for predictive maintenance of building systems (Rodriguez-Gracia et al., 2023).
Climate Control	Repair and Maintenance	Environmental and Economic	Optimise heating and cooling systems, considering occupancy patterns, weather forecasts, and building conditions to maximise energy efficiency (Rakesh et al., 2023).
Energy Performance	Repair and Maintenance	Environmental and Economic	Predict and analyse the energy performance of buildings that allow for iterative improvements (Huang, 2023).
Energy-Efficient Supply Chains	Construction	Environmental and Economic	AI optimises construction supply chains, reducing energy consumption in transportation and logistics (Naser, 2019).
Monitor Energy Consumption	Planning and Construction	Environmental and Economic	Install IoT sensors and smart meters to monitor real-time energy consumption within construction sites and buildings (Rakesh et al., 2023).
Schedule Optimisation	Planning and Construction	Environmental and Economic	AI algorithms that optimise schedule and resource allocation in which reduce energy-intensive process and enhance overall efficiency (Tatiya et al., 2018)
Building Layout	Design	Environmental and Social	Optimise building layouts, materials, and configurations for enhanced energy efficiency (Liu et al., 2015).
Intelligent Lighting Systems	Design	Environmental and Social	Implement AI-driven lighting systems that respond to occupancy patterns, adjusting lighting levels based on real-time usage data to minimise energy waste (Oyedele et al., 2021).

TABLE 8.13   Potential of AI for SDG8

Construction Activity	Stage	Sustainability Pillar	Potential of AI
Job Creation And Retention	Construction	Economic	Diversify the industry through tech-related roles, including AI developers, data scientists, robotics technicians, cybersecurity experts, and AR/VR developers (Wahbeh et al., 2020).
Productivity	Construction	Economic	Increase productivity through automation, predictive analytics, and streamlining scheduling and resource allocation (Hsu et al., 2020).
Transparent Transactions	Planning	Economic	Implement blockchain technology for transparent and secure financial transactions, ensuring accountability and reducing fraud in construction projects (Sing et al., 2023a)
Dynamic Pricing Models	Planning and Construction	Economic	Accurate and real-time cost estimation, helping construction companies remain competitive and financially sustainable (Tatiya et al., 2018).
Repetitive Tasks	Planning and Construction	Economic	Handle repetitive tasks, increasing productivity and allowing human workers to focus on more complex and value-added activities (Pradhananga et al., 2021).
Decision Making	Construction	Economic and Social	Allowing stakeholders to make informed decisions, identify trends, and optimise operations (Lin et al., 2021).
Stakeholder Collaboration	Design and Planning	Economic and Social	Enable key stakeholders work together more efficiently through generate design collaboration (Alsakka et al., 2023).
Site Safety	Construction	Social	Monitor construction sites for safety compliances, identify potential hazards, and ensure safety protocol adherence (Teferi and Newman, 2018).
Human Resource Management	Planning	Social	Streamline the hiring processes, talent acquisition, and workforce planning for improved organisational efficiency (Dickens et al., 2020).
Skilled Workforce Development	Planning	Social	Provide personalised and adaptive learning experiences to enhance skills and productivity (Xiang et al., 2022).

TABLE 8.14   Potential of AI for SDG9

Sustainable Construction	Stage	Sustainability Pillar	Potential of AI
Autonomous Construction Vehicles	Construction	Economic	AI enable autonomous operations, improving precision, safety, and efficiency in tasks such as excavation and transportation (Alsakka et al., 2023).
Manual Labour	Construction	Economic	Robotic systems with AI capabilities for tasks like bricklaying, welding, and other repetitive activities, increase construction speed and accuracy (Barbosa et al., 2023).
Construction Plans	Design and Planning	Economic	Utilise AR for visualising construction plans, providing on-site guidance, and enhancing collaboration among construction teams for improved precision (Bigham et al., 2019).
Infrastructure Development	Design and Planning	Economic	AI optimises plans and designs, reduces costs, improves quality and durability, enables predictive maintenance, mitigates environmental impact, and enhances disaster resilience (Columbus, 2017).
Project Planning	Planning and Construction	Economic	Streamlines project planning, risk mitigation, quality control and safety enhancement (Oyedele et al., 2021).
Resource Allocation	Planning	Economic and Social	AI offers predictive insights and enhanced decision-making by optimising resource allocation and ensuring the right resources are allocated at the right time (Adams, 2017).
3D Printing	Construction	Environmental and Social	Optimise the printing process, reducing material waste and enhancing the efficiency of constructing complex structures (Yaseen et al., 2020)
Smart Construction Equipment	Construction	Social	Real-time monitoring, predictive maintenance, and efficient operation reduce downtime and enhance productivity (Singh and Singh, 2023).

construction industry can leverage the following, as shown in Table 8.14, vital technologies to attain this objective:

4.2.5 SDG 11: Sustainable Cities

SDG 11 seeks to make cities inclusive, safe, resilient, and sustainable through 10 targets and 14 indicators. Cities use AI to create infrastructure that supports sustainable urban development and long-term liveability. It does so by building essential infrastructure, affordable housing, and sustainable urban spaces and promoting resource efficiency aligned with community needs (Fei et al., 2021). The construction sector is central to optimising resource use, reducing waste, and ensuring efficient construction practices to create sustainable cities (Barros et al., 2021). It extends beyond infrastructure, emphasising inclusive and resilient communities. As urbanisation surges, construction's commitment to sustainability is crucial for shaping cities' futures. The construction industry can leverage the following, as shown in Table 8.15, vital technologies to attain this objective.

4.2.6 SDG 12: Consumption and Production

SDG 12 seeks to reduce resource consumption, minimise waste generation, and promote materials recycling through 11 targets and 13 indicators. The construction industry can further support the goal by promoting responsible consumption patterns, optimising building designs for durability and energy efficiency, and collaborating with stakeholders to develop sustainable standards and practices (Gasper et al., 2019). This multifaceted approach contributes to safer and more resilient urban environments and ensures affordable housing is cost-effective, efficient, and inclusive. Through these efforts, the construction sector can foster responsible production and consumption practices for a more sustainable future (Akinosho et al., 2020). The construction industry can leverage the following, as shown in Table 8.16, vital technologies to attain this objective.

4.2.7 SDG 13: Climate Action

SDG 13 is a critical goal as it undermines the remaining 16 SDGs and seeks to mitigate greenhouse gas emissions, promote energy efficiency, enhance resilience to climate change, and foster responsible and sustainable construction practices (Opoku, 2015). It ensures responsible construction practices that comply with environmental regulations and supports substantial reductions in the carbon footprint of construction projects, fostering sustainability

TABLE 8.15 Potential of AI for SDG11

Sustainable Construction	Stage	Sustainability Pillar	Potential of AI
Efficient Resource Management	Planning	Environmental	AI-driven tools optimise resource management using materials, energy, and water efficiently (Yigitcanlar et al., 2023)
Building Design	Repair and Maintenance	Environmental	Utilise AI to optimise the energy consumption of buildings, adjusting heating, cooling, and lighting systems based on occupancy patterns and environmental conditions (Gao et al., 2020).
Smart Urban Planning	Repair and Maintenance	Environmental	Utilise AI-driven generative design algorithms to optimise urban planning, considering factors such as traffic flow, green spaces, and energy efficiency in constructing buildings and infrastructure (Goubran, 2019).
Waste Reduction	Repair and Maintenance	Environmental	AI tools enable automated sorting, streamlining processes, and conserving resources (Ajayi et al., 2020).
Smart Cities	Planning and Construction	Environmental and Social	AI optimises transportation, enhances public services, manages energy and waste and improves security (Amleida et al., 2022).
Urban Green Spaces Planning	Repair and Maintenance	Environmental and Social	Utilise AI in the planning of urban green spaces, optimising their layout for environmental benefits, aesthetics, and public well-being (Xiang et al., 2022).
Community Engagement	Repair and Maintenance	Social	AI provides accessible participation platforms and data-driven insights by fostering collaboration, informing policies, and enhancing transparency (Yaseen et al., 2020).

TABLE 8.16 Potential of AI for SDG12

Sustainable Construction	Stage	Sustainability Pillar	Potential of AI
Resource Planning	Construction	Economic	AI can analyse historical data and predict resource requirements for construction projects, allowing for better resource planning and reducing waste (Dickens et al., 2020).
Sustainable Materials	Construction	Economic and Environmental	AI can help identify sustainable sources for construction materials and track their journey through the supply chain (Berawi, 2023).
Construction Equipment	Construction	Environmental	Integrating AI into construction equipment optimises energy usage and reduces emissions (Qiu et al., 2021).
Transportation and Mobility	Design and Planning	Environmental	AI-driven transportation systems reduce congestion, improve public transportation, lower emissions through electric and autonomous vehicles, and enhance safety (Roslo, 2022).
Environmental Impact Assessment	Planning	Environmental	Predict and assess construction projects environmental impact, considering emissions, habitat disruption, and resource depletion (Onyelowe et al., 2022).
Waste Management	Planning	Environmental	Utilise AI in waste sorting processes to enhance efficiency and accuracy, ensuring that materials are appropriately categorised for recycling (Almeida Barbosa Franco et al., 2022).
Life Cycle Assessment	Repair and Maintenance	Environmental	Integrate AI into life cycle assessments, considering environmental impacts throughout their life cycle and making informed decisions to minimise resource consumption (Fnais et al., 2022).

in the industry. By implementing the five goals and eight indicators, the construction industry's role extends to developing carbon capture and storage technologies, driving innovation, and researching emissions reduction and climate resilience. These efforts position the construction sector as vital in global initiatives to combat climate change and achieve a more sustainable future. The construction industry can leverage the following, as shown in Table 8.17, critical technologies to attain this objective.

4.2.8 SDG 15: Life of Land (Biodiversity)

SDG 15 seeks to protect, restore, and promote conservation through ecosystem preservation, reforestation and land restoration, habitat conservation, ecological corridor creation, minimising land footprint, biodiversity-friendly landscaping, and sustainable land use planning (Kumar et al., 2019). By achieving the 12 targets and 14 indicators, these actions collectively bolster the preservation and improvement of ecosystems and ensure long-term sustainability. Furthermore, it facilitates informed stakeholder engagement in land use decisions, helping to balance development and environmental conservation for long-term sustainability (Boje et al., 2020). The construction industry can leverage the following, as shown in Table 8.18, vital technologies to attain this objective.

4.2.9 SDG 17: Partnerships for the Goals

SDG 17 is an important goal as it outlines the global partnerships for sustainable development through 19 targets and 24 indicators. In addition, recognising the interconnectedness of nations, industries, and communities in achieving shared goals. The construction industry can strategically collaborate with public and private organisation to address complex challenges, from infrastructure development to environmental conservation (Yigitcanlar et al., 2022). It extends beyond collaboration as it promotes the commitment to fosterer inclusive, participatory decision-making processes that empower all stakeholders to contribute to the SDG agenda (Çetin et al., 2021). The construction industry can utilise the essential technologies outlined in Table 8.19 to achieve these objectives.

4.3 Sustainable Development Goals and Artificial Intelligence in Construction Phases

4.3.1 Planning Phase

The planning phase of a construction project holds significant importance, serving as the cornerstone upon which the success of the entire project is

TABLE 8.17   Potential of AI for SDG13

Sustainable Construction	Stage	Sustainability Pillar	Potential of AI
Carbon Footprint Monitoring	Construction	Environmental	Monitor and analyse the carbon footprint of construction projects by tracking emissions from construction activities, transportation, and material production (Kanyilmaz et al.,2022).
BIM and Energy Simulation	Design	Environmental	Simulate and analyse the energy performance of buildings and reducing carbon emissions (Pan and Zhang, 2023).
Climate Modelling	Design and Planning	Environmental	AI can predict and model future climate trends, allowing for informed decisions about project design, location, and materials to ensure long-term resilience to changing climate conditions (Naser, 2019).
Environmental Monitoring	Design and Planning	Environmental	AI provides real-time data analysis, resource efficiency optimisation and emission monitoring (Wahbeh et al., 2020).
Green Building Certification	Planning	Environmental	AI-driven data analysis can support resilience planning by ensuring regulatory compliance and optimising site selection (Mahbub, 2008).
Resilience to Climate Change	Planning	Environmental	AI can analyse climate data, create predictive models, guides material selection, enables adaptive design, and assesses climate risks (Xiao et al., 2018).
Material Selection	Planning and Construction	Environmental	Identify and optimise the use of energy-efficient and sustainable construction materials, considering factors such as embodied carbon and life cycle assessments (Gasper et al., 2019).
Mitigating Greenhouse Gas Emissions	Planning and Design	Environmental	AI optimises energy use, promotes sustainable materials analyses and reduces waste (Zhang, 2021).
Renewable Energy Integration	Planning and Design	Environmental	Optimise the deployment of renewable energy sources and ensuring efficient energy production (Gao et al., 2020).

TABLE 8.18   Potential of AI for SDG15

Sustainable Construction	Stage	Sustainability Pillar	Potential of AI
Sustainable Land Use	Construction	Economic and Environmental	AI assesses and optimises land use patterns by identifying suitable areas for development (Lepczyk et al., 2017)
Invasive Species Management	Construction	Environmental	Employ AI to identify and manage invasive species that may be introduced during construction (Nagendra and Rafi, 2018).
Soil Health Monitoring	Construction and Repair and Maintenance	Environmental	Monitor soil health before, during, and after construction by assessing soil composition, nutrient levels, and erosion risks to ensure sustainable land use (Tam et al., 2021).
Habitat Planning	Design	Environmental	Utilise AI to design and plan wildlife corridors that connect fragmented habitats by facilitating the movement of species and promoting genetic diversity (Srivastava et al., 2022).
Ecological Site Assessment	Planning	Environmental	AI can analyse environmental data to identify biodiversity hotspots, critical habitats, and areas with ecological sensitivity ecosystems (Statsenko et al., 2022).
Environmental Impact Assessment:	Planning	Environmental	AI improves Environmental Impact Assessments (EIAs) by analysing biodiversity, soil quality, and environmental data (Mahbub, 2008).
Construction Impact Prediction	Planning and Construction	Environmental	Predict and assess the environmental impact of construction by analysing potential disruptions to flora, fauna, and soil structure (Boje et al., 2020).
Erosion Control	Planning and Construction	Environmental	AI provides real-time monitoring, risk assessment, and automated alerts and suggesting soil stabilisation techniques (Ma et al., 2019).
Natural Resource Monitoring	Planning and Construction	Environmental	Implement AI to continuously monitor natural resources to ensure sustainable and responsible resource management during construction projects (Elmousalami, 2020).
Vegetation Management and Restoration	Repair and Maintenance	Environmental	Optimise planting patterns, species selection, and maintenance strategies for sustainable ecosystem restoration (Singh and Singh, 2023).

TABLE 8.19 Potential of AI for SDG17

Sustainable Construction	Stage	Sustainability Pillar	Potential for AI
Public-Private Partnerships (PPPs)	Repair and Maintenance	Economic	AI can facilitate cooperation between governments, businesses, and civil society organisations (Sacks et al., 2020).
Monitoring	Repair and Maintenance	Economic	Providing stakeholders with real-time data and insights to inform decision-making and promote accountability (Schönbeck et al., 2020).
Open Data Initiatives	Construction and Repair and Maintenance	Economic and Environmental	AI can analyse and visualise data on project performance, resource utilisation, and environmental impacts (Panteleeva and Borozdina, 2021).
Supply chain optimisation	Repair and Maintenance	Economic and Social	Optimise the integration of suppliers, contractors, and subcontractors across different regions (Poprach et al., 2019).
Sharing platforms	Repair and Maintenance	Social	AI-powered platforms can facilitate the sharing of best practices, lessons learned, and innovative solutions among construction stakeholders (Kulejewski and Rosłon, 2023).
Community Engagement Platforms	Repair and Maintenance	Social	AI-powered platforms can facilitate community engagement and participation in construction projects, promoting partnerships between project stakeholders (Lepczyk et al., 2017).

built. It is during this phase that vital decisions are made regarding timelines, budgets, and quality standards, which ultimately shape the project's outcome (Huang, 2023). A meticulously developed plan sets the stage for seamless implementation, guaranteeing efficient allocation of resources and achievement of project objectives within defined parameters. By emphasising precision and comprehensiveness in the planning process, construction teams can enhance project success while reducing risks and resource inefficiencies (Gasper et al., 2019).

The literature review found 20 papers that highlighted the importance of AI and SDG in the planning phase. ML was highlighted in 8 instances, showcasing its utility in analysing extensive datasets and deriving actionable insights to inform decision-making processes during project planning (Srivastava et al., 2022). Similarly, AI was also prominently featured in 8 papers, indicating its widespread recognition as a transformative tool for optimising various aspects of construction project planning, such as resource allocation, risk management, and scheduling. Lastly, DL emerged as a focal point in 3 papers, albeit to a lesser extent compared to ML and AI. Nonetheless, its inclusion underscores the growing interest in leveraging advanced machine learning techniques to address complex challenges encountered during the planning phase of construction projects (Chen et al., 2022).

Moreover, the three SDGs most referenced in the planning phase were SDG 7 ($n = 15$), SDG 6 ($n = 9$), and SDG 13 ($n = 9$). SDG 7 stands out as a primary focus in utilising AI to improve energy planning and management strategies. AI-powered tools can optimise energy systems, improve energy efficiency, and facilitate the integration of renewable energy sources (Saeed et al., 2022). Similarly, SDG 6 is central to AI-driven efforts in water resource management and conservation. AI technologies enable real-time monitoring of water quality, prediction of water-related risks, and optimisation of water distribution systems (Cheng et al., 2010). Lastly, SDG 13, underscores the importance of integrating AI into climate resilience and adaptation planning processes. AI algorithms can analyse vast amounts of climate data, identify trends, and generate predictive models to inform decision-making to make project more sustainable (Pan and Zhang, 2021).

4.3.2 Design Phase

The design phase of a construction project is pivotal, serving as a central point upon which the project's success hinges. During this phase, crucial decisions regarding architectural concepts, structural and material specifications, and sustainability considerations are deliberated, significantly shaping the project's eventual outcome (Turner et al., 2020). A meticulously formulated design plan lays the groundwork for seamless implementation, facilitating efficient resource allocation and alignment with project objectives within specified constraints.

The examination of existing literature revealed 25 studies that highlighted the relevance of AI and SDG within the design phase. AI emerged prominently, mentioned 14 times, signifying its pivotal role in automating tasks, generating design alternatives, and simulating scenarios to optimise outcomes (Gálvez-Martos et al., 2018). ML was mentioned 7 times, highlighting its significance in processing extensive datasets, providing decision-making insights, and fostering design innovation. Despite DL receiving only 3 mentions, its algorithms remain instrumental in uncovering hidden insights and refining aspects such as structural integrity and energy efficiency in design (Chen et al., 2022).

Furthermore, the three most frequently mentioned SDGs in the design phase were: SDG 7 ($n = 12$), SDG 9 ($n = 11$), and SDG 6 ($n = 10$). SDG 7 is particularly noteworthy as a primary focus since AI can harness and optimise energy usage in building designs, using predictive modelling and energy simulation (Baduge et al., 2022). This entails analysing various factors such as building orientation, materials, and occupancy patterns. Similarly, SDG 9 was frequently mentioned for enhancing AI-driven design optimisation algorithms, enabling engineers to develop more resilient and sustainable infrastructure solutions. Additionally, BIM augmented with AI technologies enables collaborative designs and construction planning, facilitating better coordination among project stakeholders and reducing errors. Lastly, SDG 6 was frequently mentioned concerning AI integration, assisting in the design of water-efficient systems within buildings, optimising water usage through predictive analytics and smart monitoring devices. Moreover, AI algorithms can analyse large datasets from sensors and IoT devices to identify patterns and anomalies in water usage, enabling early detection of leaks and excessive water usage (Yigitcanlar et al., 2022).

4.3.3 Construction Phase

The construction phase serves as a vital link between planning and design, translating sustainability goals into tangible structures. Inefficient practices during this phase can hinder green initiatives and waste reduction efforts, limiting a project's contribution to sustainability (Cheng et al., 2010). AI enhances project management, resource allocation, and construction workflows, ensuring alignment with SDGs. It acts as a catalyst for sustainable practices, preserving goals from planning to construction (Oluleye et al., 2022).

Upon reviewing the available literature, 24 studies were identified that emphasised the significance of AI and SDGs during the construction phase. AI featured prominently, being referenced 14 times, indicating its crucial role in automating tasks, generating design alternatives, and simulating scenarios to enhance outcomes (Bang et al., 2022). ML was mentioned 7 times, underscoring its importance in processing large datasets, offering decision-making insights, and promoting design innovation. Although DL was mentioned only 3 times, its algorithms continue to play a vital role in revealing hidden insights and refining elements like structural integrity and energy efficiency in design (Sev, 2009).

In addition, the top three most frequently mentioned SDGs in the construction phase were SDG 9 ($n = 17$), SDG 12 ($n = 18$) and SDG 13 ($n = 15$). SDG 9 stands out as a primary focus in the construction stage of projects, with AI playing a crucial role in enhancing design optimisation algorithms (Yaseen, 2021). By leveraging AI-driven solutions, engineers can develop resilient and sustainable infrastructure solutions. These advancements not only streamline construction processes but also contribute to long-term environmental and social sustainability. Similarly, SDG 12 emerges as another key area where AI integration in construction holds immense potential. Through innovative applications of AI technologies, such as predictive analytics and smart monitoring devices, construction projects can optimise resource utilisation, reduce waste, and promote sustainable consumption and production practices (Baduge et al., 2022). Lastly, SDG 13 and AI-enabled solutions offer opportunities to enhance resilience to climate-related risks, optimise energy usage, and mitigate carbon emissions throughout the construction stage (Zhang, 2021).

4.3.4 Operation and Maintenance Phase

The repair and maintenance stage of a construction project is crucial, acting as a cornerstone upon which the project's durability and longevity rely. Throughout this stage, essential tasks such as identifying repair needs, sourcing quality materials, and implementing sustainable maintenance practices are carefully considered, greatly influencing the project's ongoing performance (Klarim et al., 2023). A well-thought-out repair and maintenance plan establishes a solid foundation for effective execution, enabling the allocation of resources and adherence to project goals while navigating various constraints (Zhang et al., 2021).

The review of existing literature identified 16 studies that emphasised the importance of AI and SDGs during the repair and maintenance phase, although it was the least frequently mentioned phase. AI stood out prominently, being mentioned 12 times, highlighting its crucial role in automating tasks, generating design alternatives, and simulating scenarios to enhance repair and maintenance outcomes (Fayek, 2020). ML was cited 3 times, underlining its importance in handling large datasets, offering decision-making insights, and driving innovation in repair and maintenance procedures. Lastly, DL received only 1 mention, despite its potential in revealing hidden insights and refining aspects such as structural integrity and energy efficiency during repair and maintenance tasks (Hudson, 2005).

Furthermore, the three frequently mentioned SDGs in the repair and maintenance stage were SDG 11 ($n = 12$), SDG 7 ($n = 10$), and SDG 13 ($n = 5$). SDG 11 was frequently highlighted as AI plays a crucial role in urban planning and infrastructure management to establish more inclusive, safe, resilient, and sustainable cities (Baduge et al., 2022). AI-driven solutions enable better resource allocation, optimised transportation systems, and enhanced disaster preparedness. Similarly, SDG 7 was frequently cited as AI technologies are vital in optimising energy usage within building designs, utilising predictive modelling and energy simulation (Araugo et al., 2018). Lastly, SDG 13 was often mentioned as AI integration assists in climate resilience and adaptation efforts, facilitating the development of infrastructure solutions capable of withstanding climate-related challenges. AI algorithms analyse data from sensors and IoT devices to identify climate patterns and mitigate risks (Pan and Zhang, 2023).

5 FINDINGS AND DISCUSSION

Introducing AI into the construction industry to align with SDGs holds immense potential to revolutionise the sector, particularly in reducing its environmental footprint. The built environment plays a pivotal role in shaping the quality of life for individuals and communities, spanning from the initial planning and design phases to construction, repair, and ongoing maintenance (Dizdaroglu et al., 2012). Hence, it's imperative for the industry to remain adaptable and forward-thinking to meet evolving needs.

Creating a sustainable built environment with energy-efficient infrastructure is crucial for curbing energy demand and combatting climate change. AI-powered systems offer innovative solutions to optimise energy usage, minimise waste, and streamline resource allocation, ultimately leading to reduced carbon emissions and the preservation of natural resources (Pizzi et al., 2020). These advancements align with principles of environmental sustainability and promote a responsible, cost-effective approach to development. By harnessing AI technologies, the construction industry can accelerate progress towards SDGs, fostering a more sustainable and technologically advanced approach to building and infrastructure development, benefiting both the industry and broader society.

The correlation between construction-related SDGs is clearly demonstrated, as depicted in Figure 8.7. By aligning with SDG frameworks, the construction industry has the potential to expand its impact. This interdependence underscores the pivotal role of the construction sector in advancing the global sustainability agenda. For instance, construction projects involving affordable housing, infrastructure development, and clean energy installations directly contribute to SDGs such as SDG 7, SDG 9, and SDG 11. Simultaneously, construction projects indirectly influence SDGs like SDG 5, SDG 14, and SDG 16 through factors such as labour practices, resource consumption, waste management, and the creation of safe buildings.

This interconnectedness underscores the importance of collaborative efforts, multi-stakeholder engagement, and holistic approaches in addressing global challenges. It emphasises that achieving SDG requires integrated strategies that recognise and address the interdependencies among various goals. In essence, integrating AI into the construction industry not only drives progress towards SDGs but also underscores the importance of holistic, collaborative approaches in addressing complex societal and environmental challenges.

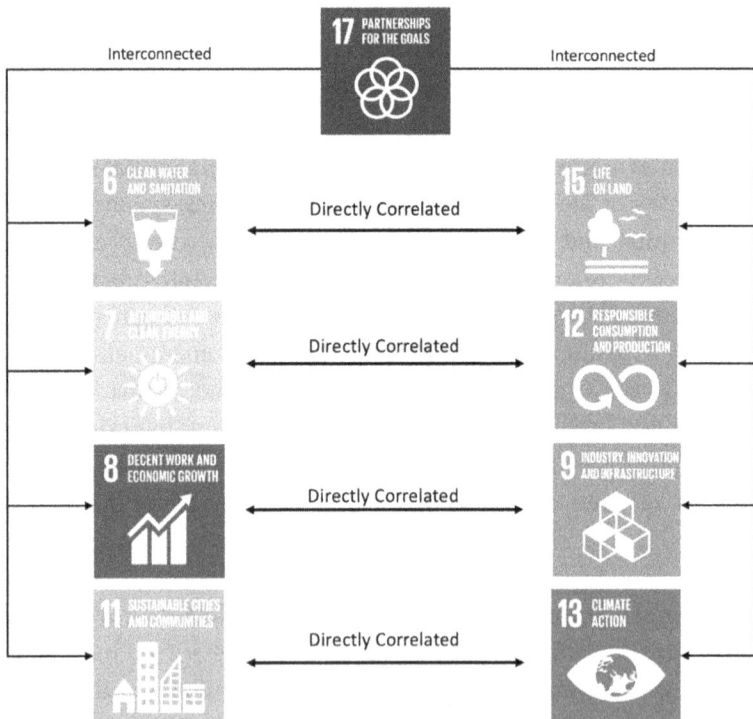

FIGURE 8.7 A direct relationship between SDG signifies a clear correlation between two specific SDGs. The importance of AI in the construction sector to achieve SDG can be broken down into several key aspects.

The convergence of SDG and AI is reshaping the landscape of the construction industry. This intersection represents a pivotal moment where traditional approaches to construction are being reimagined and redefined to align with the urgent need for environmental stewardship and resource efficiency.

Water and biodiversity management are critical aspects of sustainable development, with efficient water resource management being key to mitigating adverse effects on ecosystems and addressing land degradation, thus contributing to the objectives of SDG 6 and SDG 15. Through the integration of AI data-driven insights and real-time capabilities, the construction industry can actively contribute to environmental preservation (Froese et al., 2007). Construction and demolition processes often involve significant water usage, leading to runoff that affects water resources. By harnessing AI technology, the construction sector can reduce excessive water consumption, enhance water management practices, and mitigate environmental impacts (Opoku, 2019).

The reduction of waste and effective energy management is pivotal for advancing SDG 7 and SDG 12, which aim to facilitate clean energy access and minimise waste generation in construction activities. The integration of building information modelling (BIM), IoT, and predictive systems enhances the industry's efficiency, sustainability, and safety, pushing its capabilities forward. AI integrated into construction processes can address environmental challenges associated with waste generation, particularly from construction and demolition activities (Dickens et al., 2020). AI-driven measures play a crucial role in waste reduction, promoting material reuse, optimising resource efficiency, fostering sustainability, and mitigating environmental harm.

Innovative resource management is fundamental to the success of sustainable construction, relying on a combination of economic growth, reliable infrastructure, and innovation. Predictive analytics, supply chain optimisation, and improved communication efficiency empower organisations to increase profit margins and drive innovation (Gao, 2020). Strategic implementation of predictive analytics, AI-driven supply chain optimisation, and data-driven communication enhance sustainability efforts, aligning with SDGs 8 and 9, which promote economic growth, decent work, and technological progress.

Safety and environmental preservation are paramount considerations for construction companies undertaking large-scale sustainable projects in rapidly growing urban areas. AI is leveraged to enhance various facets of construction, resulting in increased efficiency and sustainability (Goubran, 2019). The integration of AI facilitates the delivery of essential services such as healthcare, water, and energy to communities while also reinforcing low-carbon systems within circular economies and smart cities, optimising resource utilisation in alignment with SDG 11 and SDG 13.

The importance of AI in the construction industry as a driver of SDG cannot be overstated. With its transformative capabilities, AI serves as a cornerstone in this transformative journey. It promises to reshape how projects are planned, designed, constructed, and managed (Naser, 2019). Therefore, making projects more efficient, resource-conscious, and environmentally responsible. AI's integration into the construction industry is not just a technological advancement; it's a strategic pathway to achieving SDGs (Stahl et al., 2022).

AI's impact on construction extends beyond operational enhancements; it encompasses a holistic approach towards sustainability and resilience. Through the reduction of resource consumption, minimisation of waste generation, fostering of innovation, and enhancement of safety measures, AI plays a pivotal role in steering the construction industry towards a more sustainable and resilient future, both in the short and long term (Opoku et al., 2022). Recognising AI as a critical tool in this transformative journey is imperative. It must be embraced and harnessed effectively to ensure the betterment of society and the preservation of our planet for future generations.

5.1 Key Findings

The articles acknowledge AI's pivotal role in advancing sustainability within projects and its transformative impact. AI functions not only as a catalyst but also as a guiding influence, reshaping conventional methodologies and approaches to conform with sustainable practices as seen in Table 8.20.

The literature highlighted the data-driven and predictive capabilities that AI can bring to a project, which will optimise resource allocation, streamline processes, and enhance decision-making. Using these capabilities and allowing organisations to monitor and manage tasks will contribute immensely to reaching sustainable objectives. Furthermore, AI's role in enabling projects to adapt and respond to evolving sustainability challenges (Pizzi et al., 2020). It pointed out how AI-driven technologies such as ML and natural language processing can enhance the collection and analysis of sustainability data, enabling proactive and data-informed

TABLE 8.20   Primary application of AI algorithms and its potential to revolutionise the current construction industry

Application	Description
AI-driven automation	By automating tasks like project scheduling, material procurement, and resource allocation, AI reduces human error and optimises efficiency. This streamlining not only saves time and costs but also minimises resource waste, contributing to sustainable practices (Hsu et al., 2020).
Predictive analytics	Predictive analytics leverage AI algorithms to forecast project outcomes and identify potential risks. By anticipating challenges such as delays or budget overruns, construction teams can take proactive measures to mitigate these issues, thus reducing resource waste and promoting project sustainability (Kumar et al., 2019).
Computer vision technology	Computer vision enhances monitoring and quality control on construction sites by detecting deviations from project specifications in real-time. Early identification of errors minimises rework and material wastage, ultimately leading to more sustainable construction practices (Lin et al., 2021).
Robotics and autonomous machinery	AI-driven robotics and autonomous machinery perform tasks with precision and efficiency, reducing the need for manual labour. This not only improves productivity and safety but also minimises resource consumption, contributing to sustainable construction practices (Ma et al., 2019).
Data analytics	AI-enabled data analytics optimise resource utilisation and energy efficiency by analysing data from various sources such as building sensors and weather forecasts. By identifying opportunities to reduce energy consumption and optimise resource usage, construction projects can minimise their environmental footprint and promote sustainability (Nilsson et al., 2016).
Early error detection	AI-driven technologies enable early detection of errors and deviations from project specifications, reducing the need for rework and minimising resource waste. By addressing issues promptly, construction projects can maintain their schedules and budgets while promoting sustainability (Omer and Noguchi, 2020).
Real-time insights	AI provides real-time insights into project performance, enabling proactive decision-making. By identifying inefficiencies or risks as they arise, construction teams can take timely actions to address these issues and minimise their impact on project sustainability (Opoku et al., 2022).

responses to emerging issues. AI can also improve supply chain efficiency, workforce productivity, and stakeholder engagement (Saeed et al., 2022). Therefore, contributing to broader sustainability goals and broader context of sustainable development as seen in Table 8.21.

Out of the 91 papers that were reviewed, a discernible distribution was observed based on a clear focus across different phases of the

TABLE 8.21   Key AI uses in a construction project to promote SDG derived from Chen et al. (2022)

Sustainability Pillar	Description
Environmental	• *Energy Optimisation:* Real-time monitoring lowers energy consumption and emissions. • *Material Selection:* AI minimises material waste by optimising usage. • *Water Management:* AI powered system that can identify opportunities for conservation improvements. • *Renewable energy integration:* Optimise the integration of renewable energy sources. • *Environmental impact assessment:* Identify potential risks and suggestion to minimise the harm on ecosystems. • *Green Building Design:* AI assists in designing eco-friendly structures. • *Waste Management:* AI improves waste reduction and recycling efforts.
Social	• *Safety Monitoring:* AI enhances worker safety by identifying hazards in real-time. • *Labour Optimisation:* AI optimises labour allocation for fair and efficient scheduling. • *Training and Development:* AI-driven platforms personalise learning for skill enhancement. • *Community Engagement:* AI fosters transparent communication with local communities. • *Diversity and Inclusion:* AI aids in improving workforce diversity and inclusivity.
Economic	• *Project Planning:* AI optimises schedules and resource allocation, reducing costs. • *Cost Estimation:* AI provides accurate budget forecasts, aiding financial planning. • *Supply Chain Optimisation:* AI streamlines procurement processes, minimising waste and costs. • *Productivity Monitoring:* AI tracks worker and equipment efficiency in real-time, optimising workflow. • *Predictive Maintenance:* AI predicts equipment failures, reducing downtime and repair costs. • *Performance Analytics:* AI analyses project metrics to improve productivity and profitability.

construction process. Specifically, 21 papers (23%) delved into topics pertinent to the planning phase, encompassing strategic pre-construction considerations. Similarly, 25 papers (27%) addressed the design phase, examining architectural and engineering aspects fundamental to project conceptualisation. Furthermore, 27 papers (30%) provided insights into the construction phase, exploring methodologies, technologies, and practices deployed during project execution. Lastly, 18 papers (20%) were primarily dedicated to the operation and maintenance phase, highlighting post-construction activities aimed at ensuring sustainable upkeep and longevity of built structures. This distribution reflects the diverse focus areas within the construction domain explored by the reviewed literature.

The analysed papers also provided an insight into the relevant SDGs where the construction industry can significantly contribute and make a substantial impact to improve sustainability. Out of the 169 targets outlined in SDGs, 70 (41.42%) are pertinent to the construction sector. Among these, 18.34% are directly influenced by construction projects, while 23.07% are indirectly affected by the construction industry, as seen in Figure 8.8.

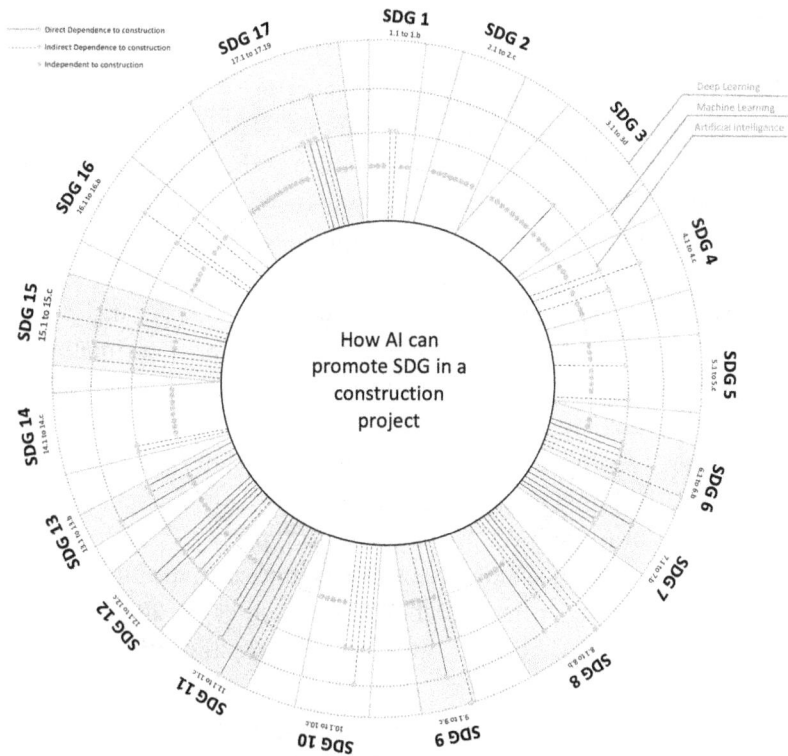

FIGURE 8.8 A visual map of the dependency of SDGs on construction activities. Targets are ordered clockwise and distributed among sub-categories of AI. Additionally, the SDGs highlighted in grey signify crucial SDGs that AI can significantly impact.

The construction industry requires different stakeholders to expand their scope of action beyond environmental attributes. Therefore, organisations that apply SDG principles to their framework must recognise substantial and positive links between targets and indicators. Organisations must also implement learning methods that embrace sustainability and innovation. Traditional approaches like post-project reviews (PPRs) and post-occupancy evaluations (POEs) in construction focus on technical aspects, overlooking the broader social, economic, and environmental impacts. These reviews should become routine and encompass non-technical aspects related to the SDGs. Additionally, companies must prioritise directing resources towards policies and programmes that can make a substantial positive impact and generate sustainable development outcomes.

5.1.1 Direct Influence on Sustainable Development

AI has emerged as a catalytic force within the construction sector, reshaping traditional practices and ushering in an era of unparalleled efficiency and sustainability. The applications of AI within this industry serve as transformative tools, revolutionising every facet of construction processes (Ma et al., 2019). At the core of AI-driven applications lies its ability to streamline workflows with unprecedented precision. By leveraging sophisticated algorithms and machine learning capabilities, these applications meticulously analyse vast datasets, identifying patterns and optimising workflows. This not only expedites project timelines but also significantly reduces inefficiencies, thereby saving time and costs (Regona et al., 2022a).

Moreover, the impact of AI extends well beyond mere efficiency gains. These technologies have become integral in enhancing project management practices. AI-powered project management tools offer comprehensive insights and predictive analytics that empower decision-makers to make informed choices (Rakesh et al., 2023). The ability to foresee potential hurdles, optimise resource allocation, and adapt plans in real time ensures projects stay on track, meeting deadlines and budgetary constraints more effectively. Safety within the construction industry has perennially been a paramount concern. AI applications are instrumental in fortifying safety protocols. Through the amalgamation of AI with IoT sensors and real-time monitoring systems, construction sites can proactively detect and mitigate safety risks. Machine learning algorithms analyse historical data to predict potential hazards, thereby fostering a safer work environment and minimising accidents (Wang et al., 2023).

Furthermore, the optimisation of resource allocation stands as a cornerstone of sustainable construction practices. AI-driven applications excel in this domain, facilitating precise project planning and predictive maintenance. Predictive analytics models anticipate maintenance needs, optimising equipment usage and reducing downtime. Real-time monitoring not only aids in tracking resource utilisation but also enables timely adjustments, contributing to reduced waste and enhanced sustainability (Bang et al., 2022). This alignment of AI technologies with SDGs is not incidental but rather a strategic synergy. By fostering precise project planning, predictive maintenance, and real-time monitoring, these technologies directly support SDG targets related to sustainable infrastructure, resource efficiency, innovation, and safe working environments (Alsakka et al., 2023).

5.1.2 Indirect Influence on Sustainable Development

The influence of the construction industry spans a vast spectrum of critical aspects that significantly impact global sustainability efforts. From resource consumption to energy efficiency, job creation, and economic growth (Gag et al., 2021). Through a concerted focus on sustainable construction practices, the adoption of eco-friendly technologies, and the steadfast adherence to responsible business ethics, the construction industry possesses the potential to significantly amplify their indirect contributions to the overarching goal of sustainable development.

Simultaneously, the influence of AI extends its reach, indirectly catalysing progress towards sustainability objectives. AI-driven innovations serve as pivotal agents in curtailing resource waste, enhancing energy efficiency, and mitigating environmental impacts. By enabling data-driven decision-making processes, these technologies empower stakeholders to make informed choices that optimise resource utilisation and minimise waste across various operational facets within the construction industry (Statsenko et al., 2022).

Moreover, the integration of AI does not just streamline processes; it also fosters a more inclusive and engaged community. Through enhanced stakeholder engagement facilitated by AI technologies, there emerges a platform for broader societal inclusion. This inclusion extends beyond economic facets, advancing educational opportunities and skill development. AI's role in promoting learning platforms and accessible educational resources contributes to a more knowledgeable and adaptable workforce, ultimately bolstering social welfare and fostering greater community resilience (Wahbeh et al., 2020).

Furthermore, AI's impact on the construction industry echoes in the development of more resilient and community-centric urban environments. By leveraging AI-driven insights, urban planning becomes more responsive and adaptable to the needs of communities. This creation of sustainable urban landscapes that prioritise the well-being of residents, promote social cohesion, and optimise resource allocation within these environments (Turner et al., 2020). In this cohesive interplay between the construction industry and AI technologies, the indirect contributions to sustainability initiatives become evident. This proactive approach not only aids in mitigating potential risks but also paves the way for the symbiotic relationship intertwines economic prosperity, ecological stewardship, and social well-being, forging a path towards a more equitable and sustainable future (Singh et al., 2023).

5.2 Research Contributions

The integration of AI into the construction industry to achieve SDGs represents a transformative approach with the potential to revolutionise project planning, execution, and management. The systematic literature review stands out in the discourse on AI's role in advancing SDGs within construction. Unlike recent studies focusing solely on specific AI technology that impacts sustainability, this literature review conducts a comprehensive examination of current AI applications and their impacts across all construction stages. It identifies various AI subsets, discussing their adaptability to construction and potential contributions to sustainable development. Moreover, it specifies the most relevant SDGs for construction, illustrating how AI can enhance sustainability considerations effectively and efficiently. Ultimately, it offers crucial guidance for researchers and practitioners aiming to harness AI for sustainable development in construction.

The findings were derived from the 91 papers and formed nine directly correlated SDGs that can help stakeholders achieve more sustainable projects. These identified SDGs promise to guide stakeholders in the construction industry towards the realisation of more sustainable projects. The technologies subject to examination are delimited to those operative within the distinct phases of planning, design, construction, and subsequent repair and maintenance processes. As AI in construction is a relatively new concept and there is limited knowledge on the importance of AI to help projects meet SDGs, the contributions of this paper are as follows:

- A new body of knowledge concerning the significance of AI in facilitating a shift from traditional approaches to construction practices that align with the SDGs.

- An understanding of existing applications that can help stakeholders achieve the nine relevant SDGs.

- Emphasises that AI will be pivotal in achieving sustainability within the construction industry.

- Provide a comprehensive and detailed insight into the application of AI, ML, and DL for achieving the identified SDGs within the construction sector.

- Acknowledges the existence of substantial research gaps in the field, emphasising that there needs to be more comprehensive investigations that specifically examine the impact of AI on SDGs within the context of the construction industry.

- A review of the opportunities and challenges that the construction industry encounters when they implement AI to achieve SDG.

- A groundwork for future research based on furthering the understanding of directly correlated SDGs.

Based on the insights derived from this chapter, future research endeavours will continue to explore the challenges related to the adoption of AI, as well as the potential opportunities it offers for the construction sector.

5.3 Research Limitations

The study has the following limitations: (a) The study relied on 91 selected papers, which may only partially represent the diverse array of global construction projects. The specific contexts and characteristics of different regions, project scales, and construction practices may not be fully captured; (b) Additional literature reviews to expand on the current findings and develop a better understanding of the opportunities and challenges of implementing AI technologies to promote SDG; (c) No data was collected. The chapter did not conduct interviews with on-site professionals to capture their insights on the application of AI throughout a building's lifecycle in support of SDG; (d) The process of identifying and correlating nine SDGs may be subject to interpretation and bias. Different stakeholders may prioritise other SDGs based on their perspectives, potentially influencing the applicability of the findings; (e) The examination of technologies within a specific phase of construction may not encompass the full range of the construction industry. Consequently, variations in practices across different sectors within the industry may need to be fully addressed; (f) Given the dynamic nature of AI in construction, specific technologies and methods may become obsolete over time. The continuous evolution of technology could impact the ongoing relevance of specific findings; and (g) While the chapter acknowledges challenges in implementing AI for SDGs in construction, a more in-depth exploration of these challenges, along with potential mitigations, could provide a better understanding. In conclusion, while the contributions of this chapter are significant, these identified limitations highlight areas for further research refinement and expansion to enhance the robustness and applicability of insights within the dynamic landscape of AI integration in the construction industry.

5.4 Research Directions

The findings of this research hold significant promise for the future trajectory of sustainable construction practices. By shedding light on the intricate interplay between technological innovation, notably AI, and the attainment of SDGs, this study lays the groundwork for advancements in the construction industry. The incorporation of AI-driven solutions holds the potential to revolutionise various stages of construction projects, ranging from initial design phases to execution and ongoing maintenance, by streamlining resource allocation, bolstering operational efficiency, and curtailing ecological footprints. Going forward, industry stakeholders can harness the actionable insights gleaned from this research to strategically deploy AI technologies, thereby expediting progress towards sustainable development objectives while fostering a culture of ingenuity and stewardship within Australia's construction sector.

6 CONCLUSION

This study has systematically reviewed the literature on AI in the construction industry using the PRISMA protocol. The study findings highlight the construction industry's importance in capitalising on the potential of AI across the entire project lifecycle, steering the industry towards practices that align with sustainability goals (Young et al., 2021). The increasing complexity of modern construction is now the main driver for developing interest in AI Given the industry's substantial energy consumption; it assumes a pivotal role in advancing sustainability (Statsenko et al., 2022). By identifying and correlating nine SDGs through the analysis of 91 articles, this chapter provides a roadmap for stakeholders to guide their efforts towards more sustainable construction projects. The contributions of this research extend beyond the identification of SDGs; it has created a new body of knowledge highlighting the significance of AI in transitioning from traditional construction approaches to practices aligned with SDGs. The chapter explores existing applications and emphasises AI's pivotal role in achieving sustainability. Governments should utilise the construction industry as a driving force by implementing effective policies and regulations. Collaborative efforts between government agencies and stakeholders are essential for integrating the SDGs into long-term business strategies and advancing sustainability (Dickens et al., 2020).

This chapter establishes a foundation for future research initiatives by acknowledging existing research gaps and emphasising the need for more comprehensive investigations specific to the impact of AI on SDGs in the construction context. It reviews current opportunities and challenges in implementing AI for SDG achievement in the construction industry. It sets the stage for continued exploration into the hurdles associated with AI adoption and the untapped potential opportunities it presents. As technology advances, the construction sector will likely see further innovations and the adoption of new tools and techniques to meet evolving challenges and requirements. As the construction industry moves forward, future research efforts should persist in delving deeper into the challenges hindering AI adoption in construction while exploring its vast opportunities. By expanding our understanding of the interplay between AI and SDGs in construction, we can contribute to a more sustainable and technologically advanced industry that aligns with global development goals. The urgent call for collective action in the construction sector is not just about achieving goals; it's about shaping a sustainable and resilient future for future generations.

ACKNOWLEDGEMENTS

This chapter, with permission from the copyright holder, is a reproduced version of the following journal article: Regona, M., Yigitcanlar, T., Hon, C., & Teo, M. (2024). Artificial intelligence and sustainable development goals: systematic literature review of the construction industry. *Sustainable Cities and Society*, 108, 105499.

REFERENCES

Adams, C. (2017). The Sustainable Development Goals, integrated thinking and the integrated report. Integrated Reporting, 1–52.

Ajayi, A., Oyedele, L., Owolabi, H., Akinade, O., Bilal, M., Davila Delgado, J.M., & Akanbi, L. (2020). Deep learning models for health and safety risk prediction in power infrastructure projects. *Risk Analysis*, 40(10), 2019–2039.

Akinosho, T.D., Oyedele, L.O., Bilal, M., Ajayi, A.O., Delgado, M.D., Akinade, O.O., & Ahmed, A.A. (2020). Deep learning in the construction industry: A review of present status and future innovations. *Journal of Building Engineering*, 32, 101827.

Almeida Barbosa Franco, J., Domingues, A., de Almeida Africano, N., Deus, R., & Battistelle, R. (2022). Sustainability in the civil construction sector supported by industry 4.0 technologies: Challenges and opportunities. *Infrastructures*, 7(3), 43.

Alsakka, F., Haddad, A., Ezzedine, F., Salami, G., Dabaghi, M., & Hamzeh, F. (2023). Generative design for more economical and environmentally sustainable reinforced concrete structures. *Journal of Cleaner Production*, 387, 135829.

Aparicio, C., Balzan, A., & Trabucco, D. (2020). Robotics in construction: Framework and future directions. *International Journal of High-Rise Buildings*, 9(1), 105–111.

Arsiwala, A., Elghaish, F., & Zoher, M. (2023). Digital twin with Machine learning for predictive monitoring of CO2 equivalent from existing buildings. *Energy and Buildings*, 284, 112851.

Baduge, S.K., Thilakarathna, S., Perera, J.S., Arashpour, M., Sharafi, P., Teodosio, B., Shringi, A., & Mendis, P. (2022). Artificial intelligence and smart vision for building and construction 4.0: Machine and deep learning methods and applications. *Automation in Construction*, 141, 104440.

Bang, S., & Andersen, B.S. (2022). Utilising artificial intelligence in construction site waste reduction. *Journal of Engineering, Project, and Production Management*, 12, 239–249.

Barbosa Júnior, I., Macêdo, A., & Martins, V. (2023). Construction Industry and Its Contributions to Achieving the SDGs Proposed by the UN: An Analysis of Sustainable Practices. *Buildings*, 13(5), 1168.

Barros, N., & Ruschel, R. (2021). Machine learning for whole-building life cycle assessment: A systematic literature review. In *Proceedings of the 18th International Conference on Computing in Civil and Building Engineering: ICCCBE 2020* (pp. 109–122). Springer International Publishing.

Baum, S., Barrett, A., & Yampolskiy, R. (2017). Modeling and interpreting expert disagreement about artificial superintelligence. *Informatica*, 41(7), 419–428.

Berawi, M.A. (2023). Smart cities: Accelerating sustainable development agendas. *International Journal of Technology*, 14(1), 1–4.

Bigham, G., Adamtey, S., Onsarigo, L., & Jha, N. (2019). Artificial intelligence for construction safety: Mitigation of the risk of fall. *In Intelligent Systems and Applications: Proceedings of the 2018 Intelligent Systems Conference (IntelliSys) Volume 2* (pp. 1024–1037). Springer International Publishing.

Boje, C., Guerriero, A., Kubicki, S., & Rezgui, Y. (2020). Towards a semantic Construction Digital Twin: Directions for future research. *Automation in Construction*, 114, 103179.

Butler, L., Yigitcanlar, T., & Paz, A. (2021). Barriers and risks of Mobility-as-a-Service (MaaS) adoption in cities: A systematic review of the literature. *Cities*, 109, 103036.

Çetin, S., De Wolf, C., & Bocken, N. (2021). Circular digital built environment: An emerging framework. *Sustainability*, 13(11), 6348.

Cheng, M.Y., Tsai, H.C., & Sudjono, E. (2010). Conceptual cost estimates using evolutionary fuzzy hybrid neural network for projects in construction industry. *Expert Systems with Applications*, 37, 4224–4231.

Columbus, L. (2017). McKinsey's state of machine learning and AI, 2017. Forbes. Available online: https://www. forbes. com/sites/louiscolumbus/2017/07/09/mckinseys-state-of-machine-learning-and-ai-2017 (accessed on 17 August 2023).

Degirmenci, K., Desouza, K. C., Fieuw, W., Watson, R. T., & Yigitcanlar, T. (2021). Understanding policy and technology responses in mitigating urban heat islands: A literature review and directions for future research. *Sustainable Cities and Society*, 70, 102873.

Dickens, C., McCartney, M., Tickner, D., Harrison, I., Pacheco, P., & Ndhlovu, B. (2020). Evaluating the global state of ecosystems and natural resources: within and beyond the SDGs. *Sustainability*, 12(18), 7381.

Dizdaroglu, D., Yigitcanlar, T., & Dawes, L. (2012). A micro-level indexing model for assessing urban ecosystem sustainability. *Smart and Sustainable Built Environment*, 1(3), 291–315.

Egwim, C.N., Alaka, H., Toriola-Coker, L.O., Balogun, H., & Sunmola, F. (2021). Applied artificial intelligence for predicting construction projects delay. *Machine Learning Applications*, 6, 100166.

Elmousalami, H. (2020). Comparison of artificial intelligence techniques for project conceptual cost prediction: a case study and comparative analysis. *IEEE Transactions on Engineering Management*, 68(1), 183–196.

Fang, X., Shi, X., & Gao, W. (2021). Measuring urban sustainability from the quality of the built environment and pressure on the natural environment in China: A case study of the Shandong Peninsula region. *Journal of Cleaner Production*, 289, 125145.

Fayek, A.R. (2020). Fuzzy logic and fuzzy hybrid techniques for construction engineering and management. *Journal of Construction Engineering and Management*, 146, 04020064.

Fei, W., Opoku, A., Agyekum, K., Oppon, J., Ahmed, V., Chen, C., & Lok, K. (2021). The critical role of the construction industry in achieving the sustainable development goals (SDGs): Delivering projects for the common good. *Sustainability*, 13(16), 9112.

Fnais, A., Rezgui, Y., Petri, I., Beach, T., Yeung, J., Ghoroghi, A., & Kubicki, S. (2022). The application of life cycle assessment in buildings: challenges, and directions for future research. *The International Journal of Life Cycle Assessment*, 27(5), 627–654.

Froese, T., Han, Z., & Alldritt, M. (2007). Study of information technology development for the Canadian construction industry. *Canadian Journal of Civil Engineering*, 34(7), 817–829.

Gálvez-Martos, J., Styles, D., Schoenberger, H., & Zeschmar-Lahl, B. (2018). Construction and demolition waste best management practice in Europe. *Resources, Conservation and Recycling*, 136, 166–178.

Gao, D. (2020). Application of computer artificial intelligence control technology in the comprehensive utilization of green building energy. In *Journal of Physics: Conference Series* (Vol. 1578, No. 1, p. 012027). IOP Publishing.

Gasper, D., Shah, A., & Tankha, S. (2019). The framing of sustainable consumption and production in SDG 12. *Global Policy*, 10, 83–95.

General, A. (2015). *United Nations transforming our world: the 2030 agenda for sustainable development*. Division for Sustainable Development Goals: New York, NY.

Ginzburg, A., Kuzina, O., & Ryzhkova, A. (2018). Unified resources marking system as a way to develop artificial intelligence in construction. In *IOP Conference Series: Materials Science and Engineering* (Vol. 365, No. 6, p. 062021). IOP Publishing.

Goubran, S. (2019). On the role of construction in achieving the SDGs. *Journal of Sustainability Research*, 1(2), e190020.

Hattacharya, R., & Bose, D. (2023). A review of the sustainable development goals to make headways through the COVID-19 pandemic era. *Environmental Progress & Sustainable Energy*, 42(4), e14093.

Hsu, H., Chang, S., Chen, C., & Wu, I.C. (2020). Knowledge-based system for resolving design clashes in building information models. *Automation in Construction*, 110, 103001.

Huang, H. (2023). Network model to optimize the process of green environmental features and urban building recognition based on lightweight image search system. *Soft Computing*, 1–11. https://doi.org/10.1007/s00500-023-08702-y

Kankanamge, N., Yigitcanlar, T., & Goonetilleke, A. (2021). Public perceptions on artificial intelligence driven disaster management: Evidence from Sydney, Melbourne and Brisbane. *Telematics and Informatics*, 65, 101729.

Kanyilmaz, A., Tichell, P., & Loiacono, D. (2022). A genetic algorithm tool for conceptual structural design with cost and embodied carbon optimization. *Engineering Applications of Artificial Intelligence*, 112, 104711.

Kulejewski, J., & Rosłon, J. (2023). Optimization of ecological and economic aspects of the construction schedule with the use of metaheuristic algorithms and artificial intelligence. *Sustainability*, 15(1), 890.

Kumar, V., & Teo, E. (2019). Towards a more circular construction model: Conceptualizing an open-BIM based estimation framework for urban mining. In *CIB World Congress* Hong Kong (Vol. 17).

Kumar, D., Singh, R. B., & Kaur, R. (2019). SDG 9: Case study–infrastructure assessment for sustainable development. *Spatial Information Technology for Sustainable Development Goals*, 157–167.

Lee, M., Yun, J., Pyka, A., Won, D., Kodama, F., Schiuma, G., ... & Zhao, X. (2018). How to respond to the fourth industrial revolution, or the second information technology revolution? Dynamic new combinations between technology, market, and society through open innovation. *Journal of Open Innovation: Technology, Market, and Complexity*, 4(3), 21.

Lepczyk, C., Aronson, M., Evans, K., Goddard, M., Lerman, S., & MacIvor, J. (2017). Biodiversity in the city: fundamental questions for understanding the ecology of urban green spaces for biodiversity conservation. *BioScience*, 67(9), 799–807.

Li, W., Yigitcanlar, T., Erol, I., & Liu, A. (2021). Motivations, barriers and risks of smart home adoption: from systematic literature review to conceptual framework. *Energy Research & Social Science*, 80, 102211.

Lin, Z., Chen, A., & Hsieh, S. (2021). Temporal image analytics for abnormal construction activity identification. *Automation in Construction*, 124, 103572.

Liu, S., Meng, X., & Tam, C. (2015). Building information modeling based building design optimization for sustainability. *Energy and Buildings*, 105, 139–153.

Ioppolo, G., Cucurachi, S., Salomone, R., Shi, L., & Yigitcanlar, T. (2019). Integrating strategic environmental assessment and material flow accounting: a novel approach for moving towards sustainable urban futures. *The International Journal of Life Cycle Assessment*, 24, 1269–1284.

Ma, L., Wang, L., Skibniewski, M., & Gajda, W. (2019). An eco-innovative framework development for sustainable consumption and production in the construction industry. *Technological and Economic Development of Economy*, 25(5), 774–801.

Mahbub, R. (2008). An investigation into the barriers to the implementation of automation and robotics technologies in the construction industry (Doctoral dissertation, Queensland University of Technology).

Mocerino, C. (2018). Digital revolution in efficient self-organization of buildings: towards intelligent robotics. In *2018 Energy and Sustainability for Small Developing Economies (ES2DE)* (pp. 1–6). IEEE.

Nagendra, S. V., & Rafi, N. (2018). Application of artificial intelligence in construction project management. *International Journal of Research in Engineering, Science and Management*, 1(12), 423–427.

Naser, M. (2019). AI-based cognitive framework for evaluating response of concrete structures in extreme conditions. *Engineering Applications of Artificial Intelligence*, 81, 437–449.

Nilsson, M., Griggs, D., & Visbeck, M. (2016). Policy: Map the interactions between sustainable development goals. *Nature*, 534(7607), 320–32.

Oluleye, B.I., Chan, D.W., & Antwi-Afari, P. (2022). Adopting Artificial Intelligence for enhancing the implementation of systemic circularity in the construction industry: A critical review. *Sustainable Production and Consumption*, 35, 509–524.

Omer, M., & Noguchi, T. (2020). A conceptual framework for understanding the contribution of building materials in the achievement of Sustainable Development Goals (SDGs). *Sustainable Cities and Society*, 52, 101869.

Onyelowe, K. C., Kontoni, D., Ebid, A., Dabbaghi, F., Soleymani, A., Jahangir, H., & Nehdi, M. (2022). Multi-objective optimization of sustainable concrete containing fly ash based on environmental and mechanical considerations. *Buildings*, 12(7), 948.

Opoku, A. (2015). The role of culture in a sustainable built environment. In: Chiarini, A. (Ed.) *Sustainable Operations Management. Measuring Operations Performance*. Springer, Cham. https://doi.org/10.1007/978-3-319-14002-5_3.

Opoku, A. (2019). Biodiversity and the built environment: Implications for the Sustainable Development Goals (SDGs). *Resources, Conservation and Recycling*, 141, 1–7.

Opoku, A. (2022). Construction industry and the Sustainable Development Goals (SDGs). In George Ofori. (Ed.) *Research Companion to Construction Economics*, (pp. 199–214). Edward Elgar Publishing.

Oyedele, A., Ajayi, A., Oyedele, L., Delgado, J., Akanbi, L., Akinade, O., ... & Bilal, M. (2021). Deep learning and Boosted trees for injuries prediction in power infrastructure projects. *Applied Soft Computing*, 110, 107587.

Pan, Y., & Zhang, L. (2021). Roles of artificial intelligence in construction engineering and management: A critical review and future trends. *Automation in Construction*, 122, 103517.

Pan, Y., & Zhang, L. (2023). Integrating BIM and AI for smart construction management: Current status and future directions. *Archives of Computational Methods in Engineering*, 30(2), 1081–1110.

Panteleeva, M., & Borozdina, S. (2021). Sustainable urban development strategic initiatives. *Sustainability*, 14(1), 37.

Pizzi, S., Caputo, A., Corvino, A., & Venturelli, A. (2020). Management research and the UN sustainable development goals (SDGs): A bibliometric investigation and systematic review. *Journal of Cleaner Production*, 276, 124033.

Poprach, S., Bolduan, T., Steuer, D., Vössing, M., & Haghsheno, S. (2019). Building the future of the construction industry through artificial intelligence and platform thinking. *Digitale Welt*, 3, 40–44.

Pradhananga, P., ElZomor, M., & Santi Kasabdji, G. (2021). Identifying the challenges to adopting robotics in the US construction industry. *Journal of Construction Engineering and Management*, 147(5), 05021003.

Qiu, Y., Wang, H., & Zhang, Q. (2021). Energy-efficient and Sustainable Construction Technologies and Simulation Optimisation Methods. In *2021 International Conference on E-Commerce and E-Management (ICECEM)* (pp. 341–351). IEEE.

Rakesh, C., Vivek, T., & Balaji, K. (2023). A Review on IoT for the application of energy, environment, and waste management: System architecture and future directions. In Govind P. Gupta, Rakesh Tripathi, Brij B. Gupta, Kwok Tai Chui (Eds.) *Big Data Analytics in Fog-Enabled IoT Networks*, 141–172.

Regona, M., Yigitcanlar, T., Xia, B., & Li, R. (2022a). Artificial intelligent technologies for the construction industry: How are they perceived and utilized in Australia? *Journal of Open Innovation: Technology, Market, and Complexity*, 8(1), 16.

Regona, M., Yigitcanlar, T., Xia, B., & Li, R. (2022b). Opportunities and adoption challenges of AI in the construction industry: a PRISMA review. *Journal of Open Innovation: Technology, Market, and Complexity*, 8(1), 45.

Regona, M., Yigitcanlar, T., Hon, C., & Teo, M. (2023). Mapping two decades of AI in construction research: A scientometric analysis from the sustainability and construction phases lenses. *Buildings*, 13(9), 2346.

Rodriguez-Gracia, D., de las Mercedes Capobianco-Uriarte, M., Terán-Yépez, E., Piedra-Fernández, J. A., Iribarne, L., & Ayala, R. (2023). Review of artificial intelligence techniques in green/smart buildings. *Sustainable Computing: Informatics and Systems*, 38, 100861.

Sachs, J. D., Schmidt-Traub, G., Mazzucato, M., Messner, D., Nakicenovic, N., & Rockström, J. (2019). Six transformations to achieve the sustainable development goals. *Nature Sustainability*, 2(9), 805–814.

Sacks, R., Girolami, M., & Brilakis, I. (2020). Building information modelling, artificial intelligence and construction tech. *Development in the Built Environment*, 4, 100011.

Saeed, Z.O., Mancini, F., Glusac, T., & Izadpanahi, P. (2022). Artificial intelligence and optimization methods in construction industry. *Buildings*, 12, 685.

Schönbeck, P., Löfsjögård, M., & Ansell, A. (2020). Quantitative review of construction 4.0 technology presence in construction project research. *Buildings*, 10(10), 173.

Sev, A. (2009). How can the construction industry contribute to sustainable development? A conceptual framework. *Sustainable Development*, 17(3), 161–173.

Singh, D., & Singh, A. (2023). Role of building automation technology in creating a smart and sustainable built environment. *Evergreen*. 10(1), 412–420.

Singh, A., Kanaujia, A., Singh, V. K., & Vinuesa, R. (2024). Artificial intelligence for Sustainable Development Goals: Bibliometric patterns and concept evolution trajectories. *Sustainable Development*. 32(1), 724–754.

Srivastava, A., Jawaid, S., Singh, R., Gehlot, A., Akram, S.V., Priyadarshi, N., & Khan, B. (2022). Imperative role of technology intervention and implementation for automation in the construction industry. *Advances in Civil Engineering*, 2022, 6716987.

Stahl, B. C., Schroeder, D., & Rodrigues, R. (2022). AI for Good and the SDGs. In *Ethics of artificial intelligence: Case studies and Options for addressing ethical challenges* (pp. 95–106). Cham: Springer International Publishing.

Statsenko, L., Samaraweera, A., Bakhshi, J., & Chileshe, N. (2023). Construction 4.0 technologies and applications: A systematic literature review of trends and potential areas for development. *Construction Innovation*. 23(5), 961–993.

Tatiya, A., Zhao, D., Syal, M., Berghorn, G.H., & LaMore, R. (2018). Cost prediction model for building deconstruction in urban areas. *Journal of Cleaner Production*, 195, 1572–1580.

Teferi, Z., & Newman, P. (2018). Slum upgrading: can the 1.5 C carbon reduction work with SDGs in these settlements? *Urban Planning*, 3(2), 52–63.

Waage, J., Yap, C., Bell, S., Levy, C., Mace, G., Pegram, T., ... & Poole, N. (2015). Governing the UN Sustainable Development Goals: interactions, infrastructures, and institutions. *The Lancet Global Health*, 3(5), e251–e252.

Wahbeh, W., Kunz, D., Hofmann, J., & Bereuter, P. (2020). Digital twinning of the built environment–an interdisciplinary topic for innovation in didactics. *ISPRS Annals of the Photogrammetry, Remote Sensing and Spatial Information Sciences*, 4, 231–237.

Wang, K., Zhao, Y., Gangadhari, R.K., & Li, Z. (2021). Analyzing the adoption challenges of the Internet of things (IoT) and artificial intelligence (AI) for smart cities in China. *Sustainability*, 13, 10983.

Wang, X., Chen, P., Chow, C. L., & Lau, D. (2023). Artificial-intelligence-led revolution of construction materials: From molecules to Industry 4.0. *Matter*, 6(6), 1831–1859.

Winge, S., Albrechtsen, E., & Mostue, B. A. (2019). Causal factors and connections in construction accidents. *Safety Science*, 112, 130–141.

Turner, C.J., Oyekan, J., Stergioulas, L., & Griffin, D. (2020). Utilizing industry 4.0 on the construction site: Challenges and opportunities. *IEEE Transactions on Industrial Informatics*, 17, 746–756.

Xiang, Y., Chen, Y., Xu, J., & Chen, Z. (2022). Research on sustainability evaluation of green building engineering based on artificial intelligence and energy consumption. *Energy Reports*, 8, 11378–11391.

Xiao, C., Liu, Y., & Akhnoukh, A. (2018). Bibliometric review of artificial intelligence (AI) in construction engineering and management. In *International Conference on Construction and Real Estate Management 2018* (pp. 32–41). Reston, VA: American Society of Civil Engineers.

Yaseen, A. (2021). Reducing industrial risk with AI and automation. *International Journal of Intelligent Automation and Computing*, 4(1), 60–80.

Yaseen, Z. M., Ali, Z. H., Salih, S. Q., & Al-Ansari, N. (2020). Prediction of risk delay in construction projects using a hybrid artificial intelligence model. *Sustainability*, 12(4), 1514.

Yigitcanlar, T., Velibeyoglu, K., & Baum, S. (2008). *Creative Urban Regions: Harnessing Urban Technologies to Support Knowledge City Initiatives Preface*. Hersey, PA: IGI Global.

Yigitcanlar, T. (2010). *Sustainable urban and regional infrastructure development: Technologies, applications and management*. Hersey, PA: IGI Global.

Yigitcanlar, T., Desouza, K., Butler, L., & Roozkhosh, F. (2020). Contributions and risks of artificial intelligence (AI) in building smarter cities: Insights from a systematic review of the literature. *Energies*, 13(6), 1473.

Yigitcanlar, T., Corchado, J., Mehmood, R., Li, R., Mossberger, K., & Desouza, K. (2021a). Responsible urban innovation with local government artificial intelligence (AI): A conceptual framework and research agenda. *Journal of Open Innovation: Technology, Market, and Complexity*, 7(1), 71.

Yigitcanlar, T., Mehmood, R., & Corchado, J. (2021b). Green artificial intelligence: Towards an efficient, sustainable and equitable technology for smart cities and futures. *Sustainability*, 13(16), 8952.

Yigitcanlar, T., Degirmenci, K., Butler, L., & Desouza, K. C. (2022). What are the key factors affecting smart city transformation readiness? Evidence from Australian cities. *Cities*, 120, 103434.

Yigitcanlar, T., Agdas, D., & Degirmenci, K. (2023). Artificial intelligence in local governments: perceptions of city managers on prospects, constraints and choices. *AI & Society*, 38(3), 1135–1150.

Yigitcanlar, T., David, A., Li, W., Fookes, C., Bibri, S. E., & Ye, X. (2024). Unlocking Artificial Intelligence Adoption in Local Governments: Best Practice Lessons from Real-World Implementations. *Smart Cities*, 7(4), 1576–1625.

oung, D., Panthi, K., & Noor, O. (2021). Challenges involved in adopting BIM on the construction jobsite. *EPiC Series in Built Environment*, 2, 302–310.

Yun, J., Lee, D., Ahn, H., Park, K., & Yigitcanlar, T. (2016). Not deep learning but autonomous learning of open innovation for sustainable artificial intelligence. *Sustainability*, 8(8), 797.

Zhang, Y. (2021). Safety management of civil engineering construction based on artificial intelligence and machine vision technology. *Advances in Civil Engineering*, 2021, 3769634.

Zhang, L., Pan, Y., Wu, X., & Skibniewski, M. J. (2021). *Artificial intelligence in construction engineering and management* (pp. 231–256). Singapore: Springer.

Index

Note: **Bold** page numbers refer to tables and *italic* page numbers refer to figures.

For Product Safety Concerns and Information please contact our EU
representative GPSR@taylorandfrancis.com
Taylor & Francis Verlag GmbH, Kaufingerstraße 24, 80331 München, Germany